Animal Cell DNA Polymerases

Authors

Michael Fry, Ph.D.
Associate Professor
Unit of Biochemistry
Faculty of Medicine
Technion-Israel Institute of Technology
Haifa, Israel

Lawrence A. Loeb, M.D., Ph.D.
Professor and Director
The Joseph Gottstein Memorial Cancer Research Laboratory
Department of Pathology
University of Washington
Seattle, Washington

CRC Press, Inc.
Boca Raton, Florida

Library of Congress Cataloging-in-Publication Data

Fry, Michael, Ph.D
Animal cell DNA polymerases.

Includes bibliographies and index.
1. Deoxyribonucleate nucleotidyltransferases.
I. Loeb, Lawrence, A. II. Title.
QP606.D46F78 1986 591.87′3282 86-6870
ISBN 0-8493-6507-4

Direct all inquiries to CRC Press, Inc., 2000 Corporate Blvd., N.W., Boca Raton, Florida, 33431.

International Standard Book Number 0-8493-6507-4

Library of Congress Card Number 86-6870
Printed in the United States

THE AUTHORS

Michael Fry, Ph.D., is an Associate Professor of Biochemistry at the Faculty of Medicine, Technion-Israel Institute of Technology in Haifa, Israel.

He was graduated from Hebrew University in Jerusalem with a B.Sc. in Microbiology in 1964, and completed his M.Sc. in 1967 and his Ph.D. in 1971 at the same university.

Between 1971 and 1973, Dr. Fry was a Fellow at the Roche Institute of Molecular Biology in Nutley, New Jersey; since 1973, he has been on the Faculty of the Technion School of Medicine.

Dr. Fry has spent two Sabbaticals (1970—80 and 1884—85) and several months most summers at the University of Washington in Seattle—an arrangement warmly anticipated by his collegues in this country.

He has published numerous papers on DNA polymerases from animal cells and on the molecular aspects of cellular aging, with particular attention to the machinery of DNA replication.

Dr. Fry is married to Iris, a student of philosophy, whose wisdom tempers his creativity; they have two daughters.

Lawrence A. Loeb, M.D., Ph.D., is Director of the Joseph Gottstein Memorial Cancer Research Laboratory, Professor of Pathology and Adjunct Professor of Biochemistry at the University of Washington in Seattle, Washington.

Dr. Loeb was graduated from New York University Medical School in 1961, and completed a Ph.D. at University of California, Berkeley, in 1967. He did an internship in Medicine at Stanford, and postdoctoral work at the National Cancer Institute.

From 1967 to 1978, Dr. Loeb was at the Institute for Cancer Research in Philadelphia; during this time, he taught at the University of Pennsylvania.

As a sought-after lecturer, Dr. Loeb has given seminars and lectures in colleges, universities and research institutes throughout the world. He has served on study sections for both the National Cancer Institute and the American Cancer Society. He has received many honors and awards, among them an Outstanding Investigator Award from the National Cancer Association. He is particularly proud of his role in writing the position paper on smoking and lung cancer for the American Association for Cancer Research.

Among Dr. Loeb's more than 200 publications are evidences of the breadth of and variety among his research interests. His major research interests include DNA replication in mammalian cells; molecular mechanisms for mutagenesis and carcinogenesis; the fidelity of DNA replication; and environmental carcinogenesis.

Dr. Loeb is married to Phyllis, a teacher, who facilitates his research in numerous small but important ways; they have three children.

In memory of Asaf Fry, whose music and science have prematurely come to an end.

TABLE OF CONTENTS

Chapter 1

GENERAL INTRODUCTION

Since the first description of a DNA polymerase in an animal cell 25 years ago,[1] an immense body of information has been accumulated on eukaryotic DNA polymerases, their classification into distinct categories, prevalence, evolution, physical and catalytic properties, and roles in DNA metabolism in vivo. The substantial number of published reviews[2-16] and collected papers[17-20] which were dedicated to eukaryotic DNA polymerases bear witness to the dynamic state of research conducted on these cardinal enzymes. The rapid pace of progress in the field of DNA polymerases of animal cells has recently accelerated. The development of an arsenal of selective inhibitors of these enzymes, the detailed analysis of DNA primase, the introduction of monoclonal antibodies against DNA polymerases, and the isolation of animal cell mutants with modified DNA polymerases are of particular importance in increasing the understanding of the enzymology and biology of DNA synthesizing enzymes in eukaryotes. Furthermore, initial works show promise of the application, in the near future, of recombinant DNA techniques to the study of animal cell DNA polymerases.

The growing body of knowledge on eukaryotic DNA polymerases and the increasing interest in these enzymes appear to warrant an attempt to summarize currently available data on polymerases and to indicate possible future directions of their study. DNA replication, repair, and recombination in animal cells are highly complex processes that involve a presently inestimable number of factors. Furthermore, variations in DNA transactions appear to exist among different types of eukaryotic cells and in cells infected with various viruses. The discourse in this review is limited by necessity, however, to a narrow aspect of DNA metabolism in animal cells. First, the discussion in this book is restricted to DNA polymerases and to cofactors that appear to interact directly with them. Hence, the plethora of enzymes and factors that are involved in DNA replication and repair is not given full exposure in this work. Second, we confined our presentation to DNA polymerases of animal cells, whereas polymerases of plant cells, bacteria, viruses, and lower eukaryotes are mentioned in selected sections throughout the book merely for comparative purposes and to delineate unique aspects of animal cell polymerases. Third, in consideration of the large number of exhaustive reviews on eukaryotic DNA polymerases, which appeared between 1974 and 1984,[2-16] in this book we mainly focus on results that were obtained in the past 5 years. Regrettably, the laudable earlier efforts by pioneering authors may not be presented in full in this review. It is indeed unfortunate that, in delineating a history of concepts of biochemistry, individual findings often become a buried archeology.

In compiling the present work, we have attempted to address two distinct audiences of researchers who have interest in DNA metabolism of higher eukaryotes. First, the book is directed to investigators who are actively engaged in the study of DNA replication and repair. For these, we have undertaken to extensively summarize and synthesize recent developments in the literature and have cataloged a large number of diverse observations. A second audience, which we believe to be a rapidly growing one, is composed of advanced graduate students who have chosen DNA replication in eukaryotes as a field of future research. For these, we have analyzed current concepts and insights, made presumptive judgments, and projected future directions of study. We hope these projections will stimulate readers to challenge our interpretations and speculations.

Writing of the main body of this book was concluded in February 1985. Selected papers that were published between February and August 1985 are reviewed in special Update sections that were added in proof to some chapters.

Many colleagues have helped us in preparing this book, and it is a pleasure to acknowledge their contributions. We thank colleagues who communicated their results to us: Drs. R. A. Bambara, E. F. Baril, R. M. Benbow, U. Bertazzoni, F. J. Bollum, A. M. de Recondo, S. L. Dresler, N. B. Hecht, J. A. Huberman, U. Hübscher, J. Hurwitz, G. Kalf, D. Korn, R. Knippers, G. Krauss, M. Lieberman, A. Matsukage, M. R. Miller, K. Ono, A. B. Pardee, E. Rapaport, S. Spadari, S. H. Wilson, T. Yagura, and S. Yoshida. We are greatly indebted to Drs. R. A. Bambara, University of Rochester Medical Center, New York; K. P. Gopinathan, Indian Institute of Sciences, Bangalore, India; T. A. Kunkel, NIEHS, Research Triangle Park, N.C.; S. H. Wilson, NCI, NIH, Md.; J. Abbotts, P. K. Liu, G. M. Martin, R. J. Monnat, T. H. Norwood, M. Reyland, and E. Snow, University of Washington in Seattle for reading parts of the book and making critical comments. Inaccuracies and ambiguities in the text remain, however, ours alone. One of us (M. Fry), was on sabbatical leave from the Technion-Israel Institute of Technology at the time of the writing of this book. He wishes to express his deep gratitude to Professor George M. Martin from the University of Washington for his support and friendship. We thank Ms. Joan Hiltner, Ms. Mary Whiting, Ms. Carol Hansen, and Ms. Sharon West for diligently and patiently putting to type the large number of drafts of this book.

REFERENCES

1. Bollum. F. J., Calf thymus polymerase, *J. Biol. Chem.*, 235, 2399, 1960.
2. Loeb, L. A., Eukaryotic DNA polymerases, *Enzymes*, 10, 173, 1974.
3. Weissbach, A., Vertebrate DNA polymerases, *Cell*, 5, 101, 1975.
4. Holmes, A. M. and Johnston, I. R., DNA polymerases of eukaryotes, *FEBS Lett.*, 60, 233, 1975.
5. Bollum, F. J., Mammalian DNA polymerase, *Prog. Nucl. Acids Res. Mol. Biol.*, 15, 109, 1975.
6. Brun, G. and Chapeville, F., Multiplicity of animal cell deoxyribonucleic acid polymerase, *Biochem. Soc. Symp.*, 42, 1, 1977.
7. Weissbach, A., Eukaryotic DNA polymerases, *Annu. Rev. Biochem.*, 46, 25, 1977.
8. Falaschi, A. and Spadari, S., The three DNA polymerases of animal cells: properties and functions, in *DNA Synthesis: Present and Future*, Molineux, I. and Kohyama, M., Eds., Plenum Press, New York, 1978, 487.
9. Sarngadharan, M. G., Robert-Guroff, M., and Gallo, R. C., DNA polymerase of normal and neoplastic cells, *Biochim. Biophys. Acta.*, 516, 419, 1978.
10. Weissbach, A., The functional roles of mammalian DNA polymerases, *Arch. Biochem. Biophys.*, 198, 386, 1979.
11. DePamphilis, M. L. and Wasserman, P. M., Replication of eukaryotic chromosomes: a close-up of the replication fork, *Annu. Rev. Biochem.*, 49, 627, 1980.
12. Scovassi, A. I., Plevani, P., and Bertazzoni, U., Eukaryotic DNA polymerases, *Trends Biochem. Sci.*, 5, 335, 1980.
13. Fry, M., Eukaryotic DNA polymerase in *Enzymes of Nucleic Acid Synthesis and Modification*, Jacob. S. T., Ed., CRC Press, Boca Raton, Fla.,1983, 39.
14. Hübscher, U., DNA polymerases in prokaryotes and eukaryotes: mode of action and biological implications, *Experientia*, 39, 1, 1983.
15. Holmes, A. M., Bollum, F. J., and Chang, L. M. S., DNA polymerases of eukaryotes, *Isozymes*, 7, 277, 1983.
16. Hübscher, U., DNA polymerase holoenzyme, *Trends Biochem. Sci.*, 9, 30, 1984.
17. Molineux, I., and Kohyama, M., Eds., *DNA synthesis: Present and Future*, Plenum Press, New York, 1978.
18. Alberts, B., Ed., *Mechanistic Studies of DNA Replication and Genetic Recombination*, Academic Press, New York, 1980.
19. deRecondo, A.-M., Ed., *New Approaches to Eukaryotic DNA Replication*, Plenum Press, New York, 1983.
20. Hübscher, U. and Spadari, S., Eds., *Proteins Involved in DNA Replication*, Plenum Press, New York, 1985.

Chapter 2

MULTIPLICITY OF DNA POLYMERASES IN ANIMAL CELLS

Variety is the mother of enjoyment.

Benjamin Disraeli
Vivian Grey

I. BASIC ENZYMOLOGICAL CHARACTERISTICS OF DNA POLYMERASES

The enzymes responsible for the accurate synthesis of the genetic information in living cells are termed DNA-dependent DNA polymerases (DNA nucleotidyl transferases EC 2.7.7.7, herein designated DNA polymerases). These enzymes are ubiquitous in cells of all prokaryotes and eukaryotes and encoded by many of their respective viruses. The universal mechanism for catalysis by DNA polymerases was initially delineated by Kornberg et al.[1] in 1956. These enzymes are responsible for catalyzing the faithful incorporation into DNA of the four deoxyribonucleoside triphosphates, dATP, dCTP, dGTP, and dTTP. The order of incorporation is directed by a complementary DNA template strand according to the Watson-Crick base pairing rules (Figure 1).[1,2] Despite considerable variance in the structure of DNA polymerases from different sources, diverse association with auxiliary proteins, and dissimilitude in catalytic behavior and in vivo roles, generally DNA polymerases have the same basic catalytic characteristics:[1,2]

1. All DNA polymerases utilize deoxyribonucleoside triphosphates (dNTPs) for the polymerization of DNA by sequential incorporation of deoxyribonucleoside monophosphates (dNMPs) with cleavage of the α-β phosphodiester bond and release of pyrophosphate. The presence of a template DNA and metal ion activator are requisites for DNA polymerase-catalyzed DNA synthesis. The order of polymerization of the deoxynucleotide substrates in the product DNA is dictated by the base sequence of the template DNA. Thus, DNA polymerases and related polymerase are singular in that the specificity of catalysis is determined by another molecule.
2. All DNA polymerases require a short segment of DNA or RNA complementary to the template and having an available hydroxyl group at the third carbon position of the deoxyribose/ribose moiety. Such a nucleic acid segment acts as a primer, and polymerization of the next nucleotide residue proceeds from its 3′-hydroxyl position.
3. All DNA polymerases catalyze DNA synthesis exclusively in the 5′ → 3′ direction of the synthesized strand.

These catalytic requirements of DNA polymerase constitute the basic theoretical and experimental framework for their structural, enzymological, and functional analysis. The participation of DNA polymerases in replicative and repair synthesis of DNA in vivo must be understood in the general context of these processes. The description of general mechanisms for DNA replication and repair in animal cells is beyond the scope of this book. Knowledge of basic rules of DNA synthesis such as semiconservative mode of replication, RNA priming, leading and lagging strand-synthesis, strand displacement synthesis, excision repair, etc. are prerequisites for understanding various parts of this book. For a masterful survey of the field of DNA replication, the reader is referred to the recent book by Kornberg.[2]

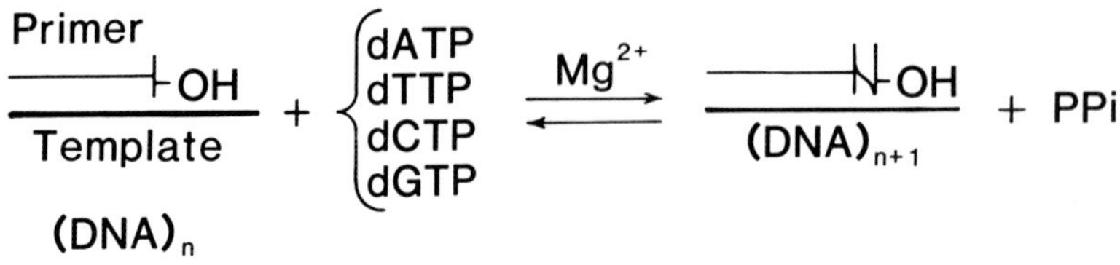

FIGURE 1. Outline of the basic mechanism of catalysis of DNA synthesis by a DNA polymerase. Details of DNA polymerase-catalyzed synthesis of DNA are described in the text. Magnesium, which is commonly the preferred metal activator of DNA polymerases, is marked in the scheme as a required component of the DNA polymerization reaction.

II. MULTIPLICITY OF DNA POLYMERASES IN ANIMAL CELLS—A BRIEF HISTORICAL PERSPECTIVE

Almost 30 years ago, in 1956, Kornberg and associates identified in *Escherichia coli* the first known DNA polymerase, later designated *E. coli* DNA polymerase I. Shortly thereafter, Bollum and Potter[3-5] demonstrated a similar catalytic activity in a eukaryotic cell. This animal cell enzyme was later classified DNA polymerase-α. In the ensuing decade, DNA polymerase activities were identified, partially purified and characterized in diverse organisms, tissues, and cells. In animal tissues the recovered activity of DNA polymerase paralleled the proliferative activity of the cells.[6] For this reason, initial efforts at polymerase purification were carried out using rapidly dividing cells yielding one predominant DNA polymerase activity.

The first extensive purification of an animal cell DNA polymerase was conducted by Yoneda and Bollum,[7] who isolated from calf thymus a high molecular weight enzyme that sedimented at ∼12 S. The requirements for catalysis of DNA polymerization of the eukaryotic enzyme were similar to those of bacterial DNA polymerase. However, two unexpected properties of mammalian DNA polymerases were noted: the major proportion of the cellular content of that enzyme was recovered in the cytoplasm,[4] and it preferentially utilized denatured DNA as a template.[7,8] In contrast, DNA polymerase purified shortly thereafter from rapidly dividing sea urchin embryos[9] was mostly localized in the cell nucleus and utilized double-stranded DNA template more efficiently than denatured DNA.[10,11] Intracellular localization and template preference of the major animal cell DNA polymerase activity were further investigated and revised in subsequent studies (cf. Table 1). That the initially described DNA polymerase activity was associated with replication of genomic DNA was clearly gleaned from studies that showed polymerase activity increased dramatically as cell division was induced in populations of quiescent cells. Human peripheral blood lymphocytes are nonproliferative cells that can be stimulated to undergo division in culture by exposure to phytohemagglutinin, a mitogenic plant lectin. Sixteen hours after stimulation, slightly before and immediately parallel with increase in the incorporation of labeled thymidine into DNA, there is an increase in the activity of DNA polymerase.[11,12] These observations are depicted in Figure 2.

For quite some time it was assumed that only a single type of DNA polymerase was present in living cells. Hence, *E. coli* DNA polymerase I was the enzyme deemed to be responsible for replication of the bacterial genomic DNA. In fact, it was even possible to demonstrate that DNA polymerase I could synthesize biologically active DNA molecules in the test tube.[13,14] Yet the catalytic properties of DNA polymerase I appeared to be inadequate for a replicative DNA polymerase.[2] A genetic approach was applied, therefore, to investigate the possibility that replication of genomic DNA is conducted in *E. coli* by an enzyme other than DNA polymerase I. Since properties of a mutant

Table 1
MAJOR DISTINGUISHING PROPERTIES OF ANIMAL CELL DNA POLYMERASES-α, -β, and -γ[a]

Property	DNA polymerase -α	-β	-γ
Cellular			
Relative activity (dividing cells)	70—90%	10—15%	1—10%
Intracellular location	Nucleus	Nucleus	Mitochondria/nucleus
Change in activity during cell cycling	Yes	No	Minimal
Physical			
Native size (kdaltons)[b]	200—600 Multimeric	40 Monomeric	180—315 Multimeric
Catalytic polypeptide[b]	150—190	40	47
Associated exonuclease[c]	No	No	No
Monospecific antibodies	Yes	Yes	Yes
Catalytic			
Preferred primer-template	Gapped DNA	Gapped DNA	Oligo(dT) · poly(rA)
Extension of RNA primer[d]	Yes	No	No
Preferred metal activator	Mg^{2+}	Mg^{2+}/Mn^{2+}	Mn^{2+}
Km for dNTPs (μM)[e]	1—15	5—30	0.2—1.0
Salt ($\geq$100 mM)	Inhibition	Stimulation	Stimulation
Processivity	Quasiprocessive	Distributive/slightly processive	Highly processive
Fidelity (nucleotide)	~1/40,000	~1/4,000	~1/8,000
Major inhibitors			
Aphidicolin (10—20 μg/mℓ)	Inhibition	No effect	No effect
N-Ethylmaleimide (0.5—10 mM)	Inhibition	No effect/slight inhibition	Inhibition
d_2TTP (d_2TTP/dTTP = 0.5—5.0)	No effect	Inhibition	Inhibition
Arylamino nucleotides	Inhibition	No effect	No effect
Phosphonacetate (10—100 μg/mℓ)	Inhibition	No effect	Inhibition
Arabinosyl-CTP	Inhibition	Lower inhibition	Lower inhibition
Inorganic phosphate (2—50 mM)	Limited inhibition	Inhibition	Stimulation
Novobiocin (0.2—0.6 mg/mℓ)	Inhibition	No effect	Partial inhibition

[a] Detailed references to experimental data on the various properties of DNA polymerases-α, -β, and -γ are given in Chapters 3, 4, and 5, respectively.

[b] Size of native DNA polymerase-γ and its constituent subunits were determined only for chick and mouse cells. Size of polymerases-α and -β and their respective catalytic polypeptides were obtained from a large number of different cell types.

[c] In an isolated case, the association of mouse-cell polymerase-α with a bidirectional $3' \rightarrow 5'$ and $5' \rightarrow 3'$ exonuclease has been reported (Chapter 3, Section IV.A). In all other tested cases, purified DNA polymerases-α, -β, and -γ were devoid of detectable nuclease activity. Analysis of polymerase-γ is at present limited to a few studies.

Table 1 (continued)
MAJOR DISTINGUISHING PROPERTIES OF ANIMAL CELL DNA POLYMERASES-α, -β, and -γ[a]

[d] Indicated is the ability of DNA polymerases to extend a natural RNA primer. Although DNA polymerase-γ is able to extend synthetic ribohomopolymeric primers, it appears not to be able to utilize natural RNA primers (Chapter 5, Section I).

[e] Km values for dNTPs vary for the various dNTPs and depend on the nature of the primer-template. Shown here is the range of Km values obtained for DNA polymerases for different sources with optimally gapped DNA and usually in experiments in which concentrations of all four dNTPs varied together.

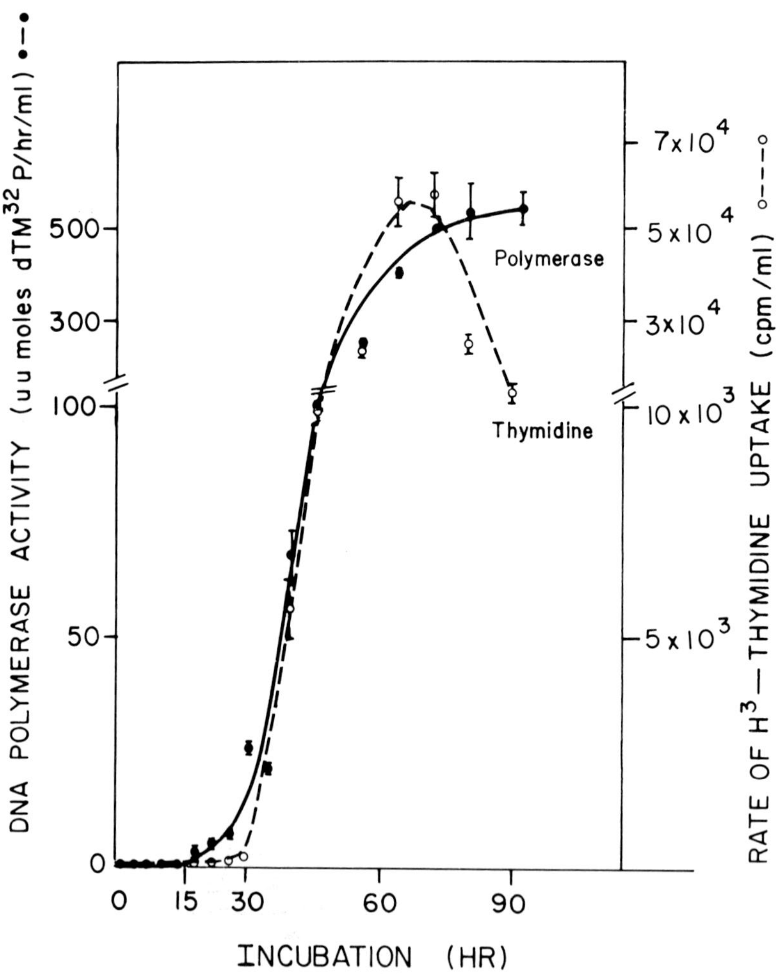

FIGURE 2. Relationship between DNA polymerase activity and rate of thymidine incorporation into DNA following phytohemagglutinin stimulation of human lymphocyte proliferation.

lacking DNA polymerase I activity could not be predicted, a large number of heavily mutagenized *E. coli* clones were quantified for cellular activity of DNA polymerase I by De Lucia and Cairns.[15] After recovering normal levels of polymerase I in 3747 clones of bacteria exposed to mutagen, the 3748th clone, designated pol A1, was found to contain 1% of the level of DNA polymerase I activity of the parental strain, yet it was viable and was able to replicate its DNA. Recovering polymerase I activity at an expected dilution in mixtures between extracts of pol A1 and parental cell strongly suggested that the mutant did not contain an inhibitor to the polymerase.[15] These results established that *E. coli* polymerase I was not absolutely required for DNA replication and provided a source for the susequent isolation of distinct *E. coli* DNA polymerase II,[16-18] and somewhat later, DNA polymerase III.[19,20] The three types of bacterial DNA polymerase differ in size, catalytic properties determined in vitro, and associated proteins.[2,20,21] Furthermore, different in vivo roles were later assigned to the three classes of *E. coli* DNA polymerases.[2,20,21]

In contrast to prokaryotes, the study of the enzymology of DNA synthesis in animal cells did not enjoy the advantage of detailed genetics, and in particular, the establishment of conditional mutants of the synthetic process. Distinguishing among different animal cell polymerases and their associated proteins had to rely completely on biochemical approaches.

About a decade after the initial isolation of DNA polymerase-α from calf thymus,[5] a distinctly different polymerase, later termed DNA polymerase-β, was isolated from several cell types.[22-24] DNA polymerase-β was distinguished from polymerase-α by virtue of its uniquely smaller molecular size, basic isoelectric point, and resistance to chemicals that react with the thiol group.[22,24,25] Shortly after polymerase-β was identified, a third type of animal cell polymerase, which was later designated DNA polymerase-γ, was isolated.[26,27] Polymerase-γ differed from α- and β-polymerases by its synthetic template requirements, preference for Mn^{2+} as a metal activator, and requirement for salt for maximum DNA synthesis in vitro, as well as its sensitivity to agents that react with protein sulfhydryl groups.[25-27]

Following their discovery, animal cell DNA polymerases of the three types were identified in a wide variety of organisms, tissues, and cells. In 1975, Weissbach et al.[25] proposed a set of basic defining criteria for the classification of known animal cell DNA polymerases under three main groups: DNA polymerases of the α, β, and γ types. At the time, a fourth category of a distinct mitochondrial enzyme was also included, but shortly afterwards the mitochondrial polymerase was shown to be identical with DNA polymerase-γ.[28,29] Although in the ensuing years some of the original parameters for the classification of eukaryotic DNA polymerases were modified and refined, the categorization of polymerases into the α, β, and γ types remains the basic working framework in this field. A property shared by all three classes of animal cell DNA polymerases is the lack of associated exonuclease activity, which is commonly found in many prokaryotic and viral DNA polymerases.[2,21,30] In 1976, an apparent fourth type of DNA polymerase that was intimately associated with $3' \rightarrow 5'$ deoxyribonuclease activity was isolated from rabbit bone marrow.[31] That activity, denoted polymerase-δ, has since been identified also in calf thymus.[32] However, it remains to be seen whether δ-type polymerase is uniquely present in selected tissues or cells, or whether it is as omnipresent as are polymerases of the α, β, and γ classes.

Following the classification of animal cell polymerases, much progress has been made in the elucidation of their structure, association with auxiliary proteins and cofactors, catalytic activity, biology, and roles in vivo. In particular, several recent methodological developments have accelerated research on important aspects of the enzymology of DNA synthesis in animal cells. First, monoclonal and polyclonal monospecific antibodies directed against DNA polymerases have become available in recent years. Monospecific antibodies have been successfully used to study the evolution of DNA polymerases-α and -β.[34-36] Monoclonal antibodies are now being used for the purification of intact molecules of polymerase-α and for the study of their intracellular localization,[34-36] subunit structure,[37-41] and association with DNA primase, an enzyme that catalyzes the synthesis of RNA primers.[38] Moreover, monospecific antibodies are now being used in studies on the biosynthesis of animal DNA polymerases and in procedures for the isolation of polymerase mRNA.[42-44] A second important recent development is the emergence of animal cell mutants that contain a mutated DNA polymerase-α.[45,46] Work with a mammalian cell mutant[46] has evolved and its DNA has recently been used for the successful transfection of a polymerase-α gene.[47] Future use of such mutants as well as utililization of the recently isolated mRNA for DNA polymerases-α and -β[43,44] may open inroads for the study of animal cell DNA polymerases by recombinant DNA technology.

III. DEFINING CRITERIA FOR CLASSIFICATION OF ANIMAL CELL DNA POLYMERASES

Experimental data accumulated in many laboratories in the last decade provide solid criteria for the rigid classification of animal cell DNA polymerases and for their distinction from prokaryotic and viral polymerases. Animal cell DNA polymerases are distinguished by cellular, physical, and enzymological parameters, as well as by their different response to selective inhibitors. Table 1 summarizes the major distinguishing features of the three principal classes of DNA animal cell polymerases. The outlined data establish a clear distinction between polymerases-α, -β, and -γ by a set of defined criteria. Unique antigenicity, typical size and subunit structure, as well as distinctive response to a wide selection of inhibitors, argue for the unequivocal singularity of each class of polymerase and for the lack of relationships between polymerases of different types. A word of reservation, however, is appropriate in regard to data summarized in Table 1. The properties of different polymerase types have been described by numerous authors for enzyme isolated from a wide variety of organisms, tissues, and cells using different conditions for enzyme isolation and detection. Naturally, therefore, a rather broad range of quantitative values for some of the parameters appear in the literature. Table 1 includes consensual criteria, and in some cases the range of quantitative values are indicated. Nevertheless, many of the parameters vary enough for the different polymerases to provide criteria for quantitative evaluation of each polymerase type in crude extracts that include more than one class of polymerase. A detailed discussion of the experimental evidence for each of the properties is presented in relevant chapters of this book.

The fact that each of the eukaryotic DNA polymerases represents a separate entity has been established most definitely by the extensive purification of the enzymes from their mixture in crude cell extracts. An example of such a purification protocol is outlined in Scheme 1. From 500 g of chick embryo tissue, Matsukage and colleagues[48] purified all three DNA polymerases, relying heavily on separation by column chromatography. This procedure is somewhat atypical, first because initial separation of DNA polymerase-β from -α is commonly achieved by chromatography on DEAE-cellulose, and second, because the three classes of polymerase were obtained by a single protocol. With the development of monoclonal antibodies, it appears likely that purification will be greatly simplified and expedited, as indeed was recently demonstrated by several groups.[38-41]

IV. ROLES OF DNA POLYMERASES-α, -β AND -γ IN DNA REPLICATION AND REPAIR

Soon after the multiplicity of types of animal cell DNA polymerases was recognized, an attempt to assign specific roles for each class of polymerase in cellular DNA metabolism commenced. In bacteria, conditional mutants have traditionally provided the most unambiguous assignment of function to enzyme. In eukaryotes, however, mutants in DNA synthetic pathways have only recently become available, and the deter-

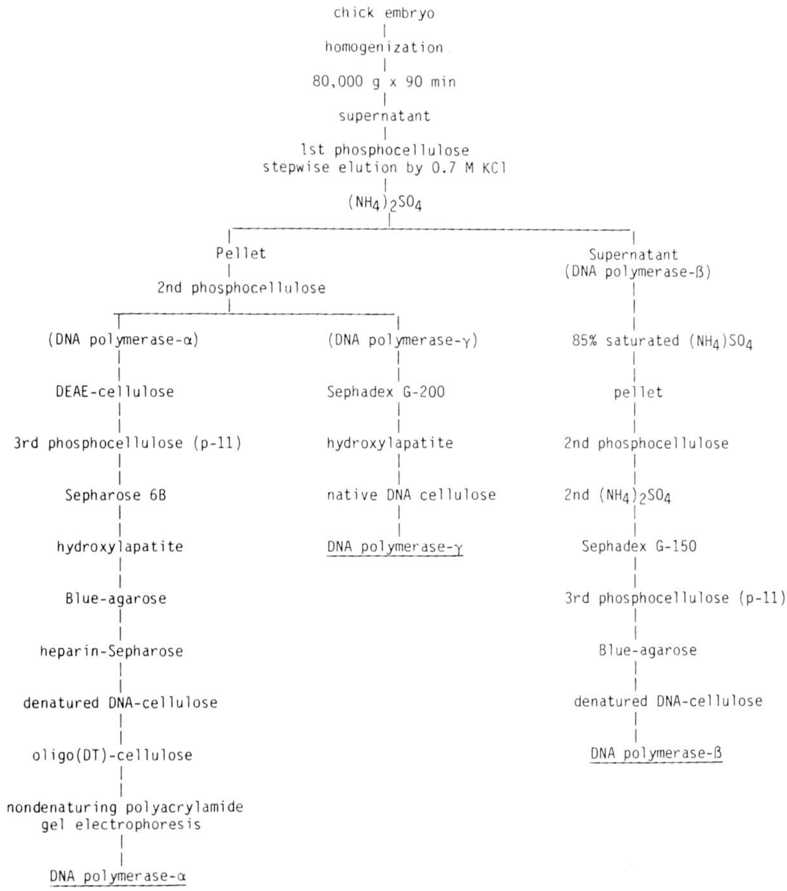

SCHEME 1. Procedure for the systematic fractionation and purification of chick embryo DNA polymerases-α, -β, and -γ. (Adapted from Matsukage, A., Yamaguchi, M., Tanabe, K., Nishizawa, M., and Takahashi, T., *New Approaches in Eukaryotic DNA Replication,* de Recondo, A.-M., Ed., Plenum Press, New York, 1983, 81. With permission.)

mination of in vivo roles for polymerases had to be based on correlative studies. Hence intracellular functions of DNA polymerases-α, -β, and -γ, as well as their subcellular location, were inferred by correlating their studies with replication of nuclear and mitochondrial DNA, and with DNA repair synthesis.

The development of an arsenal of selective inhibitors allowed correlation of the arrest of activity of any single DNA polymerase with inhibition of DNA replication and repair synthesis in vivo. However, recent developments have opened exciting inroads for studying the function of DNA polymerases in cellular metabolism of DNA. First, the availability of mutant cells that contain a mutated DNA polymerase-α allows one to correlate the mutated phenotype and the defect in DNA polymerase activity.[46,49] The application of monospecific antibodies to polymerases in subcellular systems that perform DNA replication and repair synthesis can be used to monitor the role of these enzymes in DNA transactions.[50] The construction of in vitro systems, which perform repair of defined damage to DNA by the concerted action of several enzymes, including DNA polymerases-α or -β, may also provide a more precise definition of the role of each enzyme in DNA repair.[51-54] Such studies are extensions of previous observations on unique catalytic properties of DNA polymerases, which alluded to their possible function in repair synthesis of DNA.

Evidence collected by many laboratories led to our present partial understanding of the in vivo function of DNA polymerases-α, -β, and -γ. A considerable body of mostly correlative data suggests that DNA polymerase-α is responsible for the replication of nuclear DNA. It is not clear, however, what role, if any, may be played by DNA polymerases-β and -γ in the replication of genomic DNA. Growing evidence indicates that DNA polymerase-β participates in certain types of DNA repair synthesis. Yet, convincing data also show that polymerase-α also takes part in repair synthesis of at least some types of damage to DNA. It is completely obscure whether or not polymerase-γ has any role in the repair of damaged DNA. DNA polymerase-γ is confined in the cell to both mitochondria and nuclei. By virtue of its being the only DNA polymerase present in mitochondria, polymerase-γ is deemed responsible for the replication of mitochondrial DNA. It is a complete mystery at present what function is filled by the nuclear fraction of polymerase-γ.

The complex and still largely unexplored field of functional in vivo roles of DNA polymerases-α, -β and -γ is discussed at length in Chapters 6 and 7. It should be noted, however, that considering the complexity of DNA replication and DNA repair, and the likelihood of concerted participation of a yet undefined number of factors in these processes, the attempt to assign unique function to each polymerase type could be premature and perhaps naive. Until a better understanding of the intricacies of DNA metabolism in animal cells is attained, the present assignment of roles to DNA polymerases must be regarded as tentative.

REFERENCES

1. Kornberg, A., Lehman, I. R., Bessman, M. J., and Simms, E. S., Enzymic synthesis of deoxyribonucleic acid, *Biochim. Biophys. Acta,* 21, 197, 1956.
2. Kornberg, A., *DNA Replications,* W. H. Freeman, San Francisco, 1980. (Supplement: 1982.)
3. Bollum, F. J. and Potter, V. R., Thymidine incorporation into deoxyribonucleic acid of rat liver homogenates, *J. Am. Chem. Soc.,* 79, 3603, 1957.
4. Bollum, F. J. and Potter, V. R., Incorporation of thymidine into deoxyribonucleic acid by enzymes from rat tissue, *J. Biol. Chem.,* 233, 478, 1958.
5. Bollum, F. J., Calf thymus polymerase, *J. Biol. Chem.,* 235, 2399, 1960.
6. Loeb, L. A., Eukaryotic DNA polymerases, *Enzymes,* 10, 173, 1974.
7. Yoneda, M. and Bollum, F. J., Deoxynucleotide-polymerizing enzymes of calif thymus gland. I. Large scale purification of terminal and replicative deoxynucleotidyl transferases, *J. Biol. Chem.,* 240, 3385, 1965.
8. Bollum, F. J., Thermal conversion of non-priming deoxyribonucleic acid to primer, *J. Biol. Chem.,* 234, 2733, 1959.
9. Loeb, L. A., Mazia, D., and Ruby, A. D., Priming of DNA polymerase in nuclei of sea urchin embryos by native DNA, *Proc. Natl. Acad. Sci. U.S.A.,* 57, 841, 1967.
10. Loeb, L. A., Purification and properties of deoxyribonucleic acid polymerase from nuclei of sea urchin embryos, *J. Biol. Chem.,* 244, 1672, 1969.
11. Loeb, L. A., Agarwal, S. S., and Woodside, A. M., Induction of DNA polymerase in human lymphocytes by phytohemagglutinin, *Proc. Natl. Acad. Sci. U.S.A.,* 61, 827, 1968.
12. Loeb, L. A. and Agarwal, S. S., DNA polymerase. Correlation with DNA replication during transformation of human lymphocytes, *Exp. Cell Res.,* 66, 299, 1971.
13. Goulian, M. and Kornberg, A., Enzymatic synthesis of DNA. XXIII. Synthesis of circular replicative form of phage ϕX174 DNA, *Proc. Natl. Acad. Sci. U.S.A.,* 58, 1723, 1967.
14. Goulian, M., Kornberg, A., and Sinsheimer, R. L., Enzymatic synthesis of DNA. XXIV. Synthesis of infectious phage ϕX174 DNA, *Proc. Natl. Acad. Sci. U.S.A.,* 58, 2321, 1967.
15. De Lucia, P. and Cairns, J., Isolation of an *E. coli* strain with mutation affecting DNA polymerase, *Nature (London),* 224, 1164, 1969.
16. Kornberg, T. and Gefter, M. L., DNA synthesis in cell-free extracts of a DNA polymerase-defective mutant, *Biochem. Biophys. Res. Commun.,* 40, 1348, 1970.

17. Moses, R. E. and Richardson, C. C., A new DNA polymerase activity of *Escherichia coli.* I. Purification and properties of the activity present in *E. coli* POL A1, *Biochem. Biophys. Res. Commun.,* 41, 1557, 1970.
18. Knippers, R., DNA polymerase II, *Nature (London),* 228, 1050, 1970.
19. Kornberg, T. and Gefter, M. L., Purification and DNA synthesis in cell-free extracts: properties of DNA polymerase II, *Proc. Natl. Acad. Sci. U.S.A.,* 68, 761, 1971.
20. Nusslein, V., Otto, B., Bonhoeffer, F., and Schalber, H., Function of DNA polymerase III in DNA replication, *Nature (London) New Biol.,* 234, 285, 1971.
21. Hübscher, U., DNA polymerases in prokaryotes and eukaryotes: mode of action and biological implications, *Experientia,* 39, 1, 1983.
22. Weissbach, A., Schlabach, A., Fridlender, B., and Bolden, A., DNA polymerases from human cells, *Nature (London) New Biol.,* 231, 167, 1977.
23. Baril, E. F., Brown, O. E., Jenkins, M. D., and Laszlo, J., Deoxyribonucleic acid polymerase with rat liver ribosomes and smooth membranes. Purification and properties of the enzymes, *Biochemistry,* 10, 1981, 1971.
24. Chang, L. M. S. and Bollum, F. J., Low molecular weight deoxyribonucleic acid polymerases in mammalian cells, *J. Biol. Chem.,* 246, 5835, 1971.
25. Weissbach, A., Baltimore, D., Bollum, F. J., Gallo, R. C., and Korn, D., Nomenclature of eukaryotic DNA polymerases, *Science,* 190, 401, 1975.
26. Fridlender, B., Fry, M., Bolden, A., and Weissbach, A., A new synthetic RNA-dependent DNA polymerase from human tissue culture cells, *Proc. Natl. Acad. Sci. U.S.A.,* 69, 452, 1972.
27. Bolden, A., Fry, M., Muller, R., Citarella, R., and Weissbach, A., The presence of a polyriboadenylic acid-dependent DNA polymerase in eukaryotic cells, *Arch. Biochem. Biophys.,* 153, 26, 1972.
28. Bolden, A., Pedrali-Noy, G., and Weissbach, A., DNA polymerase of mitochondria is γ polymerase, *J. Biol. Chem.,* 252, 3351, 1977.
29. Bertazzoni, U., Scovassi, A. I., and Brun, G. M., Chick-embryo DNA polymerase γ. Identity of γ polymerases purified from nuclei and mitochondria, *Eur. J. Biochem.,* 81, 237, 1977.
30. Fry, M., Eukaryotic DNA Polymerase, in *Enzymes of Nucleic Acid Synthesis and Modification,* Jacob, S. T., Ed., CRC Press, Boca Raton, Fla., 1983, 39.
31. Byrnes, J. J., Downey, K. M., Black, V. L., and So, A. G., A new mammalian DNA polymerase with 3′ to 5′ exonuclease activity: DNA polymerase δ, *Biochemistry,* 15, 2817, 1976.
32. Lee, M. Y. W. T., Tan, C.-K., Downey, K. M., and So, A. G., Further studies on calf thymus DNA polymerase-δ purified to homogeneity by a new procedure, *Biochemistry,* 23, 1906, 1984.
33. Chang, L. M. S., Plevani, P., and Bollum, F. J., Evolutionary conservation of DNA polymerase structure, *Proc. Natl. Acad. Sci. U.S.A.,* 79, 758, 1982.
34. Bensch, K. G., Tanaka, S., Hu, S.-Z., Wang, T. S.-F., and Korn, D., Intracellular localization of human DNA polymerase-δ with monoclonal antibodies, *J. Biol. Chem.,* 257, 8391, 1982.
35. Matsukage, A., Yamamoto, S., Yamaguchi, M., Kusakabe, M., and Takahashi, T., Immunochemical localization of chick DNA polymerases α and β, *J. Cell Physiol.,* 117, 266, 1983.
36. Nakamura, H., Morita, T., Masaki, S., and Yoshida, S., Intracellular localization and metabolism of DNA polymerase α in human cells visualized with monoclonal antibodies, *Exp. Cell Res.,* 151, 123, 1984.
37. Sauer, B. and Lehman, I. R., Immunological comparison of purified DNA polymerase α from embryos of *Drosophila melanogaster* with forms of the enzyme present *in vivo, J. Biol. Chem.,* 257, 12394, 1982.
38. Wang, T. S.-F., Hu, S.-Z., and Korn, D., DNA primase from KB cells. Characterization of a primase activity tightly associated with immunoaffinity-purified polymerase-α, *J. Biol. Chem.,* 259, 1854, 1984.
39. Wahl, A. F., Kowalski, S. P., Harwell, L. W., Lord, E. M., and Bambara, R. A., Immunoaffinity purification and properties of a high molecular weight DNA α polymerase, *Biochemistry,* 23, 1895, 1984.
40. Masaki, S., Tanabe, K., and Yoshida, S., Large polypeptides of 10S DNA polymerase α from calf thymus: rapid isolation using monoclonal antibody and tryptic peptide mapping analysis, *Nucl. Acids Res.,* 12, 4455, 1984.
41. Karawya, E., Swack, J., Albert, W., Fedorko, J., Minna, J. D., and Wilson, S. H., Identification of a higher Mr α-polymerase catalytic polypeptide in monkey cells by monoclonal antibody, *Proc. Natl. Acad. Sci. U.S.A.,* 81, 7777, 1984.
42. Yamaguchi, M., Matsukage, A., Takahashi, T., and Takahashi, T., Biosynthesis of DNA polymerase β in chick embryonic cells, *J. Biol. Chem.,* 257, 3932, 1982.
43. Yamaguchi, M., Takahashi, T., Yasuda, K., Shimura, Y., and Matsukage, A., Characterization of messenger RNA for chick-embryo DNA polymerase β and its translation product *in vitro, Eur. J. Biochem.,* 133, 277, 1983.

44. Detera-Wadleigh, S., Karawya, E., and Wilson, S. H., Synthesis of DNA polymerase by *in vitro* translation of calf RNA, *Biochem. Biophys. Res. Commun.*, 122, 420, 1984.
45. Sugino, A. and Nakaymara, K., DNA polymerase-α mutants from *Drosophila melanogaster* cell line, *Proc. Natl. Acad. Sci. U.S.A.*, 77, 7049, 1980.
46. Liu, P. K., Chang, C.-C., Trosko, J. E., Dube, D. K., Martin, G. M., and Loeb, L. A., Mammalian mutator mutant with an aphidicolin-resistant DNA polymerase α, *Proc. Natl. Acad. Sci. U.S.A.*, 80, 797, 1983.
47. Liu, P. K. and Loeb, L. A., Transfection of the DNA polymerase-α gene, *Science*, 226, 833, 1984.
48. Matsukage, A., Yamaguchi, M., Tanabe, K., Taguchi, Y. N., Nishizawa, M., and Takahashi, T., DNA chain elongation mechanism of DNA polymerase-α, β and γ, in *New Approaches in Eukaryotic DNA Replication*, de Recondo, A.-M., Ed., Plenum Press, New York, 1983, 81.
49. Liu, P. K., Chang, C.-C., and Trosko, J. E., Evidence for mutagenic repair in V79 cell mutant with aphidicolin-resistant DNA polymerase α, *Somat. Cell Mol. Genet.*, 10, 235, 1984.
50. Miller, M. R., Ulrich, R. G., Wang, T. S-F., and Korn, D., Monoclonal antibodies against human DNA polymerase-α inhibit DNA replication in permeabilized human cells, *J. Biol. Chem.*, 260, 134, 1985.
51. Bose, K., Karran, P., and Strauss, B., Repair of depurinated DNA *in vitro* by enzymes purified from human lymphoblasts, *Proc. Natl. Acad. Sci. U.S.A.*, 75, 794, 1978.
52. Mosbaugh, D. W. and Linn, S., Excision repair and DNA synthesis with a combination of HeLa DNA polymerase β and DNase V, *J. Biol. Chem.*, 258, 108, 1983.
53. Mosbaugh, D. W. and Linn, S., Gap filling DNA synthesis by HeLa DNA polymerase α in an *in vitro* base excision DNA repair scheme, *J. Biol. Chem.*, 259, 10247, 1984.
54. Evans, D. H. and Linn, S., Excision repair of pyrimidine dimers from simian virus 40 minichromosomes *in vitro*, *J. Biol. Chem.*, 259, 10252, 1984.

Chapter 3

DNA POLYMERASE-α

I. INTRODUCTORY REMARKS

DNA polymerase of the α category was the first eukaryotic DNA polymerase to be isolated and characterized by Bollum in 1960.[1] Subsequently, polymerases of the α class were found to be ubiquitous in proliferating cells of numerous eukaryotes, including insects (e.g., *Drosophila melanogaster*,[2-4] echinodermata (e.g., sea urchin),[5-8] amphibia (e.g., *Xenopus laevis*,[9-11] birds (e.g., chick embryo,[12-14] and mammals.[15-20] Furthermore, DNA polymerases analogous to the α type were identified in numerous unicellular lower eukaryotes such as *Ustilago maydis*,[21] *Euglena gracilis*,[22] *Saccharomyces cervisiae*,[23-25] *Tetrahymena thermophila*,[26] *T. pyriformis*,[27,28] and *Dictyostelium discoideum*[29] to name but a few. (For a review on selected aspects of DNA polymerases from lower eukaryotes, see Reference 30 and Chapter 9, Section I.)

Consequent with the identification and isolation of polymerase-α from a large variety of organisms, tissues, and cell types, evidence has accumulated indicating that polymerases of the α class play a central role in the replication of genomic DNA. The attention of a large number of investigators has thus focused on polymerase-α. Considerable effort has been invested in the purification of α-polymerase, characterization of its structural and catalytic properties, and the study of its activity and functions in vivo. The establishment of basic defining criteria for polymerase-α such as molecular size (>100 kdaltons), acidic isoelectric point, and sensitivity to salt, (>50 m*M*), inhibition by chemicals that react irreversibly with sulfhydryl groups, and by several other selective inhibitors,[31-37] has been instrumental in the earlier phases of the characterization of the α class of DNA polymerases.

However, a precise definition of the molecular structure of polymerase-α proves to be a surprisingly difficult and elusive task. Perhaps the major single barrier standing in the way of understanding the structure and function of polymerase-α is the difficulty encountered in attempts to purify the enzyme in its native form. First, α-polymerases display a high degree of heterogeneity among different species, and more importantly, they frequently appear in a number of distinct molecular forms within a single cell type.[37,38] Purification of α-polymerase is further impeded by its relatively low amount, even within dividing cells, and by its instability under conditions of conventional cell fractionation and enzyme purification. Polymerase-α appears to be exceptionally sensitive to proteolytic cleavage. The unexpected fact that cleaved polymerase fragments are still catalytically active has proven to be a major obfuscation in defining the size of the catalytic "core" subunit.[4,39,40]

In the face of these difficulties, and with admirable effort and ingenuity, polymerase-α was isolated from several cell types and purified to near homogeneity by a number of research groups. It is now recognized that α-polymerase is composed of several polypeptide subunits and that it may further be associated with additional auxiliary proteins. However, the precise number, size, and function of the various subunits of the polymerase molecule are not completely known at present and are still under debate. It is moot yet whether polymerase-α "holoenzyme", in which all the main functions required for deoxynucleotide selection and polymerization in vivo are included, is already in the researcher's hand. Furthermore, since only a few defined mutations in α-polymerase have been described,[41,42] most of the accumulated knowledge on the enzymic activity of polymerase-α was based on the development of functional in vitro assays for polymerase and activities associated with it. In vitro assays for DNA polymerization enabled the detailed description of the catalytic properties of the DNA polymerizing polypeptide in α-polymerase, i.e., the "catalytic subunit".[14,43,44] Similarly, a specific assay for the production of an RNA primer allowed the definition and characterizations of a DNA primase activity, which is intimately associated with α-polymerase.[45-49] The structure and function of polymerase-α, its subunits and auxiliary proteins, are actively investigated by the development of new functional assays,[50-53] immunochemical techniques,[55-60] selection for specific mutations in DNA replication proteins,[41,42] and by the application of DNA transfection and molecular cloning methodology.[59,61]

In this chapter we shall survey emerging current concepts on the structure of α-polymerase and its accessory proteins. The considerable body of data on the catalytic properties of polymerase-α will be reviewed, and the various selective inhibitors that distinguish polymerase-α from other polymerase classes will be described. The functional in vivo roles of DNA polymerase-α and its biology will be considered in Chapters 6, 7, and 9.

II. STRUCTURE OF POLYMERASE-α

Decay is inherent in all component things. Work your salvation with diligence.

The Pali Canon circa 500—250 B.C.
Mahaparinibbana-Sutta 6:10

A. Enzyme Purification Methods

Until the recent introduction of immunochemical isolation techniques, the structure of polymerase-α was deduced from analysis of purified enzyme preparations that were obtained by conventional techniques of protein purification. Between the mid-1970s and the beginning of the next decade, several research groups employing laborious and prolonged purification procedures obtained preparations of α-polymerase of near or apparent homogeneity.[38,62-70] As a source for polymerase-α, various research groups used proliferating tissues such as calf thymus,[38,63,64] regenerating liver,[65] bone marrow,[66] and embryonic tissues[4,67] or dividing cultured cells.[43,62,68-70] Implicit in this choice was the observation that augmented levels of polymerase-α are present in dividing cells and that the α enzyme may be involved in DNA replication (cf. Chapter 6). Polymerase-α was purified from various cellular extracts by steps of salting-out, ion-exchange and DNA cellulose column chromatographies, and gel filtration and velocity sedimentation (cf. Chapter 2). These laborious procedures yielded purified polymerase preparations at a low recovery, thus necessitating large initial quantities of cells. Early reports on the molecular size of native polymerase-α varied widely with values ranging between 70 and 1000 kdaltons.[32] Earlier procedures for polymerase-α purification did not include sufficient precautionary measures to minimize proteolysis.

In view of more recent experimental results, it can now be assumed that the lower range of molecular size (perhaps <200 kdaltons) was due to isolation of partially degraded yet catalytically active enzyme molecules.[71] The higher range of sizes obtained for α-polymerase (perhaps >500 kdaltons) may also be artifactual, resulting in part from aggregation of enzyme molecules in media of low ionic strength.[72,73] Furthermore, the presumed asymmetric nature of the polymerase-α molecule often led to discrepant apparent sizes deduced from gel filtration and from velocity sedimentation analysis.[4,18,65,73,74] Lastly, association of polymerase-α with auxiliary proteins, which are not formally subunits of the enzyme molecule, may also increase its observed size.[52,70]

Many of the shortcomings of the classical purification procedures are presently being circumvented with the advent of new techniques and approaches to the elucidation of the structure of polymerase-α. Foremost of these developments is the introduction of either monospecific polyclonal[40] or monoclonal[54-59] antibodies against α-polymerase and their implementation for enzyme purification. These immunoglobulins enabled the isolation of pure polymerase-α in procedures of very few or even a single step of either immunoprecipitation[57,59,60,75] or immunoaffinity column chromatography.[58,341] An additional technique which aided in the structural analysis of the size of the catalytic subunit of polymerases in general and α-polymerase in particular is the so-called "activity gel" electrophoresis.[39,60,76-78] This procedure allows for a direct detection of the catalytic polypeptide of a polymerase after it has been resolved by denaturing gel electrophoresis and subsequently renatured by leaching out the sodium dodecyl sulfate (SDS) from the gel and then assaying for DNA polymerization activity *in situ*.[76,77] Finally, an increasing concern about the detrimental effect of proteolysis during α-polymerase purification is leading to inclusion of a number of different protease inhibitors in each and every buffer used during purification.[4,58,64] The practice of these techniques yields a more uniform consensus on the size of the catalytic subunit of polymerase-α. However, the number and size of other subunits of α-polymerase is still uncertain.

B. Size of the Catalytic "Core" Subunit

Purification of α-polymerase has been conventionally monitored by following enzyme activity, i.e., its ability to catalyze the incorporation of radioactively labeled deoxyribonucleoside triphosphate into a newly made DNA product. Studies on extensively purified polymerase-α have yielded a plethora of divergent sizes for catalytically active polypeptides. Recent analyses of immunoaffinity purified α-polymerase increasingly indicate that it contains an exceptionally large-sized catalytic polypeptide of 180 to 200 kdaltons. Previously obtained smaller size enzymatically active polypeptides appear to be fragments generated by proteolytic cleavage. Yet it is possible that such fragments exist in vivo and that controlled proteolysis is a regulatory mechanism in the activation/inactivation of polymerase-α (cf. Chapter 10, Section III).

In the following pages we shall describe results on the subunit structure of highly purified polymerases of the α class, which were obtained from a variety of cell types. This section will be concluded by a summary of these results and an attempt to form a tentative general picture on the structure of this complex enzyme.

C. Calf Thymus DNA Polymerase-α

Due to its high content of proliferating cells and to its availability in large quantities, thymus tissue from calf has been the most widely utilized source for α-polymerase. The pioneering extensive purification of calf thymus DNA polymerase-α by Yoneda and Bollum[15] resulted in the isolation of a single catalytic entity with a sedimentation coefficient of approximately 12 S. The molecular heterogeneity of α-polymerase was most extensively documented in purification procedures employed by Holmes and collaborators.[17,38,74,79-84] These researchers purified five molecular forms of polymerase-α from thymus gland, each form having a different elution pattern on columns of DEAE cellulose and a distinct hydrodynamic behavior upon glycerol gradient centrifugation.[17,38]

The occurrence of multiple molecular forms of α-polymerase, which differ in their size and/or their chromatographic pattern, was extensively documented in a variety of cells and tissues.[2,3,13,17,19,43-45,69,87-92] Some of this microheterogeneity of polymerase-α is due to proteolytic cleavage of the enzyme during its isolation.[2-4,38-40,71,76] However, as will be discussed below, (Sections III and IV of this chapter) multiple forms of α-polymerase may have biological signficance. In the thymus, Holmes et al.[17,38] separated forms that were designated αA1 and αA2 (sedimenting at 8.0 to 8.4 S in glycerol gradients of 0.51), αB (5.2 S), αC (7.3 S), and αD (6.8 S). Each of these forms was purified

2000- to 3000-fold. Forms A1 and A2 have a similar native molecular size of 210 to 230 kdaltons and can be distinguished by their differing elution profiles on DEAE cellulose.[71,38] Denaturing gel electrophoresis reveals that partially purified form A1 consists of two major polypeptides of 150 to 170 kdaltons and 50 to 70 kdaltons.[38,79,80] Treatment of A1 with 2.4 *M* urea converts it to the C form, which contains only the 150- to 170-kdalton major polypeptide; a released basic polypeptide of 50 to 70 kdaltons can be discerned by gradient centrifugation.[38,79,83] It has been further demonstrated that generation of form C from form A1 occurred also in the presence of several protease inhibitors, suggesting that urea does not expose a hidden proteolytic cleavage in a high-molecular-weight precursor for both the 150- to 170-kdalton and the 50- to 70-kdalton polypeptides.[83] Furthermore, following removal of urea, an A type form of α-polymerase could be reconstituted from the C form and the basic 50- to 70-kdalton protein.[38,83] Since form C is an active DNA polymerase of the α type, it must be assumed that the 150- to 170-kdalton polypeptide is the catalytic subunit. However, it has been reported that forms A1 and C exhibited different template preferences[83,84] and possessed different processivity.[85] In addition, three forms of thymus α-polymerase differed radically in their sensitivity to the selective inhibitor butylanilinouracil,[86] and forms A1 and C exhibited difference in heat lability and in sensitivity to *N*-ethylmaleimide.[38,81] Hence, the 50- to 70-kdalton subunit, although devoid of enzymic activity by itself, does modify the overall properties of the polymerase.

Form B was found to contain a major polypeptide of about 103 kdaltons and several lines of evidence indicated that this form is an active fragment of α-polymerase generated by partial proteolysis.[82] The D form was found to contain a polypeptide of 140 kdaltons and to possess physical and catalytic properties that distinguish it from the A1 and C forms of α-polymerases.[82] The relationship between form D and the other subspecies of the thymus polymerase-α is not clear.[82] It was suggested that the 150 to 170 kdalton subunit constitutes the "core" catalytic component of α-polymerase, which associates with a 50- to 70-kdalton subunit to form a "holo" enzyme. The observed multiple molecular forms of α-polymerase result from either limited proteolysis or from dissociation of the catalytic and auxiliary subunits.[82]

Multiple molecular forms of calf thymus α-polymerase were also observed and purified extensively by Gross and Krauss.[64,94] A 5.7-S subspecies of the polymerase was purified to near homogeneity, and upon denaturing gel electrophoresis was found to consist of an equimolar mixture of two polypeptides of 123 and 134 kdaltons.[94] It is difficult to equate the 5.7-S form to any of the types of calf thymus polymerase-α that were described by Holmes et al.[38] However, some evidence suggested that the 5.7-S species is a product of partial degradation of a high-molecular-weight protein precursor.[94] Subsequent to the purification of the 5.7-S polymerase, the same investigators extensively purified a 9-S form of calf thymus α-polymerase, whose yield depended on the rapidity of the isolation process and increased with the inclusion of protease inhibitors in buffer solutions.[64] This form of α-polymerase has a native size of 500 kdalton and on denaturing gel displayed polypeptides of 148 to 158, 59, 55, and 48 kdaltons at a molar ratio of 1:1:1:5, respectively.[64] Although it was not formally proven, it is tempting to assume that the 148- to 150-kdalton polypeptide constitutes the catalytic core, whereas the lower-molecular-weight components are accessory subunits. The subunit structure of the 9-S enzyme as described by Gross and Krauss is generally similar to that of the initially described *Drosophila* α-polymerase[4] and rat liver polymerase-α[65,73] (see below). In addition, a third form of 7.0 to 7.5 S was isolated, but its subunit composition remained undetermined.[64] Each of the three forms of calf thymus α-polymerase, (5.7, 7.0, and 9.0 S) exhibit a different fidelity in copying synthetic templates.[91] Differences in the efficiency of extending mismatched primers were also observed with the 5.7-S and 9-S enzymes.[95] Hence, limited proteolysis and/or association with accessory subunits subtly affect the enzymic properties of the catalytic core or α-polymerase.

Yoshida and colleagues isolated a high molecular form of calf thymus polymerase, which sedimented at 10 S and which, unlike most previously described polymerase-α forms, copied at a high efficiency the ribo strand of the primer template oligo (dT)·poly (rA).[96,97] This 10-S form of the calf α-polymerase was subsequently purified to homogeneity.[98] The homogenous native enzyme, which was associated with primase activity (for discussion, see Section III.B of this chapter), had a size of 200 to 250 kdaltons and on denaturing gel electrophoresis was found to exhibit three larger polypeptides of 140, 145, and 150 kdaltons and three or four smaller subunits of 43 to 50 kdaltons.[98] If all the observed polypeptides constitute subunits of the polymerase, the molecular weight of the native enzyme suggests that they cannot all be included within the same polymerase molecules in stoichiometric amounts. Peptide mapping of the higher size subunits revealed similarities in their primary structure. These similarities as well as the in vitro interconversion among the polypeptides suggested that the microheterogeneity of the large polypeptide component results from limited proteolytic digestion.[98] Monoclonal antibodies prepared against the 10-S polymerase[56] were used to further delineate its structure. Immunoprecipitation of the purified polymerase-α led to isolation of the 140-, 145-, and 150-kdalton polypeptides, again suggesting their close relationship.[75] However, the same monoclonal antibody precipitated from fresh crude extracts of calf thymus larger polypeptides of 150, 180, and 240 kdaltons. Only the amount of precipitated 150-kdalton subunit was fully correlated with the immunoprecipitated polymerase-α activity.[75] Peptide mapping of the isolated polypeptides showed, however, great similarity between the 150- and 180-kdalton subunits.[75] It was suggested, therefore, that the 180-kdalton subunit is a precursor of the catalytic subunit and that the 150-kdalton core is derived from it. The 140- and 145-kdalton fragments are, most probably, products of partial proteolysis.[75]

An even more extensive heterogeneity of calf thymus DNA polymerase-α has been reported by other investigators. A 118-kdalton polypeptide was observed by Albert et al.[99] on denaturing gel electrophoregram of an apparently homogenous enzyme. Upon renaturation, that polypeptide regained polymerase activity, and it is thus a catalytic core unit.[99] In addition, about five polypeptides of 54, 56, 57, 58, and 64 kdaltons, all without detectable enzymatic activity, were identified.[99] The similarity between the primary structure of 118-kdalton polypeptides and the smaller subunits led to the suggestion that all those peptides may be derived from a common larger precursor molecule of >140 kdalton.[99] In fact, on activity gels, a 200-kdalton enzymatically active fragment was discerned.[99] Hübscher and colleagues[100,101] confirmed and extended the findings of Albert et al.[99] by showing that when α-polymerase activity was conventionally monitored with an activated DNA template during its purification, an enzyme of 7.4 S (200 to 250 kdaltons) was obtained.[100,101] This so-called "core" enzyme is constituted of two catalytically active polypeptides of 125 and 150 kdaltons and three inactive fragments of 50 to 60 kdaltons which could be products of degradation of the catalytic subunit.[100] However, when purification of α-polymerase was monitored with self-primed single-strand parvovirus DNA or with phage M13 DNA, which are presumed to better mimic the in vivo DNA template, a >11.3-S (>500 kdaltons) polymerase with primase activity associated with it was obtained.[100,101] The subunit composition of the "holo" enzyme was not established due to difficulties encountered in its purification.[100] However, impure preparation of the >11.3-S enzyme contained at least nine polypeptides, one of which was a catalytic unit of 125 kdaltons revealed on activity gel.[100] It should be remarked here that although activity gels usually demonstrate the presence of catalytic subunits of α-polymerase of >125 kdaltons,[39,59,60] in some cases, active fragments fail to renature and a polymerase "signal" is not obtained.[60,77] Thus the size of the catalytic polypeptide(s) may have to be confirmed by independent methods in addition to activity gel electrophoresis.

Bambara and associates[58] recently purified α-polymerase from calf thymus gland by a three-step procedure whose key feature was immunoaffinity chromatography on columns of immobilized anti-α-polymerase monoclonal immunoglobulin. A highly purified resulting enzyme had a sedimentation coefficient of 10.2 S and calculated native molecular size of 404 kdaltons. Analysis by SDS gel electrophoresis resolved the purified α-polymerase into three major polypeptides of 170, 70, and 60 kdaltons with minor bands of 180 and 65 kdaltons.[58]

Chang et al.[341] used a roughly similar immunoaffinity purification scheme to isolate α-polymerase from calf thymus. The highly purified polymerase sedimented as a double peak at 7.4 and 9.3 S corresponding to sizes of 158 and 223 kdaltons, respectively. Whereas the 223-kdalton form possessed both DNA polymerase and DNA primase activities, the 158-kdalton polymerase was devoid of primase activity. Denaturing gel electrophoresis revealed that the 223-kdalton DNA polymerase-primase complex consisted of a 160- and 68-kdalton polypeptides and of a minor component of 185 kdaltons. It has been suggested that the 160-kdalton polypeptide consitituted the polymerase catalytic subunit, whereas the 68-kdalton component may represent DNA primase.[341] The relatively simple subunit structure of the polymerase as well as some of its catalytic properties suggested that the investigated enzyme preparation did not represent a complete replication complex and that additional protein components of the "holo" enzyme may have been missed.[341]

Grummt et al.[63] purified calf thymus polymerase-α to apparent homogeneity, obtaining an enzyme that contained seven polypeptides with molecular sizes of 64, 63, 62, 60, 53, 55, and 52 kdaltons.[63] Apparently the absence of polypeptides larger than 100 kdaltons stands in clear contrast to all the studies described above. Taking into account the massive body of evidence on the sensitivity of polymerase-α to proteolysis, it is a reasonable speculation that limited proteolysis was responsible for cleavage of larger polypeptide(s) into the fragments observed. Other results on the composition of α-polymerases from cultured mouse cells[110] and human KB[109] cells, which also showed presence of only polypeptides smaller than 100 kdaltons were recently revised and subunits of 150 to 190 kdaltons have been detected (for discussion see below).

D. Rat Liver

Polymerase-α from regenerating rat liver was purified to homogeneity and characterized as to its structural and enzymic properties by de Recondo and colleagues.[65,73,102] The homogenous enzyme sedimented at 9.7 and 7.4 S in gradients of low and high ionic strength, respectively.[65] The effect of salt on the apparent size of native α-polymerase was also reported for enzymes isolated from other cells types.[4,103] Analysis of the purified α-polymerase on SDS gels yielded protein bands of 156, 64, 61, 58, and 54 kdaltons.[65,73] All the catalytic activity correlated with the 156-kdalton polypeptide, which could be resolved from all other subunits by DNA cellulose affinity chromatography. Yet, reconstitution of the 54- to 64-kdalton units with the catalytic subunit increased its activity.[65,73] A molar ratio of 0.7 to 1.2 of the liver polymerase catalytic polypeptide to the smaller consitituents argues for a general structural similarity between rat,[65,73] calf thymus,[38,75,99,101] chick embryo,[67] and *D. melanogaster*[4,40] α enzymes, i.e., a 125- to 190-kdalton catalytic core associated with additional subunits in equimolar proportions. Recent studies have shown that indeed a close structural similarity exists between α-polymerases from rat liver and *Drosophila*.[117] That the catalytic subunit of rat liver polymerase-α and the other polypeptides in the homogenous enzyme do not originate from a common precursor was demonstrated by the dissimilitude in their tryptic peptide patterns. However, partial homology existed between tryptic peptide maps of the catalytic subunits of α-polymerase from *D. melanogaster* and rat liver. Hence, it was suggested that each subunit in the rat enzyme is a distinct polypeptide and that the primary structure of the catalytic subunit is evolutionarily conserved.[117]

E. Chick Embryo

Matsukage's group purified to homogeneity DNA polymerase-α from chick embryo.[67] The apparently homogenous enzyme preparation was resolved on SDS gel electrophoregrams into a group of three or four bands of 133 to 155 kdaltons and a group of four polypeptides of 51, 54, 56, and 59 kdaltons.[7] Primary structure similarity, as revealed by tryptic peptide analysis, was demonstrated among members of the high-molecular-weight groups size proteins and among the low-size polypeptides. However, the two groups of polypeptides did not share common amino acid sequences.[67] The size of the native homogenous enzyme (200 kdaltons) suggests that basically it is composed of one large (∼150 kdaltons) and one smaller (∼55 kdaltons) subunit. Since size microheterogeneity among the polypeptides of the two groups was observed also when protease inhibitors were present, it was argued that this microheterogeneity might reflect an in vivo variegation of the basic components of α-polymearase.[67] Yet keeping in mind the extreme lability of polymerase-α to proteases, inclusion of some inhibitors of proteases does not eliminate the possibility of proteolytic cleavage by inhibitor-resistant specific protease.

F. *Drosophila*

One of the best characterized polymerase-α is that of *D. melanogaster* embryos, isolated, purified and extensively studied by Lehman and his group,[4,40,46,104-108] *Drosophila* embryos divide every 10 to 20 min and contain high levels of polymerase-α. This organism offers the potential advantage of gene identification and manipulation. Earlier attempts to characterize *D. melanogaster* polymerase were impeded by extensive proteolysis during isolation of the enzymic protein.[2,3] Proteolysis was subsequently minimized through use of techniques for a relatively rapid enzyme purification and by introduction of protease inhibitors into all purification buffers.[4,104] A homogenous preparation of the enzyme obtained sedimented at 8.6 to 9.1 S, and its calculated molecular size was 280 kdaltons. Under denaturing conditions, the homogenous α-polyerase displayed four electrophoretic bands of 148, 58, 46, and 43 kdaltons which were designated subunits, α, β, γ, and δ, respectively.[4,104] This designation should not be confused with animal cell DNA polymerases-α -β, -γ, and δ which are distinct enzyme classes (cf. Chapter 2).

The primary structure of the *Drosophila* polymerase subunits was dissimilar as deduced from staphylococcus protease peptide mapping, and they seemed not to originate by proteolytic cleavage of a common percursor molecule.[104] Glycerol gradient sedimentation in presence of 2.8 *M* urea dissociated the 148-kdalton polypeptide from the other components, and only this isolated subunit exhibited DNA polymerization activity. Although the β subunit did not possess polymerase activity, reconstitution of the α polypeptide with the β, γ, and δ subunits, increased the catalytic activity of the former.[104] As will be discussed later, (Section IV.B.1 of this chapter) primase activity was subsequently found to be tightly associated with the *Drosophila* polymerase-α — appearing to reside either with the β or the γ subunits or with both.[46,105,108] However, the extreme caution required in reaching for conclusions on subunits size in intact polymerase-α was demonstrated later when Sauer and Lehman used monospecific rabbit antibodies against the electrophoretically isolated polymerase subunits.[40] An antibody directed against homogenous α subunit (148 kdaltons specifically interacted with 166- and 185-kdalton polypeptides in the crude extracts.[40] It seems, therefore, that despite the precautionary measures taken, limited proteolysis still occurred during purification of the polymerase, reducing the size of the α subunit. This was confirmed with the development of an even more rapid procedure for the purification of α-polymerase from freshly harvested embryos in the presence of an assortment of protease inhibitors. Polymerase-α purified by this new method was shown to contain polypeptides of 182, 60, and 50 kdaltons which had cross antigenicity with the α, β, and δ subunits of the initially purified enzyme.[106] Glycerol-urea gradient centrifugation revealed that the polymerase activity coincided with 182-kdalton polypeptide. However, the previously described subunit (43 kdaltons),[4,104] was not found in the newly prepared enzyme, and it was suggested that it was a contaminant that is highly abundant in early embryo extracts.[106] Measurements of the frequency of misincorporation by the intact enzyme indicated a surprising high degree of fidelity, as would be expected, of a polymerase central to DNA replication.[108] However, the purified enzyme exhibited low processivity suggesting that additional factors must be operative during in vivo DNA replication. These observations, together with the association of primase with the homogenous polymerase-α, suggest that this polymerase preparation represents the best approximation of the DNA replicating "holo" enzyme of a eukaryote.

G. *Xenopus laevis*

Ovaries of the frog *X. laevis* contain two discernable molecular forms of α-polymerase designated α_1- and α_2-polymerase.[11,47,48] Whereas α_1-polymerase is tightly associated with DNA primase, polymerase-α_2 is devoid of primase activity.[11,47,48] Recently, Benbow and associates[345] purified to homogeneity form α_2 of the polymerase. The highly purified enzyme has a native molecular size of 180 kdaltons and consists of two polypeptides of 105 to 120 and 50 to 55 kdaltons. Activity gel electrophoresis identified the $\sim$120-kdaltons polypeptide as the catalytic subunit.[345] Data obtained recently indicate that α_1-polymerase, which possesses primase activity, consists of at least one additional polypeptide of approximately 50 kdaltons.[347] It appears that the size of the catalytic subunit of α_1-polymerase is larger than that of α_2-polymerase.[345] Whether the catalytic subunits of polymerases α_1 and α_2 are derived from a common high-molecular-weight precursor is not clear at present.

H. Bone Marrow

Purified rabbit bone marrow DNA polymerase-α exhibited two forms designated α_1 and α_2, which differed in their behavior upon molecular sieve filtration.[66] The highly purified α_1 form exhibited on SDS gel electrophoregram one band of 135 kdaltons, whereas α_2 contained in addition to that high-molecular-weight polypeptide, other components of 50 to 70 kdaltons. DNA polymerase-α activity correlated with the amount of the 135-kdalton subunit, whereas the function of the smaller subunits is now unknown.[66]

I. Mouse Myeloma

The structure of polymerase-α from mouse myeloma cells has been investigated by Wilson and group.[69,110] A high-salt extract of these cells contains three relatively abundant molecular species of α-polymerase designated α_1, α_2, and α_3. In an initial study, forms α_1 and α_2 were purified to homogeneity.[69] Both α_1 and α_2 polymerases contained two polypeptides of 52 to 54 and 42 to 48 kdaltons.[69] At that time it was argued that these polypeptides were unlikely to be products of limited proteolysis, and a proposed terameric structure (54 kdaltons)$_2$·(47 kdaltons)$_2$ was presented.[69] Interestingly, one of these forms of mouse myeloma polymerase contained a $3' \rightarrow 5'$ and $5' \rightarrow 3'$ bidirectional exonuclease, an unusual feature of animal cell polymerases (see Section IV.A in this chapter). In a later publication, it was shown that activity gel electrophoresis of myeloma cell extracts revealed ∿120-kdalton and 76-kdalton catalytic polypeptides.[110] Purification to homogeneity of the third species of mouse myeloma α-polymerase yielded a preparation which upon denaturing gel electrophoresis contained predominantly a 120-kdalton polypeptide.[110] It was argued that an α enzyme may be devoid of any smaller accessory subunits and may contain only two catalytic subunits of >120 kdaltons.[110]

Structural heterogeneity of different molecular species of mouse myeloma α-polymerase and/or limited proteolysis may provide explanations for the apparent contradictory subunit structures obtained in the above described studies. In this context, it is pertinent to note that Wilson and colleagues have studied recently the catalytic subunit of BSC-1 cultured monkey cells by use of precipitating anti-α-polymerase monoclonal antibodies.[60] The immunoprecipitate obtained from fresh crude extracts was shown to contain a ∿190-kdalton polypeptide that upon electroelution from SDS polyacrylamide gels and renaturation displayed catalytic properties of α-polymerase.[60] By contast, direct activity gel analysis of the crude extracts failed to demonstrate the presence of the ∿190-kdalton polypeptide. In those activity gels, only polypeptides of 40, 70, and 115 kdaltons were discerned.[60] These results demonstrate that under proper isolation conditions a polypeptide of 180 to 200 kdaltons can be identified as the site of catalytic activity in α-polymerase as also shown recently for calf thymus,[75] KB cultured human cells,[109] *D. melanogaster,*[40,106] and mouse ascites cells.[68] Furthermore, the results of this study illustrate one limitation of activity gel analysis—it may fail to detect active polypeptides that cannot renature and regain enzymic activity under the conventional experimental conditions.

J. Cultured Human Cells

Korn's group conducted pioneering studies of the purification, structure, and enzymic properties of polymerase-α from cultured human KB cells.[43,44,62,93,109] These workers were likely to be the first to present results on a near homogenous preparation of any type of α-polymerase.[62] In their initial study, Korn and colleagues resolved on DEAE cellulose two forms of α-polymerase that, under the extraction conditions that were employed, were in the cytoplasm and a third form that remained tightly bound in the nuclei.[43,62,93] The major cytoplasmic enzyme designated α was the first to be purified to near homogeneity.[62] That polymerase had a molecular size of 140 kdaltons in its native form, and on SDS gels displayed subunits of 76 and 60 kdaltons that were present at equimolar proportions.[62] In a later study it was shown that all three forms of KB cell polymerase-α possessed a similar apparent size of 140 kdaltons, and all contained a quartet of polypeptides of 70, 65, 59, and 55 kdaltons.[93] Indirect experiments, which could not rule out selective proteolytic cleavage of the polymerase, indicated that gross proteolysis was undetectable in the KB cell extracts.[93]

These results must, however, be revised in light of recent structural data obtained on a primase-containing KB cell α-polymerase, which was isolated by immunoaffinity adsorption to Protein A-monoclonal anti-α-polymerase IgG-agarose.[109] Purified polymerase that was obtained by this method contained polypeptides of 180 and 165 kdaltons, trace amounts of 140- and 125-kdalton fragments, and three smaller peptides of 77, 55, and 49 kdaltons.[109] Use of a spectrum of monoclonal antibodies directed against KB cell α-polymerase demonstrated that a set of polypeptides of 125 to 180 kdaltons as well as 77 kdaltons and a cluster of polypeptides of 49 to 60 kdaltons were present in all the immunoaffinity purified polymerase-α preparations. The three groups of polypeptides were unrelated to each other as revealed by peptide mapping.[109] These results reinforce again the growing body of evidence which indicates that polymerase-α contains a 180- to 200-kdalton polypeptide that is most probably the catalytic subunit and a cluster of additional smaller subunits.[40,68,75,106,109]

That polymerase-α purified to homogeneity by certain procedures may include accessory proteins not seen in other preparations was shown by Baril and group.[70] Polymerase-α that was purified by these investigators from HeLa cells, contained three chromatographically distinct forms of enzyme designated α_1, α_2, and α_3. Form α_2, which was purified to homogeneity by a series of chromatographic steps, appeared as a protein of 600 kdaltons that contained polypeptides of 140, 69, 55, 57, and 24.5 kdaltons. Form α_3, however, was a protein of 220 kdaltons only and contained two polypeptides of 140 and 65 kdaltons.[70] Interestingly, two protein cofactors (C1 and C2), which enabled α_2-polymerase to copy efficiently denatured DNA, could be released from the enzyme by two chromatographic steps.[70] Cofactor C1 was a 96-kdalton tetramer of the 24.5-kdalton polypeptide that was detected in the "holo" enzyme, and C2 was the 51- to 55-kdalton polypeptide.[70] As will be discussed later, (Section IV.C.2 of this chapter) these cofactors which were recently described in detail for cultured monkey cells,[52,53] serve as primer-recognition auxiliary proteins for polymerase-α. Whether these polypeptides can be regarded as integral subunits of the polymerase or as dissociable cofactors is a matter of definition of the composition of the "holo" enzyme. The constitution of the "holo" polymerase should be unraveled in the near future with the employment of immunoaffinity purification procedures.[58,60,75,109,341]

The considerable effort by many researchers to elucidate the structural details of polymerase-α is beginning to yield a coherent picture. Table 1 summarizes the main elements of results described heretofore. In cases that conclusions on the structure of an α-polymerase investigated by any one group were modified in the course of study, only revised concepts are presented in the table.

Despite the large variety of sources for polymerase-α and the often confusing and contradictory results that were obtained, some general conclusions on the structure of polymerase-α may already be drawn. First, an increasing body of data indicates that when isolated by appropriately rapid and nondestructive procedures, polymerase-α has a catalytic polypeptide core of >120 kdaltons and most probably of 180 to 200 kdaltons.[17,38,40,60,64,68,75,106,109,241] Interestingly, recent results on the structure of homogenous DNA polymerase of the lower eukaryote *S. cerevisiae* also showed that the size of a major polypeptide in that enzyme is 120 to 170 kdaltons.[111-113] The evolutionary manifestations of these recent studies and other germane observations are discussed in Chapter 9, Section I.A). The controversy of few years ago on the question of whether or not polypeptides of >120 kdaltons exist in polymerase-α[37] seems to have been affirmatively resolved by consensual results. A related question, however, regarding the size microheterogeneity of the larger polypeptides in α-polymerase is still unresolved. Some researchers obtained by a variety of isolation methods a family of larger polypeptides,[60,67,68,75,98,101] whereas others detected a single large polypeptide.[17,38,65,70,106] Whether the observed microheterogeneity, as well as the multiplicity of molecular forms of α-polymerase reflect an in vivo heterogeneity of α-polymerase as argued by some authors,[67,71] remains moot. The number and sizes of polymerase-subunits other than the catalytic core polypeptide is another open question.

Multiple polypeptide chains other than the catalytic core was detected in purified preparations of α-polymerase (Table 1). However, only a demonstration of a specific function in DNA polymerization for any of these polypeptides will provide a satisfactory proof for their being subunits of the polymerase molecule. At present only DNA primase was shown to be associated with the replicative enzyme. In some cases, primase activity was physically separable from the catalytic polypeptide, whereas in others, both DNA polymerase and DNA primase seemed to reside in the same polypeptide chain (see Section IV.B.1 of this chapter). In addition to primase, some accessory proteins, notably C1 and C2, may also associate with the catalytic core.[52,53,70] In summary, although our understanding of the structure and functional properties of the

Table 1
SIZE AND SUBUNIT COMPOSITION OF HIGHLY PURIFIED PREPARATIONS OF POLYMERASE-α

Source of polymerase	Molecular form[a]	Specific activity[b] (fold purification)	Size of native enzyme (kdaltons)	Size catalytic subunit (kdaltons)	Size other subunit (kdaltons)	Number other subunits	Remarks[c]	Ref.
Calf thymus	A1,A2 (8.0—8.15)	(2000—3000)	200 ± 15	150—170 (1)	50—70 (1)	1		17, 38
	B (5.2 S)	(2000—3000)	103 ± 8	[d]				38, 82
	C (7.3 S)	0.15 × 10^5 (2000—3000)	165 ± 10	150—170 (1)				17, 38,83
	D (6.8 S)	0.1 × 10^5 (3000)	144 ± 10	[d]				82
	9 S	0.6 × 10^5 (12,000)	Approx. 500	148 or 158	48, 55, 59			64
	5.7 S	1.0 × 10^5 (20,000)	Approx. 250	123 and 134				94
[e]	10 S	1.0 × 10^5 (25,000)	200—250	140, 145, 150	45—50	3—4		98
[e]	10 S			180, 150 (240)	43—50		Immun.	75
[e]	10.2 S	0.35 × 10^5 (5,000)	404	170, 180	60, 65, 70	3	Immun.	58
				118 (>140) + 54—64 (5)				99
	7.4 S		200—250	125, 150				100, 101
[e]	>11.3 S		>500	125		11		100, 101
[e]	9.3 S	1.05 × 10^5 (5,000)	223	160 (185?)	68	1	Immun.	341
			404—413		52—64	7		63
Rat liver		0.45 × 10^5 (×3,000)	155—250	156	54—64	4		65, 73
Chick embryo		0.45 × 10^5 (49,000)	200	130—155 (3—4)	51—59	4		67
Drosophila[e]		0.5 × 10^5 (×7,500)	280 (Ref. 4)	166, 185	85, 60		Immun.	40,
				182	60, 50			106
Xenopus laevis[e]	"α_2"[f]	2.9 × 10^5 (×41,600)	180	120	55			345

Table 1 (continued)
SIZE AND SUBUNIT COMPOSITION OF HIGHLY PURIFIED PREPARATIONS OF POLYMERASE-α

Source of polymerase	Molecular form[a]	Specific activity[b] (fold purification)	Size of native enzyme (kdaltons)	Size catalytic subunit (kdaltons)	Size other subunit (kdaltons)	Number other subunits	Remarks[c]	Ref.
Rabbit marrow	"α_1"	0.3—2.0 × 10^5	215	135	54—66	4		66
Monkey BSC-1 cell				Approx. 190 (40,70,115)			Immun.	60
Mouse myeloma				120(76)				110
Human KB cells[e]				125—180(4)	77,49—60		Immun.	109
Mouse ascites	10 S	0.3 × 10^5		125—178(3)	48—70	6		68
Human HeLa cell	α_2[g]	2.6 × 10^5	600	140(1)	24—69	4		70
	α_3	Homogenous	220	140(1)	65	1		70

Note: Data are presented in the order of their description under Section II of Chapter 3.

[a] Molecular form is specified when one of several distinct molecular forms of α-polymerase was purified.
[b] Specific activity is expressed in units per milligram purified enzyme protein. Unit of activity is according to the author's definition.
[c] Purification by immunoaffinity is specified as "immun." All other enzyme preparations were obtained by conventional protein purification procedures.
[d] Forms B and D are products of partial proteolytic digestion of form C (References 38, 82). In parenthesis: number of polypeptides with catalytic activity.
[e] An enzyme preparation which contains DNA primase activity.
[f] Form α_2 is devoid of DNA primase activity.
[g] An enzyme preparation which contains cofactors C2 and C2 (Reference 70).

catalytic polypeptide of α-polymerase is becoming clearer, the number, sizes, and roles of other subunits of the polymerase are yet to be elucidated.

III. IS POLYMERASE-α MODIFIED POSTTRANSLATIONALLY?

The microheterogeneity of both polymerase-α molecules and of their catalytic core have been extensively documented (Section II of this chapter). Although this heterogeneity has been shown to be due to a large extent to artifacts of the preparative procedures employed, some size and electric charge variance among the polymerase-α molecules appears to remain. An attractive possible explanation for this heterogeneity is that the polymerase molecules are modified postsynthetically.

Foremost of such possible posttranslational processes may be proteolytic cleavage of a large precursor α-polymerase molecule into a smaller molecule. The proteolytic conversion may modify the polymerase activity or the mode of its interaction with other subunits and auxiliary proteins and may serve, therefore, as an enzyme activation-inactivation mechanism in vivo.[71] Viral, bacterial, and animal cell hormones and secretory proteins have to be cleaved proteolytically in order to perform their functions in the cell. It may be, therefore, that some of the proteolytic interconversions among the molecular forms of polymerase-α may be due to genetically programmed postsynthetic proteolytic activation of the polymerase. Microheterogeneity in the size of the α-polymerase catalytic core, which was observed in the presence of various protease inhibitors, somewhat supports such contention. In addition, relatively low sizes of the catalytic polypeptide were obtained despite the absence of detectable gross proteolysis in cell extracts. (Cf. Reference 4 with References 40 and 106, and Reference 93 with 109.) These observations may argue for specific proteolytic cleavage of larger precursor polypeptides of the polymerase. Hübscher has offered a model for the activation of polymerase-α by selective proteolysis, which renders the enzyme fit for association with other subunits and accessory proteins in the replicative complex.[71] However, at the time of this writing, we find that convincing experimental evidence for programmed proteolytic activation of α-polymerase is missing. This intriguing possibility will certainly be put to critical tests when a cloned gene for the α-polymerase catalytic core becomes available.

The possibility of posttranslational modification of α-polymerase through glycosylation has been experimentally explored by Basu et al.[118-121] Polymerase-α isolated from mouse IMR-32 human neuroblastoma cells displays three chromatographically distinct forms; α_1, α_2, and α_3.[118-119] These forms possess different primer-template preferences and more interestingly, they are differentially inhibited by ricin B, a plant toxin that binds specifically to terminal β-linked galactose in glycoconjugates.[119] Furthermore, concanavalin A, a glycoprotein-binding plant lectin, has been found to also inhibit polymerase-α, and there is direct binding of α-polymerase to concanavalin A-Sepharose columns.[118,119] Form α_2 is strongly inhibited by ricin 1B, a subunit of intact ricin. However, when α_2-polymerase was purified from cells that were treated with tunicamycin, an inhibitor of protein glycosylation, the enzyme was found to become resistant to ricin 1B.[120] Moreover, polymerase-α_2 isolated from untreated cells and from cells that were exposed to tunicamycin in displayed altered heat lability.[120] These indirect experiments raise the tantalizing possibility that polymerase-α is modified and perhaps regulated by glycosylation.[121]

Other possible postsynthetic modifications of α-polymerase or one of its accessory proteins may be mediated by phosphorylation-dephosphorylation reactions. Danse et al.[187] demonstrated that the activity of partially purified chick cell polymerase-α was stimulated following phosphorylation by a cAMP-independent protein kinase. Due to the impurity of the polymerase used in that study, it is not clear whether the increased activity was due to direct phosphorylation of the enzyme or rather to a modification of an accessory protein. In this context it should be mentioned that the activity of an α-polymerase stimulatory protein from mouse cells was found to be modulated by phosphorylation (see Section IV.C.3 of this chapter).[185] Cell cycle-dependent modulations in the activities of nuclear protein kinases[188] may be coupled to phosphorylation of components of the replicative complex, one of which may be α-polymerase.

Glycocylation as well as phosphorylation, ADP-ribosylation acetylation, and any other of the possible postsynthetic protein modifications are, therefore, viable possible sources for heterogeneity within the polymerase-α molecule population. Further study of posttranslational modifications of polymerase-α may shed light on possible routes for the modulation of its activity in vivo.

IV. AUXILIARY ACTIVITIES ASSOCIATED WITH DNA POLYMERASE-α

The structural complexity of DNA polymerase-α described heretofore suggests that in the holoenzyme, the catalytic core of the polymerase is associated with a number of other polypeptides that play a role in the highly intricate replication process. Much effort had been put into the identification of functions other than DNA polymerase which may be linked to α-polymerase. It is clear that most of these polymerase-α-associated components are not present in extensively purified enzyme preparations. However, they might function in close association with α-polymerase in the course of DNA replication in vivo. In this section, results of these studies will be described in the order of activities looked for or identified.

A. Exonuclease

Early studies on prokaryotic DNA polymerases established that DNA polymerase I from *Escherichia coli* is able to conduct several enzymatic functions as a single polypeptide of 109 kdaltons. In addition to DNA polymerization in the 5′→3′ direction, polymerase I is able to carry out also the exonucleolytic hydrolysis of DNA in the 3′→5′ direction ("proofreading") and also hydrolyze DNA in the 5′→3′ direction ("nick-translation").[122,123] Similar exonucleolytic activities have been identified in DNA polymerases of various bacteria, bacteriophage, and animal viruses such as herpes simplex virus.[122,123] As DNA polymerases began to be described in eukaryotes, investigators searched for deoxyribonucleases which may be tightly associated with these enzymes. In lower eukaryotes, exonucleolytic activities copurify with an α-like DNA polymerase. Hence, exonucleases were identified in partially purified α-like polymerases from *U. maydis*,[124] *Euglena gracilis*,[22] *T. pyriformis*[27] and *S. cerevisiae* [pol B (II) but not pol A (I)].[125] The association of DNA polymerase from lower eukaryotes with exonuclease and its evolutionary implications were reviewed by McLennan and Keir.[30] Although early reports indicated that in some cases exonucleases copurified with or were part of α-polymerases from higher eukaryotes, later studies on more highly purified DNA polymerases-α demonstrated that in the large majority of cases, exonucleolytic activity could not be detected in purified polymerase-α from sea urchin,[5] *D. melanogaster*,[4] regenerating rat liver,[73] calf thymus,[64] and human KB cells,[44] to mention but a few examples. Two exceptions should be noted, however. First, one of the two highly purified mouse myeloma polymerase-α forms exhibited exonucleolytic activity, which hydrolyzed DNA processively in the 5′→3′ direction and distributively in the 3′→5′ direction.[69] A second, more extensively documented case, is that of DNA polymerase-δ. Polymerase-δ, isolated from rabbit bone marrow cells and from calf thymus, displays properties generally similar to those of α-polymerase, but the highly purified enzyme contains a 3′→5′ exonculease activity.[126] The properties of polymerase-δ are discussed in detail in Chapter 5. In conclusion, in most cases DNA polymerase-α from animal cells is not associated with traceable activity that hydrolyzes DNA. Few exceptions to this general phenomenon were, however, noted. It remains to be determined whether or not 3′→5′ exonuclease activity associated with any eukaryotic DNA polymerase functions in proofreading as to increase the fidelity of DNA synthesis.

B. DNA Primase

Discontinuous synthesis of nascent strands of DNA in both prokaryotes and eukaryotes and in their respective DNA viruses is initiated with the formation of short oligoribonucleotides, which are subsequently elongated by DNA polymerase.[122,123,127] In *E. coli*[123,128-130] and its phages T7[131-133] and T4,[134-136] unique RNA polymerases have been characterized which catalyze the synthesis of RNA primers. These enzymes are generically designated "DNA primase" although each possesses unique structural and catalytic properties. Extensively documented evidence indicates that nascent DNA chains, which are isolated from normal and virus-infected animal cells, contain transient oligoribonucleotides of variable sequence which are covalently linked to the 5′ end of the newly made DNA.[137-145] The possible association of α-polymerase with these short RNA oligomers was first suggested by the finding that the α enzyme was the only class of eukaryotic DNA polymerases that was able to extend RNA primers[66,69,105,146,148] (see Section VI.B of this chapter).

The first clue that α-polymerase may be associated with DNA primase was provided by evidence that the polymerase was able to copy unprimed synthetic homopolymers such as poly(dC) or poly(dT) in an ATP-requiring reaction.[148,149] In the last several years, several laboratories reported on the purification of DNA primase from a variety of animal cells. In different cell types, primase was found to be either inseparable from the catalytic subunit of α-polymerase,[150,151] to constitute distinct subunit(s) in the oligomeric α-polymerase molecule,[108,341] to be associated with a discrete molecular form of the α-polymerase,[47,48,92,152,153] or to constitute an entity physically separable from polymerase-α.[154] These findings and some of the catalytic properties of DNA primase from animal cells are described in this section.

1. Structural Relationships Between Primase and α-Polymerase

Seno and colleagues described a DNA primase in Ehrlich ascites mouse cells.[45,92,152,153,155] Two molecular forms of α-polymerase, α_1 and α_2, can be resolved from these cells by virtue of either their chromatographic behavior and/or different sizes.[45,92] The larger of these two polymerase species, α_1, was found to be associated with DNA primase which enabled it to copy unprimed poly(dC) or poly(dT).[45,152] Furthermore, primase activity was involved in ribonucleotide triphosphates-dependent copying of circular single-stranded phage fd DNA by α_1-polymerase.[152,155] The efficient synthesis of RNA primers on poly(dT) and fd DNA was found to require a 130-kdalton stimulating protein.[45,152] That stimulatory factor increased the affinity of the polymerase-α-primase complex ("DNA replicase") for the template and enhanced the rate of DNA synthesis. However, the factor did not affect either the length of either the primer RNA (8 to 10 ribonucleotides) or the DNA product (300 to 1000 deoxynucleotides long).[156]

These observations were extended: two forms of DNA polymerase-α could be resolved by ion exchange chromatography from a variety of animal cells and only one form, which was designated replicase, was associated with DNA primase.[153,157] In almost all cases, the size of replicase was slightly higher than that of the α-polymerase form which was devoid of primase activity.[153] The stimulating protein obtained from mouse ascites cells enhanced both the activities of polymerase and primase in replicases obtained from various vertebrate cells.[153] The tight association between α-polymerase and DNA primase within the replicase molecule was demonstrated by coprecipitation of both activities by monoclonal antibodies directed against KB cell polymerase-α.[158] However, it was noted that one monoclonal neutralizing antibody (SJK 132-20) strongly inactivated DNA polymerase activity without affecting the activity of primase.[158] Furthermore, aphidicolin, a specific inhibitor of α-polymerase (see Section VI.B), inhibited polymerase-α activity on fd DNA, but had only a slight effect on synthesis of oligoribonucleotides by DNA primase.[158] That the ATP analog, 2′-deoxy-2′-azidoadenosine-5′-triphosphate (azido ATP), inhibited copying of poly(dT) by replicase, but was less effective in inhibiting utilization of preprimed poly(dT)-oligo(rA), suggested that polymerase and primase activities reside at disparate catalytic sites.[159] It is not clear, however, whether these two distinct sites occupy different regions of a single polypeptide or whether they are located on two separate polypeptides.

A similar nondissociable primase-polymerase-α complex was purified by immunoaffinity separation from human KB cells.[109] Although physical dissociation of the primase and polymerase activities was not achieved, primase was only partially inactivated by two neutralizing monoclonal antibodies that completely inactivated α-polymerase. Furthermore, primase was found to be more heat labile but more *N*-ethylmaleimide resistant than polymerase-α.[109] Hence, in KB cells, as in mouse ascites cells, although the primase and polymerase-α functions could not be physically resolved, these two activities seem to constitute catalytically and structurally distinct entities.

That primase activity may indeed be associated with a discrete molecular form of α-polymerase was also demonstrated in the frog *X. laevis*. Shioda et al.[47] and Riedel et al.[48] reported that a DNA primase activity, which enabled *X. laevis* α-polymerase to copy poly(dT) and single-stranded phage M-13 DNA, was exclusively associated with only one of two forms of α-polymerase that were separated from the frog eggs. Primase copurified with α_1-polymerase through six column chromatography steps and cosedimented with that form of the polymerase in glycerol gradients.[47] Furthermore, it was reported that *X. laevis* DNA primase comigrated with the >100-kdalton polymerase polypeptide on activity gel electrophoregrams.[160]

Intimate association between α-polymerase and primase has been observed for calf thymus α-polymerase, and an argument was raised for the coexistence of both α-polymerase and primase catalytic sites within the same single polypeptide.[150] Primase and α-polymerase could not be separated by velocity sedimentation through gradients that contained 1.7 *M* urea. The two activities were not resolved from each other on activity gel electrophoregrams and comigrated as a single band of 125 kdaltons.[150] Polymerase-α and primase copurified through several chromatographic steps and maintained constant activities ratio throughout the purification procedure, and both activities were retained in a nearly homogenous preparation of calf thymus 9-S α-polymerase.[162] However, primase could be selectively inactivated during electrophoresis in the presence of urea, whereas polymerase-α was more sensitive to the sulfhydryl group-reactive agent, *N*-ethylmaleimide.[162] Despite their clearly different response to urea and *N*-ethylmaleimide, attempts to physically separate the polymerase from primase were not successful. Yoshida and associates demonstrated that primase activity was present only in the 10-S form of the calf thymus polymerase-α and not in the possibly fragmented 6.5-S subspecies.[49] Primase was inseparable from the 10-S polymerase-α through several steps of purification, and both activities coprecipitated when interacted with anti 10-S polymerase-α monoclonal antibody.[49] Hence, in this study too, attempts for physical separation between calf thymus α-polymerase and primase were not successful. Chang et al.[341] purified primase-containing α-polymerase from calf thymus by employing an immunoaffinity column chromatography purification procedure. Two forms of α-polymerase with sizes 158 and 223 kdaltons were obtained, and only the latter exhibited DNA primase activity. The polymerase-primase complex of 223 kdaltons consisting of two polypeptides of 160 and 68 kdaltons and DNA polymerase and DNA primase functions, respectively, were tentatively assigned to these two components.[341] The absence of primase activity in the 160-kdalton form of polymerase contrasts with previous observations on the inseparability of primase and polymerase in calf thymus 125-kdalton polypeptide.[150]

Philippe et al.[147] studied the association between rat liver polymerase-α and DNA primase. Both activities were not detected in the mature liver, and both were induced with similar kinetics during regeneration of the organ. Furthermore, the two activities could not be dissociated by either velocity gradient sedimentation or by column chromatography. Also studied by these researchers was the relationship between DNA polymerase-α and DNA primase in a conditional temperature sensitive mutant of BALB/C-3T3 cell line, which failed to enter S phase at a nonpermissive elevated temperature.[147] Primase was physically inseparable from polymerase at the permissive temperature, but at the restrictive temperature, as cells accumulated at the G_1-S phase boundary, primase could be dissociated from polymerase upon chromatography on a column of heparin-Sepharose. This observation suggests that in 3T3 cells the two enzymic activities reside on two different polypeptides and that when DNA polymerase-α cannot be occupied in active replication. Primase tends to dissociate from polymerase.

Similar tight association between primase and α-polymerase has been recently found in HeLa cells.[166] The human cell primase copurified with α-polymerase through five steps of purification without any apparent dissociation between the two activities. However, primase could be distinguished indirectly from α-polymerase by its preferential rate of heat inactivation at 41 or 45°C.[166]

In contrast to the above described findings, assignment of the primase activity to subunit(s) other than the catalytic polypeptide in the oligomeric polymerase-α molecule has been achieved for *D. melanogaster,* human lymphocytes, and mouse hybridoma. Lehman and associates described in *Drosophila* a primase activity that was part of the homogenous α-polymerase.[163] Using their most updated purification procedure, these authors obtained a primase-containing homogenous α-polymerase, which consisted of a catalytic α subunit (182 kdaltons) and two accessory subunits, β and γ whose sizes were 60 and 50 kdaltons, respectively.[106] Glycerol gradient velocity sedimentation in presence of 2.8 *M* urea separated the α catalytic subunit from the comigrating β and γ subunits[107] (Figure 1). The isolated subunit α was devoid of primase activity, whereas it was detected in a mixture of β and γ subunits.[108] The association of *Drosophila* primase with the β/γ subunit(s) and its absence from the catalytic α subunit are in contrast to the reported association of both polymerase and primase activities with the catalytic subunit of calf thymus polymerase-α.[150]

Complete dissociation between polymerase-α and primase was reported by Tseng and Ahlem.[154,165] Using either human lymphocytes[165] or mouse hybridoma cells,[154] these researchers purified a DNA primase that either contained very low activity of α-polymerase[165] or was completely devoid of polymerase activity.[154] The mouse cell DNA primase was purified to homogeneity by several chromatographic steps. During the last steps of purification, primase was dissociated from α-polymerase, and a homogenous preparation of a polymerase-free DNA primase, which consisted of two polypeptides of 56 and 46 kdaltons in a 1:1 stoichiometry, was obtained.[154] The readily dissociable primase of mouse hybridoma cells stands in contrast to primase, which is tightly associated with α-polymerases of calf thymus,[49,150,162] Ehrlich ascites mouse cells,[158] and HeLa cells.[166]

In attempting to summarize the structural relationship between DNA primase and α-polymerase, one notes that several different interrelationships between these two activities have been reported. In some experimental reports, primase seems to reside with the catalytic polypeptide of α-polymerase.[150,160] In other instances, polymerase and

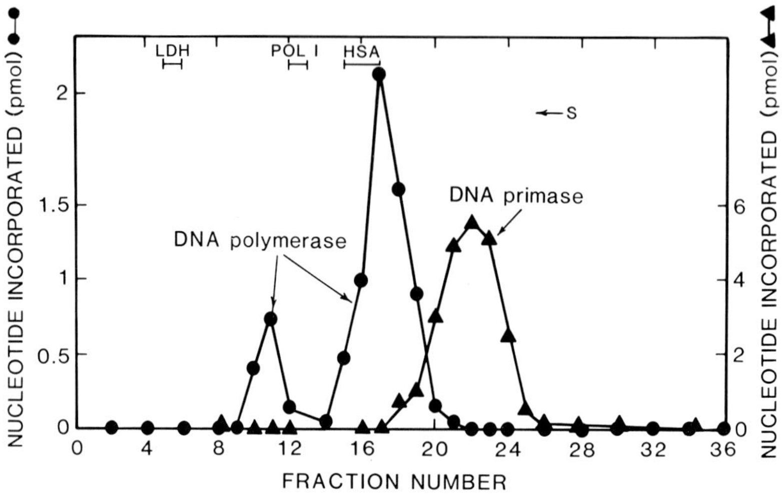

FIGURE 1. Dissociation of DNA primase from the catalytic subunit of DNA polymerase-α of *Drosophila melanogaster.* Highly purified DNA polymerase/primase enzyme complex was treated at 4°C for 4 hr with 3.4 *M* urea and sedimented at 41,000 rpm in 10 to 30% glycerol gradient containing 2.8 *M* urea, for 72 hr in a Beckman® SW 41 rotor. Fractions of the gradient were collected and assayed for activities of DNA polymerase(•—•) and primase(▲—▲) as described in Reference 107. (Adapted from Kaguni, L. S., Rossignol, J.-M., Conaway, R. C., Banks, G. R., and Lehman, I. R., *J. Biol. Chem.,* 258, 9037, 1983. With permission.)

primase are physically inseparable, but each possessed distinguishing structural (antigenic) or catalytic features.[109,158,162,166] In the case of *Drosophila,* polymerase and primase were convincingly assigned to different subunits in the oligomeric α-polymerase molecule,[107] and in mouse hybridoma cells, primase was isolated free of polymerase activity.[154] It is not clear at the time of this writing whether in some cells primase can act independently of polymerase whereas in others these two activities must act in concert, or whether a unifying model for the structural and catalytic relationship between these two enzymes will be found. Some of the main results on animal cell DNA primases are summarized in Table 2.

2. Major Enzymic Properties of DNA Primase

DNA primases from animal cells can be identified as unique activities that cannot be classified under one of the three known types of RNA polymerases of animal cells. DNA primases from various cells and tissues are not inhibited by α-amanitin, an inhibitor of RNA polymerases II and III.[45,47,109,157,161-163,168] Similarly, rifamycin AF/013, an inhibitor of RNA polymerases I to III, does not affect primase activity in SV40-infected BSC-1 cells.[168] The ATP analog azido ATP, which strongly arrests the activity of mouse ascites cell primase, is less effective toward RNA polymerases I and II.[159] Last, the considerable size of the catalytic subunits of RNA polymerases I, II, and III excludes the possibility that they constitute a component of the relatively small primase-polymerase-α complex.[45]

Several investigators analyzed the position and size of the oligoribonucleotide product resulting from primase catalysis. Transfer of α-^{32}P label from dAMP in DNA to neighboring 3′, 2′-rAMP residue serves to prove the formation of covalent bond between the RNA primer at the 5′ terminus of DNA and the nascent DNA chain.[152,163,165,166] The presence of a triphosphate group at the 5′ terminus of the oligoribonucleotide 5′ end has also been demonstrated.[152,166] Using electrophoresis for sizing of newly made primer RNA and DNA after their isolation by DNase and alkali

Table 2
SOME PROPERTIES OF DNA POLYMERASE-α-ASSOCIATED DNA PRIMASE

Source	Association with [a] subspecies of α-polymerase	Dissociated from α-polymerase catalytic core	Size of primase polypeptide (kdaltons)	Ribonucleotides[b] in product	Remarks	Ref.
Xenopus laevis [c]	Yes ("α_1","A")	No	> 100 (Ref. 160)	Approx. 10		47, 48, 153, 160
Drosophila melanogaster		Yes	50/60	8—15		107, 164
Salmon[c]	Yes	No		8—10		153
Chick[c]	Yes	No		8—10		153
Mouse ascites[c]	Yes (α_1, replicase)	No		8—10	Immun.[d]	45, 92, 152, 155 ,158
Mouse hybridoma[c]		Yes	56 + 46 (5.5 S)	9—10		154
Rat thymoma[c]	Yes	No				157
Calf thymus		No	125			150
	Yes (9 S)	No		8—15		162
	Yes (10 S)	No		Approx. 10	Immun.	49
	Yes (9.3 S)		68	<8 to 10	Immun.	341
Human HeLa cell	Yes (Ref. 153)	No		8—15		153, 166
Human KB cell		No		6—10	Immun.	109, 167

[a] Association of DNA primase with one of several molecular forms of DNA polymerase-α is indicated and specified if reported.
[b] Number of ribonucleotides polymerized is the minimal number and refers to the canonical priming reaction product according to Hu et al.[167]
[c] DNA polymerase-primase reaction stimulated by a 130 kdalton protein factor isolated from mouse ascites cells.[45,152]
[d] Immunoprecipitation with anti-α-polymerase monoclonal antibody of polymerase-α-primase complex is indicated (immun.) when performed either to purify the complex or in attempt to dissociate primase from polymerase.

treatment, respectively, it was shown that the primer consists of 6 to 15 ribonucleotide residues (Table 2), and that α-polymerase extends in vitro the RNA primer by several hundreds to several thousands deoxyribonucleotide residues.[47,48,155,156,166] Either in the absence of deoxyribonucleotide triphosphate substrates or with low or absent α-polymerase activity, the primer RNA is not extended onto a DNA chain, but rather becomes longer than its initial canonical size of 6 to 15 nucleotides. Thus, for example, the polymerase-free primase from mouse hybridoma cells synthesizes on natural DNA templates 50 residues ribonucleotide chains, which are primary multiples of the initial primer length of 9 to 10 nucleotides.[154,169] However, in the presence of α-polymerase, only the basic unit length of 9 to 10 residues is synthesized.[154]

Korn and associates investigated the mode of action of α-polymerase-primase complex from human KB cells in the presence of variable concentrations of deoxyribonucleoside triphosphate.[167] These authors found that the size of the RNA made in vitro was strongly dependent on the concentration of deoxyribonucleoside triphosphates in the system. Thus, in the absence of deoxy substrates, the α-polymerase-primase complex synthesized a family of oligoribonucleotide molecules consisting of 24 to 36 residues each. At low concentrations, (0.08 to 0.8 μ*M)* of deoxyribonucleoside triphosphates, the primase-polymerase complex catalyzed the synthesis of mixed oligomers, which were composed of alternating tandem arrays of oligo ribo- and oligo deoxyribonucleoside units.[167] At high concentrations of deoxy-substrates (5 μ*M)* the product of the primase-polymerase complex was a single RNA oligomer (6 to 10 ribonucleoside residues), which was linked to a DNA chain.[167] A model was proposed which suggested that primase and polymerase-α constitute two separate active sites in the enzyme molecule, and their relative activity is determined by ambient deoxyribonucleoside triphosphates. Hence, primase quasi-processively synthesizes a canonical oligomer of 6 to 10 ribonucleotide residues. When DNA polymerase-α is not activated by ambient deoxynucleotide substrates, the primase will continue to perform two or three rounds of semiprocessive synthesis of RNA. However, if deoxyribonucleoside triphosphate activates the polymerase-α site, the RNA primer will be extended with deoxyribonucleotides. When substrates for the polymerase are at a limiting level, both polymerase and primase act tandemly. At high substrate levels, however, the polymerase-α mode is "locked" immediately after a canonical-length RNA is polymerized.[167] Primase activity requires four ribonucleoside triphosphates, but significant activity of the enzyme is observed in the presence of ATP alone.[47,155,162,163] This limited stringency for ribo-substrates could be due to a "relaxed" requirement for the synthesis of RNA that is entirely complementary to the template.[165] Evidence also suggests that ATP is required by the primase-polymerase-α complex both as a substrate for RNA synthesis and as a necessary component for DNA elongation past secondary structure barriers in the template DNA[160] (see Section V.F.2 of this chapter).

An in vitro system to analyze the initiation of DNA synthesis by mouse hybridoma cell primase that is free of α-polymerase has been identified recently by Tseng and Ahlem.[170] These authors examined the sites of RNA initiation on isolated SV40 DNA, which acts in vivo as a single replicon and in which a 65-base pairs region has been genetically defined as a *cis* acting essential origin of replication.[170] The in vitro synthesis of RNA by primase on separated strands of cloned SV40 DNA was found to be strongly site specific and to correspond closely with the replication origin region and with sites previously mapped as RNA-DNA junction sites.[170] However, all the initiation sites observed in this model system were on the leading strand away from the origin band. No initiation sites were found for the lagging strand. The possibility was raised that factors additional to primase are required for initiation of DNA synthesis on the lagging strand.[170]

C. Primer-Template Recognition Proteins

Several laboratories reported in the past few years the isolation of proteins from assorted cell types that modify the interaction of polymerase-α with the primer-template. Whether some of these proteins are subunits of an oligomeric polymerase-α or are rather auxiliary protein, which interact with the polymerase at the more intricate organizational level of the replicative complex, is unclear at the present time. Ancillary proteins that modulate the interaction of polymerase-α with primer and/or template will be classified in this section according to their observed or proposed roles. We have arbitrarily divided proteins that affect the activity of DNA polymerase-α into: (1) proteins which have been shown to destabilize double helical DNA, (2) primer recognition factors, and (3) assorted DNA binding proteins.

1. Helix Destabilizing Proteins

Herrick and Alberts[171-173] isolated from calf thymus a series of DNA binding proteins. These proteins were designated helix destabilizing proteins on the basis of their capacity to melt duplex DNA and DNA binding proteins on the basis of their preferential binding to immobilized DNA. One of these proteins, termed unwinding protein 1 (UP1), was purified to homogeneity and its size was determined to be 24 kdaltons.[171] UP1 melted double-stranded nucleic acids as monitored by its ability to enhance hyperchromicity for double strand DNA and for the synthetic hybrids poly[d(A-T)] and poly[r(A-U)].[172] That protein also melted fold-back hairpin structures from denatured DNA at temperatures lower than the DNA melting point.[172] The UP1 protein bound to phage fd single-strand DNA, saturating it at a ratio of one UP1 molecule per seven DNA bases, and maintained it in a rigid and extended conformation.[173] The activity of partially purified calf thymus polymerase-α was stimulated by UP1 in copying on the templates: oligo(dG)·poly(dC), heat-denatured DNA, and most notably, double-stranded DNA with single-strand template tails.[173] Amounts of UP1 in excess to that required to occupy all template sites inhibited α-polymerase, whereas the activities of β-polymerase and of rabbit bone marrow polymerase-α were not affected by UP1 at any amount tested.[173] Calf thymus DNA polymerase-α has been reported to catalyze *de novo* synthesis of poly[d(A-T)].[174,175] Addition of UP1 to the reaction mixture decreased the lag time for polymerization and increased its rate.[174] Analysis of the kinetics of synthesis of poly[d(A-T)] by α-polymerase indicated that UP1 stimulated the reaction by melting the heteropolymer and providing the enzyme with a single-stranded template. It cannot be ruled out, however, that the *de novo* synthesis observed in these studies was due to a few contaminating DNA molecules, which served as initial primer-templates. With this interpretation, stimulation by UP1 could be due to addition of more contaminating DNA with the protein, thus diminishing the time lag required for extensive synthesis. It also remains to be seen if highly purified polymerase-α is stimulated by UP1 as partially purified enzyme is or whether additional proteins are required for stimulation.

Cobianchi et al.[114] purified from calf thymus a single-strand DNA binding protein of 18 to 20 kdaltons which appeared to correspond to a selected basic subclass of a variety of synthetic and natural DNA templates, whereas polymerase-β activity was only slightly affected in its presence. Valentini et al.[176,192] surveyed a family of antigenically related single-strand DNA binding proteins from calf thymus. Although the different binding proteins differed in their ability to stimulate α-polymerase, it was found that in general, maximum stimulation was attained at a saturating protein to DNA ratio, suggesting stretching of the template by stoichiometric binding of the protein.

A helix unwinding protein from regenerative rat liver and an analogous molecule from nonregenerating liver were investigated by de Recondo and group.[177-182] A single-strand DNA binding protein, designated helix destabilizing protein 25 (HD25), was purified to homogeneity from regenerating rat liver.[178] HD25, which was found to be a tetramer of 25 kdalton subunits, melted poly[d(A-T)] and destabilized supercoiled SV40 DNA, presumably by melting (dA)·(dT)-rich regions in it.[178] Both activities of polymerase-α and -β were stimulated by HD25. The stimulation of α-polymerase activity was strictly dependent on the molar ratio of HD25 to DNA; at ratios higher than 1.7, HD25 became inhibitory for the α enzyme.[177] Stimulation of β-polymerase activity was evident only with DNA containing short gaps, becoming measurable at primer to template ratios >0.5.[177] A protein closely related to HD25, which appeared as a monomeric 25 kdalton polypeptide and which was designated S25, was isolated from normal rat liver. Protein S25 differed from HD25 by charge, its inability to form an oligomeric protein complex, and by two peptide differences upon comparative peptide mapping of the homogenous proteins.[180-183] In contrast to HD25, the monomeric S25 failed to melt DNA and did not stimulate α-polymerase under any of the conditions tested.[180,181] Preferential interaction of S25 with supercoiled DNA and reduction in the DNA linking number suggested that this protein is a structural element of chromatin.[181,182] It has been demonstrated recently that S25 is identical to the previously recognized nonhistone chromatin protein HMG1.[182] The close structural similarity between S25 (HMG-1) and HD25, and yet, their very different effects on DNA melting and on the activity of DNA polymerase, raises the possibility that limited posttranslational modifications of one of these proteins leads to drastic alterations in its biological activity.

Purification of a helix destabilizing protein from mouse cells was reported by Planck and Wilson.[183] That protein, designated HD1, enhanced the processivity of mouse cell DNA polymerase-α.[184]

In summarizing data on the functional relationship between α-polymerase and helix destabilizing proteins, it becomes evident that the primary site of action of these proteins is on the template DNA and that the activity of polymerase-α is affected mainly as a result of changes in the template. This conclusion is sustained by reports on the binding between helix destabilizing proteins and DNA and the physical changes which that binding induces in the DNA.[172,176,181,183] It is reasonable to assume that the primary in vivo function of at least some of these proteins is not concerned with polymerase-α. Hence, some helix destabilizing proteins display high binding affinity to ribohomopolymers;[172] one of these proteins, HD25, is not specific for α-polymerase and also affects polymerase-β activity,[179] and a related protein (S25) is an established structural component of chromatin.[182] Helix destabilizing proteins may be viewed, therefore, as a family of nucleic acid binding and modifying proteins with diverse biological functions. If these proteins merely exert their effect on polymerase-α through changes which they induce in the template, it is difficult to explain observations on tissue specificity of the responding α-polymerase.[173] As in the case of phage T4 replicating complex, specific interaction between some of these proteins and α-polymerase should be explored by immobilizing either one on a solid matrix and using it to identify associated proteins that will bind to the immobilized component.

2. Primer Recognition Protein Factors

Recent reports from the laboratories of Baril and DePamphilis demonstrated that animal cell α-polymerase may associate with proteins that serve as primer recognition factors by decreasing the K_m of the enzyme for the 3′-hydroxyl group primer terminus.[52,53,70,197] A single-strand DNA binding protein designated C1 was isolated from HeLa cells.[197] The protein factor, which was associated with α-polymerase in the crude cell extract, was separated from the enzyme by DNA cellulose chromatography. Reconstitution of α-polymerase with partially purified C1 increased 15- to 30-fold the activity of the enzyme with denatured DNA template.[197] The product DNA synthesized by the α·C1 complex was covalently linked to back-folded hairpin structure on the sparsely self-primed denatured DNA. Subsequently, three species of α-polymerase were separated and purified from HeLa cells.[70] One of the three molecular forms, an enzyme of 600 kdaltons which was designated polymerase-α_2, was associated with two protein cofactors C1 and C2, which were, respectively, a tetramer of a 24-kdalton polypeptide and a monomeric chain of 51 kdaltons.[70] The C1C2 cofactors preferentially enhanced the utilization of denatured DNA by polymerase-α_2 and could be dissociated from the enzyme by exposure to the nonionic detergent Triton®.[70] Analogous C1C2 factors have been purified from CV-1 cultured monkey cells.[52] The single observed effect exerted by C1C2 on α-polymerase was the considerable decrease in the K_m of the enzyme for the 3′-hydroxyl terminus of the primer stem.[53] Essentially all monkey cell polymerase-α is complexed with C1C2 in crude extracts, and the purified core enzyme can reassociate with the dissociated factors. It seems, therefore, that C1C2 cofactors serve as accessory subunits, which enable α-polymerase to replicate long stretches of DNA which are relatively sparsely primed as is the case in the advancing replicating fork.[52,53,127]

A less well-defined protein factor that may be analogous to C1C2 has been purified from calf thymus.[50] This 70-kdalton protein stimulates the utilization by α-polymerase of ϕX174 single-strand DNA singly primed with a complimentary restriction fragment. It has been suggested that the protein factor increased the affinity of the α-polymerase to the primer stem.[50]

3. DNA Binding Proteins

Different laboratories reported on DNA binding proteins, which modulate the activity of polymerase-α. The source of these proteins, their properties, possible mechanism of action, and roles in vivo are varied. Even though not documented, many of these proteins are likely to have helix destabilization activity.

A single-strand DNA binding protein of 30 to 35 kdaltons has been purified from mouse ascites cells.[185] That protein stimulates the activity of homologous mouse cell α-polymerase with denatured DNA template, but does not affect the activity of polymerase-β or of bacterial DNA polymerase. Maximum stimulation is attained with amounts of protein that saturate the template, and at higher levels, the protein factor becomes inhibitory to α-polymerase.[185] A requirement for salt to minimize dissociation of the enzyme and factor from template DNA suggested that a ternary complex is formed between these components.[186] Interestingly, the stimulatory factor could be phosphorylated by a chromatin-associated protein kinase, and the phosphorylated form of the protein was devoid of stimulatory activity.[185]

A 10- to 20-kdalton DNA binding protein factor, which stimulated the activity of homologous DNA polymerase-α on denatured DNA template, was recently isolated from mouse liver.[189,190] That protein, designated factor D, was found in association with chromatin and remained at a constant level during the cell cycle in the regenerating liver.[189] Purified factor D did not affect the activity of homologous β-polymerase and stimulated α-polymerase activity only with selected templates. Hence, in addition to copying of (dT)-rich tracts in denatured DNA, factor D stimulated the in vitro copying by α-polymerase of poly(dT) primed with either oligo(dA) or poly(dA). By contrast, the factor had no effect on the utilization by α-polymerase of activated DNA or of complementarily primed poly(dC), poly(dG), or poly(dA).[190,191] This intriguing template specificity was found, however, not to be limited to α-polymerase. Factor D did not stimulate the activities of mouse cell polymerase-β and of hamster cell polymerase-α, but it did exert template specific stimulation on calf thymus α-polymerase, γ-polymerase from HeLa cell and calf thymus, and DNA polymerase I from *E. coli.*[191] It is not clear, therefore, whether the primary target of factor D within the cell is polymerase-α. Factor D was found to decrease the K_m of responsive polymerases for sparsely primed heteropolymeric DNA and for primed poly(dT), but it did not affect their affinity for primed poly(dA) or dNTP substrates.[191] In contrast to its ability to bind to DNA, factor D was found to be unable to bind significantly to polymerase-α.[191]

Additional DNA binding proteins, which stimulate DNA polymerase-α activity, were detected in some lower animal cells. A single-strand DNA binding protein of 28 kdalton was isolated from *X. laevis*.[193] This 28-kdalton protein was not specific for α-polymerase and stimulated to a similar limited extent polymerases-β and -α. Two DNA binding proteins, which stimulated homologous polymerase-α by directly interacting with it, were isolated from sea urchin.[194,195] One of these factors, a 220-kdalton protein designated SF1, increased the V_{max} of α-polymerase without altering its K_m for deoxynucleoside substrates.[195] Analysis by gel filtration indicated that a complex was formed between the polymerase, SF1 and ATP and that ATP was hydrolyzed in the partially purified complex to AMP and pyrophosphate[195] (see also Section IV.C.4 of this chapter for discussion on polymerase-α stimulation by ATPase-catalyzed hydrolysis of ATP). Three probably related sea urchin replicative stimulatory factors of different molecular size were recently described.[196] These factors were isolated from sea urchin embryos, but were not detected in unfertilized eggs. Since DNA polymerase-α is present as a preformed enzyme in sea urchin embryos,[344] it is conceivable that one or more of these factors is involved in the initiation of the rapid DNA replication after fertilization.

4. Other Proteins Which Affect α-Polymerase

Several laboratories reported on proteins from assorted cell types that enhanced, depressed, or modified the activity of DNA polymerase-α.

Kalf and collaborators showed that the activity of polymerase-α declined sharply in giant rat trophoblasts toward the final stages of embryonic development.[198] Mixing experiments suggested that the decay in enzymatic activity was due to disappearance of an α-polymerase stimulatory activity. Subsequently, an 85-kdalton stimulatory protein was isolated from trophoblasts and was found to enhance the in vitro activity of the polymerase four- to sixfold on various natural and synthetic templates.[199]

A factor that may regulate the activity of α-polymerase via its interaction with deoxynucleoside triphosphates was isolated. In the presence of a 70-kdalton protein, which copurified with calf thymus polymerase-α, the copying in vitro of oligo(dT)·poly(dA) was inhibited by dGTP and dCTP with parabolic-competitive and linear-competitive kinetics.[200] Dissociation of the protein factor by gel filtration canceled the inhibitory effect of these substrates. The interesting possibility that ambient levels of deoxyribonucleoside triphosphates may regulate the rate of DNA synthesis by direct interaction with α-polymerase auxiliary protein(s) was raised.[200]

Méchali and de Recondo demonstrated that highly purified polymerase-α from regenerating rat liver was tightly associated with a protein which was responsible for binding of the polymerase to double-stranded SV40, which was devoid of 3′-hydroxyl termini.[201] After a nucleoprotein complex was formed between the SV40 DNA and α-polymerase, the binding protein could be dissociated from α-polymerase by sucrose gradient centrifugation. It has been suggested that a subunit of the rat enzyme is responsible for the DNA binding event which is functionally and physically separable, therefore, from catalysis of DNA synthesis.[201]

A role for DNA-dependent ATPase as a stimulator of DNA polymerase-α was documented. Cobianchi et al.[114] purified a DNA-dependent ATPase from human cells, which stimulated DNA polymerase-α to copy poly[d(A-T)] or supertwisted BK virus DNA that contained single-strand sequences primed by *E. coli* RNA polymerase. The

ATPase could unwind terminal portions of duplex DNA by binding to a single-strand region and moving in the 5′→3′ direction in respect to the bound DNA strand.[114] Boxer and Korn[115] purified to near homogeneity a 75-kdalton DNA-dependent ATPase from human cells. DNA unwinding activity could not be demonstrated for the KB cell ATPase,[115,116] but in the presence of ATP, it stimulated copying of nicked duplex DNA by both α- and β-polymerase.[115] It should be noted, however, that a DNA-dependent ATPase, which was purified from Novikoff hepatoma, specifically stimulated β-polymerase without affecting α-polymerase (cf. Chapter 4, Section III.B).

Finally, partially defined proteins, which specifically inhibit the in vitro activity of α-polymerase, were described in phytohemagglutinin-stimulated lymphocytes,[202] cultured rat hepatoma cells,[203] and the extracellular medium of calf thymus cells grown in culture.[204]

D. Polymerase-α and Diadenosine-5′,5‴-P¹-P⁴-tetraphosphate-Binding Proteins

Several lines of experimental evidence suggest that there might be links among the following: intracellular level of the purine nucleotide diadenosine 5′,5‴-P^1-P^4-tetraphosphate (Ap_4A) and positive regulation of DNA replication in animal cells and DNA polymerase-α.

The unusual purine nucleotide Ap_4A is a naturally occurring product of the back reaction of the amino acid activation step in protein synthesis.[205,206] Several authors noted a close inverse correlation between the intracellular levels of Ap_4A and replicative activity of cells. Rapaport and Zamecnik[207] found that intracellular levels of Ap_4A in several cultured cell lines varied considerably at different proliferative states. Whereas fluctuations in intracellular concentration of ATP were not related to the division time of cells, levels of Ap_4A ranged between 0.1×10^{-4} of the ATP concentration in slowly growing cells to 20×10^{-4} of its concentration in rapidly dividing cultures.[207] Furthermore, inhibition of cellular protein synthesis by selective antibiotics as well as arrest of entry into S phase by hydroxyurea led to 50- to 100-fold decrease in the intracellular level of Ap_4A.[207]

Following these observations, Grummt et al.[208] demonstrated that mitogenic stimulation of cultured 3T3 and BHK cells led to a 1000-fold increase in the intracellular concentration of Ap_4A. As these cells progressed through the G_1 phase into the S phase, the concentration of Ap_4A increased from 0.01 to 10 μM. When entry of serum-stimulated cells into S phase was blocked by hydroxyurea, intracellular levels of only 1 μM Ap_4A were attained.[208] Furthermore, V79 lung fibroblasts, which progress into S phase directly following mitosis, contained high basal levels of Ap_4A and its concentration increased only tenfold between mitosis and S phase.[208] The correlation between intracellular levels of Ap_4A and DNA replication was further documented for mouse Ehrlich ascites cells, which cease to replicate their DNA under anaerobic conditions. Two hours after anaerobiosis commenced, intracellular levels of Ap_4A decreased by about 50-fold, whereas ATP, ADP, and AMP were maintained at a virtually unaltered concentration.[209] In other experiments,[210,211] the exposure of permeabilized G_1 hamster cells to Ap_4A showed that within minutes after the addition of this purine nucleotide, the cells incorporated labeled thymidine into DNA at about half the rate of S-phase cells. The stimulatory effect of Ap_4A could not be mimicked by adenosine nucleotides and pyrophosphate, and it was abolished by inhibitors of DNA replication indicating that Ap_4A stimulates replicative rather than DNA repair synthesis.[210,211] Furthermore, electron microscopy revealed an increase in the number of replication eyes in DNA of quiescent cells which were exposed to Ap_4A, and pulse-chase experiments showed that DNA segment with size of Okazaki fragments of DNA were synthesized in these cells.[210,211]

A possible link of α-polymerase to the positive correlation between Ap_4A levels and

DNA replication has been experimentally explored.[63] Equilibrium dialysis of nearly homogenous calf thymus α-polymerase and Ap_4A demonstrated a strong binding between the two. The binding was highly specific to Ap_4A, as was shown by dialysis competition experiments with assorted analogs of the nucleotide. Furthermore, Ap_4A binding activity declined with the diminishing activity of α-polymerase in differentiating neurons. Lysates of mature, terminally differentiated neurons that contain only β- and γ-polymerases did not bind Ap_4A as determined by equilibrium dialysis.[63] Of the seven polypeptides which were contained within the apparently homogenous α-polymerase, only a 57-kdalton subunit was found to bind Ap_4A specifically. The binding of Ap_4A to the polypeptide was more heat labile and more *N*-ethylmaleimide resistant than the catalytic activity of α-polymerase.[63] It should be noted, however, that the subunit composition and size, which were reported in this study for calf thymus α-polymerase, differ considerably from the structure of that enzyme that was deduced by several other laboratories (see Table 1). Strong affinity between purified DNA polymerase-α and Ap_4A was also found for human DNA polymerase-α.[212] Ap_4A bound to 660- and 145-kdalton forms of HeLa cell polymerase-α at molar ratios of 0.5 and 0.08, respectively. The preferential association of Ap_4A with the 660-kdalton polymerase was due to its linkage to an Ap_4A binding protein, which could be separated from the polymerase by hydrophobic column chromatography.[212,213] In a recent work, the Ap_4A binding protein from HeLa cells was isolated and defined as a dimer of two 47-kdalton polypeptides.[214] In experiments on the calf thymus Ap_4A binding proteins, about half the binding activity copurified with a 450-kdalton α-polymerase, whereas the remaining binding activity was resolved as a 100-kdalton polymerase-free protein.[215] The polymerase-α-associated binding activity could be dissociated from the enzyme by hydrophobic chromatography. Both the polymerase-bound and free Ap_4A binding protein were found to be composed of a 54-kdalton polypeptide.[215] The free and polymerase-α-associated Ap_4A binding proteins differed only in that the former displayed an activity of Ap_4A phosphohydrolase.[215]

What can be the source of Ap_4A that tightly binds to the α-polymerase-associated protein? Tryptophanyl t-RNA synthetase copurified together with HeLa DNA polymerase-α and the aminoacyl t-RNA charging enzyme could be dissociated from the polymerase, together with the Ap_4A binding protein, by hydrophobic chromatography.[213,214] Whether the association of α-polymerase with that t-RNA activating enzyme is serendipitous or bears a functional in vivo significance is not clear. It should be noted that a defined subspecies of plant cell α-like DNA polymerase also copurified with tryptophanyl t-RNA synthetase, but it did not have a detectable Ap_4A binding activity.[216]

What could be the mechanism of modulation of α-polymerase activity by Ap_4A? It has been shown that Ap_4A could serve as an efficient primer for the in vitro synthesis of poly(dA) on a poly(dT) template by α-polymerase.[213] Furthermore it has been demonstrated that Ap_4A could prime the in vitro copying by α-polymerase of a synthetic double stranded octadecamer, which contained a part of the origin of replication of SV40 DNA, and it was experimentally demonstrated that the primer Ap_4A was covalently linked to the product DNA.[217] Whether the priming capability of Ap_4A in an in vitro polymerase-α directed DNA synthesis reflects its role in vivo is a matter of speculation. It still remains to be demonstrated whether the priming Ap_4A molecule is the same one that binds to the polymerase-α associated protein.

The correlation between intracellular level of Ap_4A and initiation of DNA replication and between the binding of Ap_4A to a polymerase-α associated protein are highly intriguing in their potential to offer a mechanistic model for the regulation of polymerase-α activity and perhaps for DNA replication in animal cells. Some major questions concerning the Ap_4A system remain, however, open at this time. First it is not clear how Ap_4A is able to transform quiescent cells into cells which start to actively replicate their genome several minutes after exposure to the nucleotide.[210,211] Theoretical considerations as well as various experimental systems to investigate the cell cycle emphasize the long preparatory G_1 period that preceded cell entry into S phase. Even if Ap_4A is the terminal positive pleotypic regulator that allows cells to enter into S phase, it is difficult to envision immediate G_1 to S phase transition without an adequate time period allotted for macromolecular synthesis. One such step may be indicated by the massively documented data showing that the activity of α-polymerase itself increases several fold prior to S phase (see Chapter 6). A second question concerns the association between α-polymerase and the Ap_4A binding protein. It is not clear whether the binding protein represents an integral subunit or accessory protein of the α-polymerase, or rather it is a fortuitously copurified protein. Similar questions concern the aminoacyl t-RNA synthetase. The significance of Ap_4A binding to polymerase-α is obscure. Although the ability of Ap_4A to serve as a primer in α-polymerase directed in vitro synthesis of DNA has been demonstrated,[213,217] its possible role in priming DNA replication in vivo is still a matter of speculation. It should be noted that Ono et al.[218] were unable to detect any stimulatory effect of Ap_4A on DNA polymerases-α, -β, and -γ from Ehrlich ascites cells, and the purine nucleotide failed to prime DNA synthesis by terminal deoxynucleotidyl transferase. A similar lack of stimulatory effect of Ap_4A on mammalian DNA polymerase was reported by others,[219] and Ap_4A inhibited the activity of DNA primase associated with calf thymus polymerase-α and did not act as a primer on phage M13mp7 single-stranded DNA.[220]

In summary, the tantalizing observations on a possible functional association between Ap_4A production and level in cells, DNA replication, and the activity of DNA polymerase-α do not provide at present a comprehensive mechanistic model for the involvement of Ap_4A in the regulation of α-polymerase activity and perhaps DNA replication in vivo.

V. CATALYTIC PROPERTIES OF DNA POLYMERASE-α

"Where order in variety we see, and where, though all things differ, all agree."

Alexander Pope
Windsor Forest

DNA polymerase-α, like all known prokaryotic and eukaryotic cell DNA polymerases, replicates polynucleotide templates by the sequential addition of deoxyribonucleoside monophosphate residues to the 3′-hydroxyl terminus of a primer stem.[122] The catalytic properties of DNA polymerase-α are defined, therefore, by its interaction with template, primer, deoxyribonucleoside triphosphate substrates, and divalent cation activator. The kinetic analysis of α-polymerase action is made extremely complex because of the large variety of templates and primers, which the enzyme utilizes, and because of the effect of protein cofactors and small molecules such as salts ATP and Ap_4A on the polymerization reaction. Furthermore, much of the kinetic data that were collected on polymerase-α in numerous studies were obtained either for enzymes, which contained only a part of the full complement of subunits and/or for polymerases that were partially degraded. In most cases, therefore, findings on kinetic parameters of purified DNA polymerase-α pertain mostly to the catalytic core of the enzyme. The possibility that the catalytic properties of an intact multmeric polymerase-α that interacts with other cofactors may differ from those of the isolated core should, therefore, be kept in mind.

A. Interaction of Polymerase-α with the Template

The relative rate at which different natural and synthetic DNA templates are copied in vitro by DNA-polymerase-α depends on the source of the α type polymerase, on its subunit composition, the presence of cofactors, and on specific conditions of the polymerization reaction. Examples for the effects of cofactors such as primase, C1C2, and other proteins on the utilization of unprimed and sparsely primed templates are given throughout this chapter.[45-48,50,52,53] Other works cited under various sections demonstrate different template utilization efficacies by different molecular forms of polymerase-α.[38,83,90,97,98] In general all of the templates that α-polymerases utilize are polymers of deoxyribonucleosides. During the early stages of research on animal cell DNA polymerases, it was found that DNA polymerases-β and -γ copied at a high efficiency the ribo strand of oligo(dT) · poly(rA) (see Chapters 4 and 5). It was initially assumed that α-polymerase cannot copy the ribo strand in such primer template. However, accumulated evidence indicated that α-polymerases from sea urchin,[8] *D. melanogaster*,[3] human KB cells[221], and human IMR cells[119] succeeded in copying poly(rA) primed with oligo(dT) albeit at a relatively low efficiency. It has been established that the 10-S form of calf thymus polymerase-α can utilize polyriboadenylic acid as template under defined specific reaction conditions.[96,97] However, to our knowledge, there are no definitive studies suggesting that α-polymerase is able to utilize natural RNA as a template.

One of the more significant catalytic properties of α-polymerase is its preferential interaction with single-stranded DNA. In principle, heat-denatured DNA is a relatively poor template presumably due to its low content of fold-back primer stems. Yet early observations of α-polymerase from *D. melanogaster* which was probably partially degraded during its isolation, indicated that the enzyme copied partially denatured activated DNA even better than activated DNA.[3] This result was interpreted as indicating that the polymerase required stretches of single strands for binding to the template. Copying of denatured DNA at a measurable rate was also observed for homogeneous *D. melanogaster* α-polymerase[4] and polymerase-α from calf thymus cells.[222] The ability of α-polymerase to utilize the single-stranded denatured DNA template was attributed to binding of the enzyme to single-strand stretches which, in turn, enable the elongation of primers formed by intrastrand base pairing between the 3′ terminus of the template and a complementary sequence within the template.[223] A detailed mechanistic study on the interaction of α-polymerase with the template DNA and with its other reactants was conducted by Korn and group.[43,44,221,223-227] In these studies, a nearly homogeneous KB cell DNA polymerase was used, which was probably partially degraded and did not possess primase activity (See Reference 57 and Section II of this chapter). A detailed analysis of the interaction of the enzyme with various templates was performed by combining kinetic competition between templates[44,223,224,227] and assessing directly enzyme-template binding on glycerol gradients.[224] Competition experiments, as well as direct velocity sedimentation analysis of template enzyme binding revealed that homogeneous polymerase-α did not interact with supercoiled or relaxed DNA duplex. Furthermore, the polymerase did not interact with 3′-hydroxyl ends in single-stranded M-13 DNA. It was concluded therefore, that the interaction of α-polymerase with the template requires initial binding of the enzyme to single-strand region.[44,223,224] Furthermore, KB polymerase-α was found to bind preferentially to (dT)-containing homopolymers and heteropolymers.[44,227]

Similar preference of mouse cell polymerase-α for binding to (dT)-rich sequences in DNA has also been described.[228] These observations have been interpreted as reflecting preference by the polymerase for (dT) rich sequences in the template.[227] It was further hypothesized that (dA)-(dT)-rich regions in the genome melt, and the polymerase binds to these single stranded tracts to commence replication.[224,228] Evidence has been presented that interaction of KB cell polymerase-α with single-stranded DNA resulted in allosteric activation of a second template binding site within the enzyme molecule.[44,228] A model was proposed by Korn and associates, which described the events of polymerase-α binding to the template.[44] According to that model, α-polymerase identifies single-stranded regions on the template, binds to them, and as a result of the binding event, a second template binding site in the enzyme is dynamically activated, which consequently leads to activation of the primer binding site.[44] The model is based mainly on kinetic studies and should be confirmed by physical identification of binding sites in the α-polymerase molecules. Using denatured and native DNA as competitors for active DNA with fish α-polymerase, others[346] failed to observe cooperative binding of DNA. It has been suggested that the fish enzyme contains only a single DNA binding site. It was further speculated that the cooperativity obtained with KB cell enzyme could be due to aggregation of the polymerase-α molecule.[346]

B. Interaction of Polymerase-α with the Primer Stem

DNA polymerase-α interacts in vitro and extends both deoxyribonucleotidyl and ribonucleotidyl primer stems. Studies on α-polymerase from various cells types indicated that they, in contrast to β- and γ-polymerases, are able to elongate in vitro synthetic and natural RNA primers.[66,69,105,146,148,230-234] The failure of polymerases-β and -γ to utilize RNA primers and the ability of α-polymerases to extend such primers constituted an early indication to the possible involvement of α-polymerase in RNA primed discontinuous synthesis of DNA (see Section IV.B of this chapter).

A detailed kinetic analysis of the interaction of KB cell polymerase-α with both ribo and deoxyribonucleotidyl primer stems was conducted by Fisher and Korn.[44,225] Using hook polymers that were terminated with either ribo- or deoxyribonucleotides, it was demonstrated that α-polymerase extended both types of primers with similar K_m and V_{max}. The minimum length primer that was required by the KB cell polymerase-α to initiate DNA polymerization was found to be an octamer. This minimum size was invariable with changing reaction temperature or with base composition of the primer.[44,225] Furthermore, a single mismatch at the end of the primer stem prevented binding of the polymerase to the primer terminus.[44,225] This was determined by sedimentation on sucrose gradients of enzyme and template assaying for free and bound DNA polymerase activity. Lastly, although presence of H, OH, or PO_4 groups at the 2′ terminal position of the primer did not affect binding of the α-polymerase, the existence of the 3′-phosphoryl group at the primer terminus diminished polymerase-primer association.[225]

In a recent study on the in vitro replication of phage M13 DNA by calf thymus 9-S polymerase-α, oligo(dA) and oligo(rA) primers of variable length were used to initiate DNA synthesis.[220] A minimal primer length of four residues was required, optimum DNA synthesis was attained with hexameric primer, and oligo riboadenylic primers were slightly more efficient than deoxyriboadenylic initiators.[220] In a detailed study on the effect of partial mismatch between primer and template on DNA elongation by calf thymus DNA polymerase-α, Reckman et al.[95] used primers of the type $(dT)_8dG$, $(dT)_8dC$, and $(dT)_{10}dGdT$ to initiate synthesis on a poly(dA) template. Both the 5.7-S and 9-S forms of α-polymerase extended these mismatched primers at a much lower rate than they did with a perfectly complementary primer. The rate-limiting step in elongating the mismatched primer was the addition of the first nucleotide after the mismatch. Only after five to six perfectly paired residues were added to the primer did the rate of polymerization return to its maximum level.[95] The low rate of α-polymerase binding to a mismatched primer terminus and its inefficient elongation may have mechanistic significance vis-a-vis the fidelity of DNA synthesis.

C. Interaction of Polymerase-α with Metal Activators

Catalysis of DNA polymerization by DNA polymerase-α depends on the presence of metal activator. Maximum activation of α-polymerases is exerted by magnesium ions, whereas manganese ions are severalfold less efficient in activating polymerases of the α class. The major function of divalent metal ions in catalysis by DNA polymerases is likely to involve coordination between the incoming nucleotide substrates and the active catalytic site on the polymerase molecule. Use of electron spin resonance methodology with *E. coli* DNA polymerase I led to the identification of one tight metal binding site on the enzyme which appears to be involved in nucleotide binding. The occupancy of the single site on polymerase I is sufficient to activate DNA polymerase reaction. The metal coordinates a site on the enzyme to the β and γ phosphate in the triphosphonucleotidyl substrate, thus facilitating the departure of pyrophosphate in the polymerase reaction. Unfortunately, available amounts of purified α-polymerase have been insufficient to conduct similar studies.

Additional aspects of α-polymerase interaction with magnesium was investigated by Korn and associates.[44,225] They showed that magnesium ions enhanced binding of the enzyme to single-stranded DNA and to polypyrimidines, but not to polypurines. In addition to its effect on template binding, magnesium was found to be involved in the binding of the primer stem by α-polymerase. Kinetic data indicated that magnesium competitively inhibits replication of $(dT)_{15}$·poly(dA) whereas it does not affect enzyme bindings to poly(dA). The requirement for a primer of minimum length of eight nucleotides as well as the lack of structural similarity between competing metal and primer led to the proposal that normal binding of an octameric primer stem by polymerase-α occurs through a magnesium primer complex which involves the coordination of four magnesium ions with 7-phosphate residue in the phophodiester backbone of the primer. Furthermore, multiple magnesium binding sites in α-polymerase were suggested by kinetic analysis of the effect of metal ion on purine-pyrimidine hook template-primer. It was postulated, therefore, that KB cell α-polymerase contains four magnesium primer binding subsites.[44,225] Since octameric primers possess only seven phosphodiester bonds and yet served as optimal length initiators for KB cell α-polymerase, it was suggested that one of the four magnesium ions might be available for polymerase translocation along the template, subsequent to the insertion of the next deoxynucleoside monophosphate residue.[44,225] The effect of magnesium on the binding of α-polymerase to template and primer have been shown by direct sedimentation analysis to be independent of the presence or absence of deoxynucleoside triphosphate substrates. Hence, magnesium ions are required for the α-polymerase nucleic acid interaction independently of catalysis of DNA polymerization.[225] The low intracellular concentration of magnesium and the very high affinity of deoxyribonucleotide triphosphates to the metal raised the possibility that the deoxynucleotide substrates bring magnesium to the enzyme, and in turn, the metal ions are cycled during the polymerization.[225]

The central role of metal ions in affecting intrinsic mutagenesis by polymerase-α and animal cell polymerase of the β and α categories is discussed in Chapter 8.

D. Nucleotide Substrates for DNA Polymerase-α

In addition to the four common deoxyribonucleoside triphosphate substrates, DNA polymerases-α, as well as β-polymerases, are able to utilize deoxyuridine triphosphate (dUTP) as a substrate for DNA polymerization in vitro.[235,236] However, the nucleotide dUMP is not present in DNA made in vitro in isolated nuclei.[237] Hence, either dUTP is hydrolyzed prior to its incorporation or it is excised from the DNA product by uracil *N*-glycosylase.[237]

Studies on the competitive inhibition between the antibiotic aphidicolin and dCTP and dTTP utilization by α-polymerase, but not dATP and dGTP utilization, can be interpreted to indicate that the polymerase-α molecule includes at least two substrate binding sites, one for purine and the other for pyrimidine nucleotides (cf. Section VI.B of this chapter).

E. The Initial Ordered Interaction of Polymerase-α with its Reactants

The initial interaction of polymerase-α with its reactants,i.e., template, primer, and deoxynucleoside triphosphate substrates was investigated by several research groups.[184,226,238] The various studies of this problem resulted in a similar conclusion that in contrast to equilibrium-random mechanisms, which have been suggested for several polynucleotidyl enzymes, polymerase-α reacts with its substrates in a truly ordered fashion. Tanabe et al.[238] conducted substrate kinetic experiments which indicated an ordered mechanism for the interaction of mouse cell α-polymerase with activated DNA primer template and nucleotide substrates. Wilson and associates[184] studied the steady-state kinetics of template and substrate binding of α-polymerase and constructed a kinetic model for analysis of initial velocity data and termination probabilities. The proposed model described in detail the sequential interaction of polymerase-α first with the primer templates followed by nucleotide substrate. In addition, the model distinguished between the initial polymerization step and elongation. It was indicated that the rate-limiting step occurs at the transition between initiation and elongation and that this step is facilitated at higher levels of primer template.[184]

A detailed steady state-kinetic analysis of KB cell α-polymerase, which was conducted by Fisher and Korn,[226] established the order of the initial events of the interaction of the polymerase with its reactants. Using steady-state kinetic analysis in conjunction with velocity sedimentation analysis of polymerase-substrates binding, the steps of enzyme binding to its reactants were delineated and ordered. An innovative approach was the use of 2′,3′-dideoxy-blocked primer stem, which could not be elongated and thus permitted analysis of enzyme-substrate complex formation without DNA polymerization catalysis. With such blocked primers, the formation of an abortive complex of template-polymerase-primer-complementary nucleotide substrate could be demonstrated.[226] Both kinetic studies and direct assessment of complex formation corroborated a model that specifies initial binding of polymerase-α to the template, followed by primer binding with deoxyribonucleotide triphosphate being the third bound reactant. Specification of which of the four deoxyribonucleoside triphosphates is added to the enzyme is determined by base sequence of the template.[226] The following scheme summarizes this model for the ordered interaction of polymerase-α with its reactants.

$$E \underset{K_2}{\overset{K_1}{\rightleftharpoons}} E \cdot \text{template} \underset{K_4}{\overset{K_3}{\rightleftharpoons}} E \cdot \text{template} \cdot \text{primer} \underset{K_6}{\overset{K_5}{\rightleftharpoons}}$$

$$E \cdot \text{template} \cdot \text{primer} \cdot \text{dNTP} \overset{K_7}{\rightleftharpoons} E + \text{products}$$

This scheme is a partial steady-state kinetic model for the sequential interactions of KB cell α-polymerase (E) with its reactants. The scheme is a simplification of the model proposed by Fisher and Korn on the basis of their kinetic analysis.[226] Concentrations of reactants that may vary and thus affect the kinetics of polymerase interaction with the reactants are not presented (for detailed description of experimental and theoretical considerations see References 44 and 226).

F. Elongation by DNA Polymerase-α

1. Processivity

A basic problem concerning the mechanism of DNA chain elongation catalyzed by any DNA polymerase is the question of processivity of the reaction. It is inquired how many phosphodiester bonds are formed by a polymerase following its asociation with the template and preceding dissociation of enzyme from template. A polymerase that is completely nonprocessive or distributive will bind to the primer-template, add only a single deoxynucleoside monophosphate residue to the primer terminus, and dissociate from the template. A quasi-processive polymerase will add few residues to the growing DNA chain following each enzyme-template association event. A processive enzyme, on the other hand, will be able to catalyze the polymerization of a large number of residues following each template-binding event. Considering the low rate of collision between two macromolecules, which is limited by rate of diffusion, it must be assumed that rapid synthesis of DNA in vivo depends on a completely processive mechanism. Thus, if purified DNA polymerases are not processive, it is most probable that ancillary cellular factors render them such *in situ.*

In a pioneering study, Chang[239] devised general methods for the assessment of the processivity of DNA polymerases. In one method, calf thymus polymerases-α or -β copied excessive oligo(dT)·poly(dA) primer-template, and during the reaction the enzyme was challenged with oligo(dG)·poly(dC). The assumption was that utilization of the challenging template will be immediate only for distributive enzyme, whereas a processive polymerase will remain associated with the initial oligo(dT)·poly(dA) template for some time, and the new oligo(dG)·poly(dC) template will be copied only after a lag period.[239] Since polymerase-α did not display a lag in the copying of the challenging template, it was concluded that it is a nonprocessive enzyme.[239] That conclusion was sustained by observing synchronous extension of excessive primer molecules and by sedimentation analysis of the rate of increase in length of the product DNA chains.[239] Using the same method of template-challenge, McKune and Holmes[84] concluded that both the A and C forms of α-polymerase from calf thymus also appear to be distributive.

A more sensitive method for the measurement of processivity, which can determine processiveness values ranging between one and several hundred residues added per polymerization cycle, was introduced by Bambara et al.[240] The assessment of processivity by that method involved comparison of rates of copying of heteropolymeric natural DNA in the presence of a full vs. partial complement of nucleotide substrates. Basically, the more distributive a polymerase is, the higher will its relative rate of catalysis be in the presence of a partial complement of substrates. By devising a way to reduce recycling of the polymerase on the same primer molecule, the sensitivity and reliability of this method was demonstrated.[240] Unlike the template challenge technique, this method is independent of considerations of differences between the affinities of various DNA polymerases to different primer templates. In studies with human KB cell homogenous DNA polymerase-α and with activated DNA, it was demonstrated that the polymerase was quasi-processive incorporating an average of 11 ± 5 nucleotides per binding event and that that value was not affected by the size of the DNA gap filled.[221] This low level of processiveness was well below the limits of detectability of the template-challenge method.[239] In studies on the processivity of homogenous α-polymerase from *D. melanogaster,* it was demonstrated that the "holo" enzyme, which consisted of three or possibly four different subunits, copied phage M-13 single-stranded DNA with a processivity of 14.2 ± 3.1, whereas the catalytic subunit alone displayed a lower processivity of 5.0 ± 1.3.[105] These results support the contention that isolated α-polymerase is semiprocessive, and furthermore, they suggest that accessory subunits of the enzyme may increase the inherently low processiveness of the catalytic subunit *(vide infra)*. Nevertheless, the processivity of the holoenzyme also appears to be too low in considering requirements for a replicative polymerase. In experiments using the same method[240] for the determination of processivity, fish polymerase-α yielded a value of 5.6 ± 1.3 nucleotides polymerized per enzyme binding event.[346] In the presence of spermine at 0.5 m*M*, processivity increased modestly to 8.2 ± 1.6. Hence, the possibility that ambient small molecules might also affect processivity should not be overlooked.

That isolated polymerase-α indeed extends DNA in a semiprocessive manner was also demonstrated by Das and Fujimura.[241] These authors developed a method for assessing the processivity of copying synthetic homopolymeric DNA. Oligo(dT)·poly(dA) was copied in the presence of [^{3}H]dTTP under conditions that minimized repetitive utilization of the same primer by the polymerase. Product poly(dA) was isolated by oligo(dT)-cellulose chromatography and its length determined by differential enzymic digestion, which yielded the ratio of residues within the synthesized DNA segment to the number of 3′-hydroxyl termini extended. A value of about ten residues polymerized per template-α-polymerase encounter event was obtained at both 21 and 37°C.[241] A modification of this method, which adapted it to the use of a heteropolymeric phage ϕX174 template, was introduced by Grossberger and Clough.[242] These authors found that human lymphocyte DNA polymerase-α was quasi-processive, polymerizing 19.5 ± 4.5 nucleotides per polymerization cycle.[242]

Using poly(dT) as template and oligo(rA) as a primer, Wilson and associates[184] could directly measure lengths of product DNA made in a single cycle of mouse polymerase-α binding and synthesis. Excess of primer-template over enzyme as well as the preference of the enzyme for (rA) over (dA) primer termini, ascertained that the size of DNA that was measured electrophoretically reflected the length of products of a single cycle of polymerization. The chain length distribution of polymerase-α products indicated that the probability of chain termination decreased sharply between the first and fourth residues added by α-polymerase and subsequently increased linearly up to the tenth residue at which it reached a value of 0.45.[184]

The quasi- to low-processivity of polymerase-α does not satisfy theoretical requirements from an enzyme that is regarded as being responsible for DNA replication in vivo (see Chapter 6). Several investigators focused on factors which may increase or modulate the processivity of DNA polymerase-α. First, the subunit composition of the enzyme was considered. As described in Section II of this chapter, forms A1 and A2 of calf thymus polymerase-α can be converted by mild urea treatment to form C, which is devoid of a smaller subunit of the enzyme. Hockensmith and Bambara[85] found that form A1 possesses higher processivity than form C. That conclusion was in contrast to results of a previous similar comparison made with the less sensitive template-challenge method.[84] Several agents such as spermine, spermidine, nalidixic acid, and pyrophosphate were found to increase the termination probability after the first nucleotide residue was added to the primer terminus by mouse cell polymerase-α.[84] By contrast, the helix destabilizing protein UP1 failed to increase product length on natural DNA template, which was copied by calf thymus polymerase-α form A2, implying that it did not affect processivity.

Several authors investigated the effect of ATP on processivity. Initially, Bambara and his group[243] found that ATP stimulated the activity of calf thymus polymerase-α and increased its processivity three- to sixfold on poly(dA) primed with oligo(dT). These researchers also reported that although ATP exerted its stimulatory effect on polymerase-α by increasing its binding to the primer-template, it did not affect the processivity of the enzyme on gapped fd RF DNA template.[219] Knippers and associates[244] found that in the presence of ATP, polymerase-α from *X. laevis* synthesized longer product DNA molecules. Although formal quantitative analysis of processivity was not presented, the authors interpreted their results as indicating that ATP stabilized the interaction between α-polymerase and template DNA and consequently, increased the processivity of the enzyme.[244] A strong argument in support of the notion that the inherently low processivity of polymerase-α can be modulated by accessory proteins was provided by a study of the effect of *E. coli* single-strand DNA binding protein (SSB) on polymerase-α from *D. melanogaster*.[105] It was found that whereas the processivity of isolated α-polymerase on multiply primed M13 DNA was 14 ± 3.1, the addition of SSB to the system increased the processivity to 285 ± 4.5.[105] Interestingly, SSB affected only the processivity of an oligomeric enzyme whereas it failed to exert any effect on isolated catalytic subunit.

2. Barriers to DNA Elongation

Several laboratories have addressed the problem of structural signals in the template DNA which act as barriers to the advancement of polymerase-α. Weaver and DePamphilis[245] designed a comprehensive method for the determination of the sequences in natural DNA template that slow or terminate DNA synthesis by polymerase-α. Phage ϕX174 single-stranded circular DNA was primed at defined positions with restriction fragments which were terminally labeled at their 3′ ends. The number of product molecules could be assessed by their radioactivity, and the precise location and base composition of their termination site was determined by sequencing methods. Polymerases-α from CV-1 and HeLa cells from calf thymus were all arrested at the same sites along the template. The arrest loci were distributed nonuniformly on the template and included high proportion of G-C bases and low ratio of A-T rich sequences.[245] About one third of the stop signals possessed computer-predicted hairpin secondary structure, and the α-polymerase usually halted proximally to its stalk. However, in most cases the advancement of α-polymerase was stopped up to 25 bases upstream to any predicted hairpin structure. Moreover, results obtained in that study did not support the possibility that arrest sites are formed by long-range interactions between remote regions on the template.[245] It should be noted that computer predictions of DNA secondary structure are based on differences in energies between different base pairings, the values of which were derived mainly from ribo- rather than deoxyribonucleotide interactions.

Villani et al.[105] used singly or multiply-primed phage M-13 single-strand DNA as a template for homogenous, α-polymerase from *D. melanogaster* to study the pattern of DNA synthesis in vitro. It was found that DNA elongation in that system was terminated after several hundred nucleotides, but that addition of *E. coli* SSB increased the length of the products by about twofold.[105] These results were interpreted as an indication for the ability of SSB to relax secondary structures in the template that may act as barriers for the advancement of α-polymerase. Kaguni and Clayton[246] inserted the heavy strand origin of mouse mitochondrial DNA into the duplex replicative form of phage M13 DNA. With this insert serving as a template for *Drosophila* polymerase-α, it was found that products of discrete size were synthesized in vitro. Since DNA polymerase III from *E. coli* paused at most of the same sites as *Drosophila* α-polymerase, it was suggested that arrest of DNA synthesis at particular regions is a property of the template DNA rather than of the polymerase. A search for common structural features of the stop signals revealed that although some of those sites had a potential to form hairpin structures, many of the stopping sites did not possess a potential for secondary structure.[246] In studies with homogenous and intact *Drosophila* α-polymerases, it was found that the enzyme dissociated repeatedly at specific sites on singly primed ϕX174 DNA.[108] That these sites are kinetic pause sites rather than obstructions to polymerization could be demonstrated by the production of longer DNA molecules with extended incubation times and by the appearance of fully replicated molecules when polymerase-α was added in excess to template.[108] Similarly, calf thymus 9-S polymerase-α paused at several discrete sites along M13mp7 single-stranded template DNA, and the addition of *E. coli* SSB permitted the polymerase to proceed through some of the pausing loci.[220]

In another study, computer-predicted secondary structures in M13mp8 single-stranded DNA could be correlated with pausing sites for calf thymus α-polymerase.[247] Results of this analysis indicated that all replication barriers could be correlated with secondary structures predicted by computer calculations, which allowed for long-range base pairing, G-T mispairs, and looping-out of bases. Calf thymus 5.7-S and 9-S polymerase-α were found to stop at a position two bases in front of the hairpin structure, and from there on they polymerized few more nucleotides in a fully distributive manner.[247] Although *E. coli* SSB facilitated copying over the pause sites, it did not eliminate them completely. The full correlation between pause sites and secondary structures in the templates as described in this study, stands in contrast to earlier reports that α-polymerase paused also at loci with no apparent secondary structure.[245,246] This discrepancy may be due to differences in stringency of the rules for secondary structure formation that were put forward by the different authors.[247] Whether secondary structures are at all formed in vivo and whether they pose barriers for the advancement of the replicative complex in the cell is completely unknown at present. In vitro studies may, however, be useful for the detection of α-polymerase associated factors that enable the enzyme to proceed through replicative obstructions in the template.

A possible permissive role for ATP in copying over pause sites has been suggested.[160] DNA polymerase-α from *X. laevis* stopped at discrete sites on fd circular single-stranded DNA template. In the presence of ATP, however, longer products were obtained due to copying over stop sites by the polymerase.[160] Subsequently, it was found that ATP increased the binding of α-polymerase to singly-primed fd DNA and allowed it to copy over replication barriers.[244] A parallel study on in vitro copying of MVM parvovirus single-stranded DNA by Ehrlich ascites cell DNA polymerase-α demonstrated premature termination of synthesis in the absence of ATP or alternatively, GTP.[51,68] In contrast to these results, Reckman et al.[247] failed to observe any effect of ATP on the pattern of pausing of 9-S calf thymus α-polymerase on an M13 DNA template.

G. Gap Filling and Long-Stretch DNA Synthesis by α-Polymerase

A catalytic parameter, which is used to characterize DNA polymerases, is their preference for gaps of discrete size in duplex DNA and their ability to fill such gaps to completion as measured by sealability of the patched gap by DNA ligase. This parameter may reflect in vivo activities of DNA polymerases in short and long patch DNA repair and thus could serve as an in vitro criterion for involvement of these enzymes in different modes of DNA repair (see Chapter 7). Unfortunately, the disparity of results obtained with α-polymerases from different cell types does not allow for a simple unifying concept on its involvement in certain aspects of DNA repair.

Korn and colleagues[43,221,250] found that whereas KB cell DNA polymerase-β was most reactive on short gaps of about 10 nucleotides and was able to fill them to completion, polymerase-α was unreactive with gaps smaller than 30 nucleotides and failed to fill to completion gaps of 30 to 60 nucleotides, which it did utilize. It was reported that CV-1 monkey cell polymerase-α also failed to fill to completion gaps of 40 nucleotides. However, calf thymus and HeLa cell polymerase-α were capable of filling such gaps.[248] Human lymphoblastic polymerase-α was found to be capable of filling single nucleotide gaps in DNA.[338] By contrast, HeLa cell polymerase-α was incapable of filling one nucleotide gap, whereas β-polymerase did utilize such nicks.[339] In studies with partially depurinated phage PM2 DNA, gaps of discrete size were created in the DNA by the combined action of HeLa apurinic/apyrmidinic endonuclease II and HeLa DNase V.[340] It was found that HeLa cell polymerase-α could not utilize gaps smaller than 15 nucleotides, whereas gaps of 20 to 63 nucleotides supported limited DNA synthesis which was terminated after the gap size was reduced to 15 nucleotides.[340] This limited ability of α-polymerase to fill gaps to completion contrasted with the efficient utilization and filling to completion of gaps of all sizes by DNA polymerase-β.[340] Some of the contrasting evidence on the ability of polymerases-α from different sources to utilize gaps of certain size may perhaps be explained by differences in the subunit composition of enzymes that were used in the various studies. In accord with this presumption is the finding that form A1 of calf thymus DNA polymerase-α utilized gaps of 65 nucleotides more efficiently than form C, which lacked the 50- to 70- kdalton subunit.[85]

The ability of polymerase-α to extend primers opposite long stretches of single-stranded template DNA was investigated by several authors. Hurwitz and associates[234] found that DNA polymerase-α succeeded in elongating RNA primers on linear ϕX174 DNA template only for 230 to 390 residues. Addition of polymerase-β to the reaction system led to completion of 4500 base pair long RF structure. Similarly, Matsukage and group[14,249] reported that α-polymerases from chick and mouse cells synthesized in vitro relatively short DNA chains, about the size of in vivo Okazaki fragments, which could then be further extended by β-polymerase. Numerous studies have demonstrated, however, that under proper conditions, polymerases-α isolated from *D. melanogaster,*[105] *X. laevis,*[160] Ehrlich ascites mouse cells,[51,68] and calf thymus[220] were able to extend RNA or DNA primers by up to several thousand nucleotides.

VI. SELECTIVE INHIBITORS OF DNA POLYMERASE-α

There has been a continuing and extensive interest in searching for selective inhibitors of animal cell DNA polymerases and in particular DNA polymerase-α. It was hoped that selective arrest of DNA polymerase-α, the putative major replicative enzyme, will serve in chemotherapy of malignant cell growth. Research on discriminatory inhibitors of any of the several classes of eukaryotic DNA polymerases falls into two broad categories: (1) identification of inhibitors, determination of their inhibitory potency toward each specific type of DNA polymerase, and elucidation of their mechanism of action, and (2) use of inhibitors of defined selectivity and potency to unravel the in vivo role of each of the disparate types of DNA polymerase in repair and replication of nuclear, mitochondrial, and viral DNA. In this section we review research on inhibitors of α-polymerase, which is limited to the first area of investigation. The use of selective inhibitors to construe the rôles of α-polymerase in replication and repair of DNA is discussed in Chapters 6 and 7, respectively.

A. Sulfhydryl Group Reactive Reagents

Reagents that react irreversibly with sulfhydryl groups such as *N*-ethylmalemide (MalNet), *p*-chloromercuribenzoate (pCMB), and *p*-hydroxymercuribenzoate (pHMB) interact with any available thiol group on a protein molecule and thus have low selectivity. Nevertheless, reagents of this type were proved to be most useful in distinguishing the highly sensitive polymerase-α from the resistant β-class polymerase.[16] Whereas partially purified HeLa cell β-polymerase was resistant to low concentrations of pCMB, polymerase-α was completely inactivated.[16] Survey of polymerases of the α class from a large variety of cells and tissues established their sensitivity to MalNet at concentrations that commonly ranged between 0.1 and 10 m*M*.[32,33,251,252] In almost all the cases that were examined, DNA polymerase-β was found to be totally resistant to MalNet (see Chapter 4, Section V.A). Although pCMB and pHMB are also potent

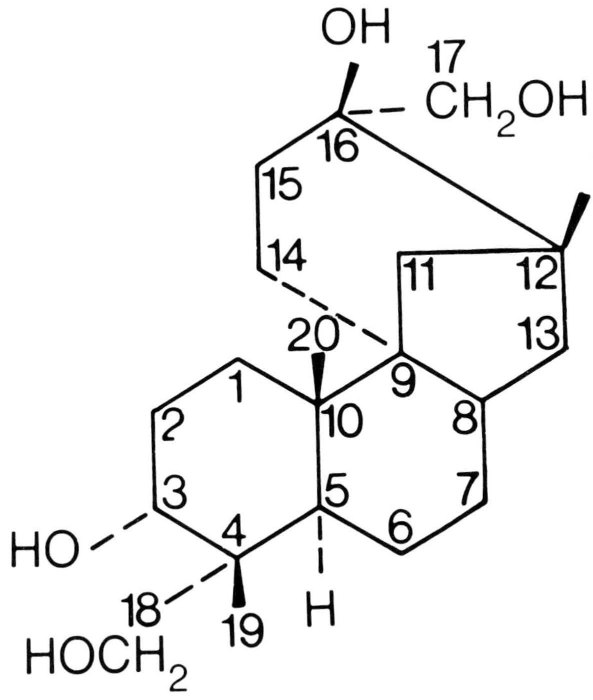

FIGURE 2. Structural formula of aphidicolin: a mycotoxin from cultures of *Cephalosporum aphidicola* that specifically inhibits DNA polymerase-α activity.

inhibitors of polymerases of the α class, their discriminatory power is lower than that of MalNet. DNA polymerase-β is resistant to low concentrations of these agents, but is inactivated at higher concentrations (Chapter 4, Section V.A). DNA polymerase-γ is as sensitive to sulfhydryl group blocking reagents as is α-polymerase (Chapter 5, Section I.C.1). Iodoacetamide was also found to be inhibitory to DNA polymerases from rat liver in the order of potency $\alpha > \gamma > \beta$-polymerase.[253] The intimate involvement of thiol groups of the α-polymerase molecule in its catalytic activity is implied by the inactivation of the enzyme by sulfhydryl group blockers. So far, this knowledge has not been put to use in investigations on the molecular details of the mechanics of catalysis. Involvement of thiol group inactivation in the mechanism of inhibition of rat liver α-polymerase by certain carcinogens was suggested by Chen and Becker.[253] The carcinogens methylnitrosourea (MNU), acetoxyacetylaminofluorene (AAF), and *N*-methyl-*N'*-nitro-*N*-nitrosoguanine (MNNG) were all found to be potent inhibitors of α-polymerase, less effective inactivators of polymerase-γ, and least inhibitory toward β-polymerase. Interestingly, inhibition of α-polymerase by these carcinogens could be reversed by the thiol group reducing agent dithiothreitol — suggesting direct interaction between the carcinogens and exposed sulfhydryl groups on the α-polymerase molecule.[253]

B. Aphidicolin

Aphidicolin, a highly specific inhibitor of animal cell DNA polymerases-α, is an antibiotic which is isolated from the fungus *Cephalosporium aphidicola* and from other fungi.[34,254] Aphidicolin is a tetracyclic diterpenoid tetraol[255] (sec Figure 2), which was found to arrest the replication of certain DNA viruses.[256] Selective inhibition of animal cell DNA replication by aphidicolin and its specific inhibitory activity on DNA polymerase-α were first described by Ikegami at al.[257] These researchers demonstrated that aphidicolin inhibited mitotic, but not meiotic, division in sea urchin embryos without having any detectable adverse effect on nondividing cells. Furthermore, in addition to being a potent inhibitor of DNA replication in vivo, aphidicolin was found to specifically inhibit the in vitro activity of isolated α-polymerase.[257] Subsequent extensive experimentation in numerous laboratories established that aphidicolin at 0.04 to 2.0 μg/mℓ arrested nuclear DNA replication in a variety of cells without affecting protein and RNA synthesis or mitochondrial DNA synthesis.[34,254,257-262,263] Moreover, α-polymerase purified from a large number of different eukaryotic cell types was found to be strongly inhibited by aphidicolin at 1 to 40 μg/mℓ, whereas activity of polymerases-β and -γ remained unaffected by the drug.[257-259,263-265] In addition, aphidicolin was found to inhibit the activity of DNA polymerases of vaccinia[265,267] and herpes simplex virus[266,267] and to block the replication of SV40 minichromosomes.[268-269] Furthermore, aphidicolin is a potent inhibitor of plant cell α-like DNA polymerases[270-272] and of yeast cell α-like polymerases I and II.[273] By contrast, aphidicolin does not exert any inhibitory effect on terminal deoxynucleotidyl transferase, viral reverse transcriptase, and bacterial and bacteriophage DNA polymerases.[34,254,257,258,265,266] Recently, two aphidicolin-related compounds, 3-deoxyaphidicolin and aphidicolin-17-monoacetate, were isolated from culture filrates of *Phoma beta,* an incitant of leaf spot disease of sugar beet.[274] The two antibiotics inhibit DNA synthesis in vivo and α-polymerase activity in vitro through mechanisms that seem to be indistinguishable from those of aphidicolin.[274]

The complex mechanism of inhibition by aphidicolin has been a subject of extensive research. As indicated by its structural formula (Figure 2), aphidicolin is not a structural analog of a nucleotide. Furthermore, this antibiotic is nonpolar and hydrophobic and thus cannot be expected to act directly as a substrate for DNA polymerase. Yet, using activated DNA as primer-template for purified polymerase-α, it was initially observed that aphidicolin acts as a competitive inhibitor with respect to dCTP, whereas it is noncompetitive with the other three substrates and uncompetitive with regard to the template DNA and magnesium activator.[34,263,267,271,275,276] Inhibition of α-polymerase activity by aphidicolin could be attenuated by varying of concentration of dCTP.[41,42,267,271,275] These kinetic data led to the proposition that aphidicolin interacts hydrophobically with DNA polymerase-α at loci in the vicinity of or directly at the dCTP binding site, which is distinguishable from binding site(s) for the other three substrates.[254,276] Investigating the mechanism of inhibition by aphidicolin of α-polymerase on synthetic homopolymers revealed, however, that the drug may interfere with the polymerization of substrates other than dCTP.

Oguro et al.[275] showed that aphidicolin had no effect on the incorporation of dAMP and dGMP using oligo(dA)·poly(dT) and oligo(dG)·poly(dC) templates, respectively, but it was a competitive inhibitor for the polymerization of dTMP on oligo(dT)·poly(dA) template. It was subsequently demonstrated that aphidicolin was a mixed-type inhibitor of dTMP incorporation by α-polymerase on activated DNA template.[276] Krokan and associates[277] also found that aphidicolin inhibited competitively the incorporation of dTMP opposite a complementary homopolymer and showed that the drug did not affect K_m values of purified α-polymerase for dATP and dGTP, but it did increase the enzyme's K_m for dTTP. Examining the mode of inhibition of purified calf thymus α-polymerase by aphidicolin, Holmes[278] noted that the drug inhibits competitively incorporation of dTMP on oligo(dT) · poly(dA) template and on activated DNA in the absence of dCTP. A plausible current concept is that the conformation, and thus sensitivity to aphidicolin, of α-polymerase varies depending on the combination and concentration of primer-template and nucleotide substrates.[263,276] Furthermore, kinetic results indicated that aphidicolin acts as an allosteric inhibitor of polymerase-α, which displays negative cooperativity among at least three substrate binding sites.[276] Differences in ambient deoxyribonucleoside triphosphates, template, and enzyme may explain, therefore, contrasting reports on the mechanism of action of aphidicolin in isolated nuclei vs. purified DNA polymerase-α. Oguro et al.[279] reported that aphidicolin inhibition of DNA synthesis in isolated nuclei differed from inhibition of purified polymerase-α in that the former was competitive when all substrates were varied together, but noncompetitive for individual substrates. By contrast, Krokan et al.[277] found that the antibiotic was a competitive inhibitor of dCMP incorporation both with isolated α-polymerase and nuclei. These discrepant results are difficult to interpret since the relative concentrations of aphidicolin and deoxynucleotide substrates at the nuclear site of DNA replication are not known.

It is most plausible to hypothesize that the selective arrest of DNA replication in vivo by aphidicolin is due to its specific inhibitory effect on α-polymerase activity. Yet it has been noticed that concentrations of aphidicolin that are required for the inhibition of purified α-polymerase activity are 10- to 1000-fold higher that those needed to suppress DNA synthesis in vivo.[257,265,273,280,281] It was pointed out, however, that similar concentrations of aphidicolin lead to killing of V79 hamster cells and to arrest of α-polymerase activity if the latter is assayed in the presence of adequately low concentrations of dCTP.[42] It has been proposed that interaction in vivo of the α-polymerase holoenzyme with other proteins enhances its sensitivity to aphidicolin.[273,281] The demonstration of increased sensitivity of yeast α-like polymerase upon its reconstitution with crude extract proteins sustained this hypothesis.[273] Also, a changing mode of aphidicolin inhibition of nuclear polymerase-α activity upon removal of nuclear proteins by salt extraction has been noted.[281] It might be, therefore, that both the potency and mode of action of aphidicolin depend on the interaction of α-polymerase with associated nuclear proteins. Yet the concentration of free aphidicolin at the site of DNA replication in nuclei is unknown, and the nucleus may concentrate the drug in vivo to levels comparable to those required to inhibit purified polymerase-α.

The strongest argument in support of the proposition that α-polymerase is the target for aphidicolin inhibition of DNA synthesis in vivo has been provided by the isolation of aphidicolin-resistant cell mutants which also display either altered α-polymerase protein or possess elevated intracellular levels of polymerase. Selection for aphidicolin-resistant mutant cells results in most cases in the isolation of cell strains that contain elevated levels of intracellular pools of deoxyribonucleoside triphosphates.[282-285] Such mutant cells are usually also resistant to inhibition of growth by arabinofuranosyl analogs of the deoxyribonucleoside substrates by virtue of the augmented intracellular concentration of deoxyribonucleotides.[284,285] Taking into account the inhibitory interaction of aphidicolin with substrate binding sites in the α-polymerase molecule, the existence of these mutants confirms only indirectly the supposition that α-polymerase is the in vivo target for aphidicolin.

A better correlation between cell resistance to aphidicolin and α-polymerase was obtained by the isolation of mutant cells of *D. melanogaster*[41] and mouse mammary carcinoma,[286] which were relatively resistant to the antibiotic and displayed elevated intracellular amounts of polymerase-α activity. However, the closest correlation between inhibition of in vivo DNA replication and aphidicolin effect on α-polymerase was demonstrated through the isolation of cell mutants that possessed an altered α-polymerase molecule. Sugino and Nakaymaera[41] selected a mutant of *D. melanogaster* embryonic cell that was tenfold more resistant to aphidicolin than wild type cells. Polymerase-α, which was isolated from the mutant strain, was also tenfold more resistant to the antibiotic than wild type cell α-polymerase.[41] Liu and colleagues[42,287-289] selected a Chinese hamster V79 cell mutant that was more resistant to aphidicolin than the parental cell. This mutant cell was also characterized by slow growth, UV-light sensi-

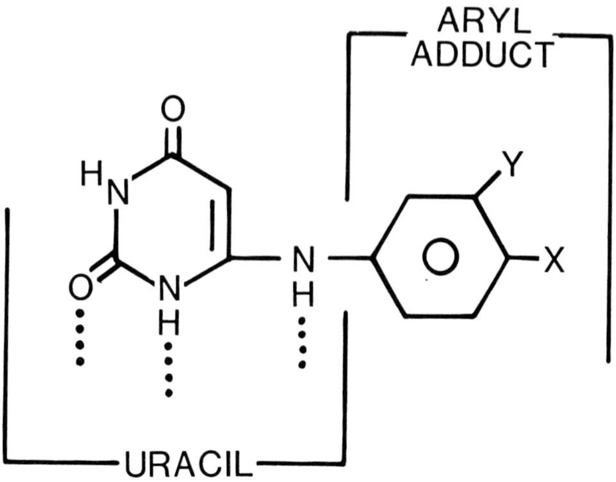

FIGURE 3. Schematic structure of synthetic uracil-based, dGTP-specific DNA polymerase inhibitor. (Adapted from Wright, G. E., Baril, E. F., Brown, V. M., and Brown, N. C., *Nucl. Acids Res.*, 10, 4431, 1982. With permission.)

tivity, and spontaneous and UV-induced hypermutability.[287-289] Most interestingly, polymerase-α, which was purified from the mutant cell strain, displayed increased resistance to aphidicolin and its K_m for dCTP was tenfold lower than that of α-polymerase of the parental cell.[42] The properties of the mutant cell and of its polymerase confirm the contention that polymerase-α is the in vivo target for aphidicolin. In addition, they have interesting implications in regard to the mutagenicity of α-polymerase and to its role in DNA replication and repair (cf. Chapter 7, Section III.B).

Last, two instances in which polymerases-α with unusual resistance to aphidicolin was identified will be mentioned. Wilson and associates[290] observed a subspecies of CV-1 monkey cell polymerase-α that was unusually insensitive to aphidicolin. Scovassi et al.[291] reported that α-polymerase from Med fly was tenfold more resistant to aphidicolin than mammalian polymerase-α.

C. 6-(Arylamino)uracils and N^2-Arylaminopurines

A promising new venue of custom-designing of selective inhibitors of DNA polymerases in general and for polymerase-α in particular was opened through the work of Brown and Wright and associates.[292-295] Appropriately constructed 6-substituted uracils of the type shown in Figure 3 are potent and selective inhibitors of replicative DNA synthesis in prokaryotic or eukaryotic cells. The inhibitory activity of these substituted uracils is based on their structural similarity to dGTP; both compounds possess one and two ring moieties, and the 6-amino group on the base domain has the potential for forming hydrogen bonds with the template cytosine.[292,293] The specificity of analogs of this type strongly depends on the nature of the substituents at the X and Y positions on the aryl moiety. Whereas small substituents such as $=CH_3$ or Cl render the analog inhibitory to *B. subtilis* DNA polymerase III but not to HeLa polymerase-α, a bulky substitution such as *p-n-* butyl at position X makes the drug inhibitory for HeLa polymerase-α, but not for the bacterial polymerase.[292] Hence, butylanilinouracil was found to block HeLa cell DNA polymerase-α with K_i 60 μM whereas β- and γ-polymerases were unaffected at 1 mM analog. The inhibitor suppressed replicative DNA synthesis in HeLa cells, and its effect on isolated polymerase-α was reversed by excess dGTP. The selectivity of butylanilinouracil was limited, however, and whereas some

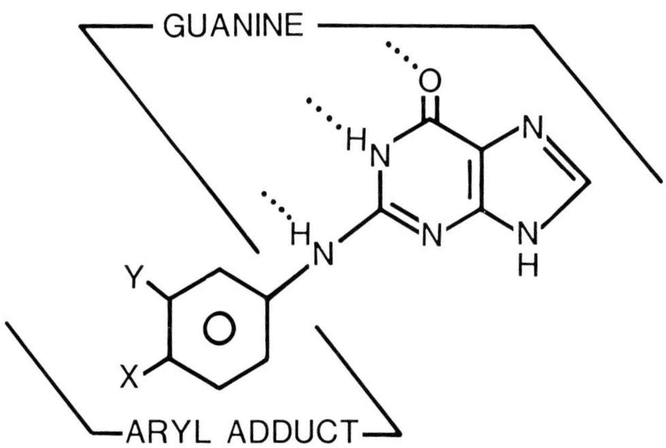

FIGURE 4. Schematic structure of synthetic guanine-based dGTP-specific DNA polymerase inhibitor. (Adapted from Wright, G. E., Baril, E. F., Brown, V. M., and Brown, N. C., *Nucl. Acids Res.*, 10, 4431, 1982. With permission.)

molecular forms of calf thymus α-polymerase were strongly inhibited by it, others were less sensitive.[292]

Related nucleotide derivatives with improved potency and specificity, N^2-*p*-*n*(butylphenyl) guanine (BuPG) and N^2-(3′,4′-trimethylphenyl)guanine (TMPG), have been introduced recently.[293] These guanine analogs, whose general structural formula is presented in Figure 4, act as competitors of dGTP to inhibit HeLa cell α-polymerase at K_i 6 to 9 μM for BuPG, but they have no effect on the activity of polymerases-β and -γ.[293] Like butylanilinouracil, BuPG was found to inhibit replicative DNA synthesis in a variety of animal cell lines, but its potency differed for different cell types.[294] Khan et al.[295] reported that phosphorylated derivatives of BuPG, BuP-dGMP, BuP-dGDP, and BuP-dGTP, exhibit greatly augmented inhibitory activity. In particular, adding phosphate groups at the 5′α and 5′β positions increased the potency of the analog toward α-polymerase several 100-fold. Thus K_i < 10 μM was obtained for BuP-dGTP.[295] This triphospho-analog of dGTP was found to be highly inhibitory for α-polymerases from various cell types and for yeast cell polymerase I, but it did not diminish the activity of polymerases-β and -γ and polymerase I from *E. coli*[295] The phosphorylated analog was not incorporated into newly made DNA chains, and unlike BuPG which inhibited only copying of cytosine-containing templates, BuP-dGTP was effective also for replication of homopolymers that did not contain cytosine. Hence, the mechanism of action of BuP-dGTP is not identical to that of BuPG. Last, as indicated by these researchers, the unusually high potency and specificity of BuPG-dGTP might be useful in future designing of specific probes for α-polymerase.[295]

D. Arabinose Analogs of Deoxynucleotides

Arabinofuranosylnucleoside analogs were the earliest described selective inhibitors of α-polymerase. The arabinose analog of cytosine, 1-β-D-arabinofuranosylcytosine (ara C), and its active phosphorylated derivatiave, 1-β-D-arabinofuranosylcytosine 5′-triphosphate (ara CTP), were found to be potent inhibitors of DNA synthesis in intact cells and in isolated nuclei, respectively.[296,297] The dCTP analog ara CTP is an effective inhibitor of eukaryotic and viral DNA polymerases.[298,299] The extent of inhibition of human lymphocyte polymerases-α and -β by ara CTP varies with different primer templates and substrates, yet, copying of activated DNA was inhibited with K_i = 2.8 μM and K_i = 13.3 μM for α- and β-polymerases respectively.[300] Almost complete resistance of rat liver and carcinoma cell polymerases-β to ara CTP was observed under conditions in which inhibition of α-polymerases is demonstrable.[301,302] Calf thymus α-polymerase was inhibited more effectively by ara CTP (K_i = 4 μ*M* than thymic β-polymerase (K_i = 32 μ*M*.)[298] Copying of activated DNA by CV-1 monkey cell polymerase-α was found to be inhibited 20 to 30 times more effectively by ara CTP than copying of the same template by β-polymerase.[268] In the same vein, it was found that ara CTP strongly inhibited HeLa cell polymerase-α, whereas β-polymerase was almost unaffected by the analog.[303] Comparing the sensitivity of polymerases-α and -β from several cell types to ara CTP, Dube et al.[235] found that, as a rule, polymerase-β was less sensitive to the analog than polymerase-α. Discordant results were reported by DiCicoccio and Srivastava[299] who found that ara CTP was inhibitory to a similar extent for DNA polymerases-α and -β from several normal and leukemic cell types.

That polymerases of the α type from different cells may possess different sensitivity to ara CTP was reported recently by Tanaka and Yoshida.[304] Based on the observation that patients with acute lymphoblastic leukemia (ALL) are less affected by chemotherapeutic treatment with ara C than individuals with acute myeloblastic leukemia (AML), the sensitivity of the respective α-polymerases to ara CTP were compared. It was found that α-polymerases from ALL patients were more resistant to ara CTP (K_i = 16.9 ± 6.9 μ*M*) than polymerase-α from AML patients (K_i = 3.8 ± 0.5 μ*M*) whereas both groups had a similar K_m for dCTP.[304] Hence, it may be that the considerable scatter of sensitivity to ara CTP of α-polymerases derived from different sources is due to differences in the affinity of the enzymic protein to the analog.

Discrepant results were obtained in studies on the selectivity of the dATP analog 9-β-D-arabinofuranosyladenine 5′-triphosphate (ara ATP) toward α- and β-polymerases. It has been reported that ara ATP inhibited α- and β-polymerases from several cell types with similar efficiency.[299] In another study, ara ATP displayed a slightly higher inhibitory effect on quail oviduct polymerase-β than on α-polymerase.[305] In contrast, mouse[306] and calf thymus[307] polymerases-α were found to be much more sensitive to ara ATP than the respective β-polymerases.

The dGTP analog 9-β-D-arabinofuranosyl 5′-triphosphate (ara GTP) was found to inhibit mouse cell polymerases-α, -β and -γ.[306] The mode of inhibition and the inhibition constant, (K_i) varied with types of substrate and polymerase, but α-polymerase was reported to be more sensitive to the inhibitor than β- and γ-polymerases.[306]

The UTP analogs 1-β-D-arabinofuranosyluracil 5-triphosphate (ara UTP) and its 5-alkylated[308] and 5-halogenated[309] derivatives inhibit both viral and animal cell DNA polymerases. Polymerase-α and -β were found to be inhibited by ara UTP, whereas polymerase-α activity was inhibited by ara UTP containing longer 5′-alkyl side chains.[308] Halogenated ara UTP derivatives inhibited α-polymerase more effectively than β-polymerase, but the kinetic constants strongly depended on the nature of the template used.[309] The alkylated derivatives of ara UTP were recently found to inhibit the activity of terminal deoxyribonucleotidyl transferase, which does not require a template.[310] It might be, therefore, that the inhibition exerted by these analogs of DNA polymerase-α is mediated through a substrate-site and does not require interaction of the enzyme with template-primer.

E. Phosphonacetic Acid

Phosphonacetic acid was originally identified as a potential antiviral drug when it was found to inhibit herpes virus replication without interfering with growth of the human host cell. Subsequently, phosphonacetic acid was shown to inhibit the activity of isolated DNA polymerases of several animal cell viruses.[252] Although eukaryotic cell DNA polymerases were initially considered to be resistant to phosphonacetate,[311] it was found later that polymerase-α[312-315] and -γ[312,315] were inhibited by the reagent at concentrations of 10 to 100 μg/mℓ that were equal to or greater than those required to inhibit viral polymerases. In contrast to that inhibition, polymerase-β was resistant to phosphonacetate at these concentrations.[312,314-316] The inhibition of herpes virus DNA polymerase by phosphonacetic acid is due to interaction of the agent with the pyrophosphate binding site of the enzyme.[314] It may be that the resistance of β-polymerase to inhibition by phosphonacetic acid results from the reported inability of that enzyme to conduct pyrophosphate exchange or pyrophosphorolysis.[317] Unlike the previously described inhibitors of DNA polymerase-α aphidicolin, ara C, 6-(arylamino)uracils, and BuPG, phosphonacetic acid is not a general inhibitor of replicative DNA synthesis in animal cells since its uptake into some cell types is very limited.[318]

F. 9-(2-Hydroxyethoxymethyl)guanine (Acyclovir) and 2′-Nor-2′-deoxyguanosine (2′-NDG)

The two guanosine analogs, acyclovir (acycloguanosine) and 2′-NDG, are efficient inhibitors of herpes virus-induced DNA polymerase and arrest herpes virus growth in infected cells.[319-323] The relevance of these two antiviral drugs to the subject of this section is that both agents also inhibit host cell polymerase-α but not polymerase-β. After being phosphorylated to its triphosphate form, acyclovir inhibits herpes virus HSV-1 DNA polymerase (K_i = 0.03 to 1.4 μ*M* by being incorporated into DNA in place of dGMP, but since it lacks 3′-hydroxyl group, the analog acts as a chain terminator.[319] DNA polymerases-α from different cell types are also inhibited by triphosphoacyclovir at K_i = 0.18 to 4.0 μ*M* depending on primer-template, whereas polymerase-β appears to be resistant to the drug.[319,320] The guanosine analog 2′-NDG is a more potent antiherpetic agent than acyclovir due to its more efficient phosphorylation by virus-induced thymidine kinase and GMP kinase to its triphosphate form.[321,322] Triphospho-2′-NDG like acyclovir inhibits HSV-1 DNA polymerase and less efficiently α-polymerase, but it does not affect the activity of β-polymerase.[323]

G. Novobiocin

The antibiotic, novobiocin, which paradigmatically acts as an inhibitor of DNA topoisomerase, is also an inhibitor of α-polymerase. Novobiocin at 4 μ*M* was found to completely abolish the activity of human carcinoma cell polymerase-α, whereas β-polymerase activity remained unaffected.[302] Similarly, Edenberg[324] reported that 0.2 mg/mℓ novobiocin abolished half the activity of α-polymerase, whereas β-polymerase remained unaffected. Hence, the inhibitory effect of novobiocin on cellular DNA replication does not provide unequivocal evidence for the participation of DNA topoisomerase in replication since polymerase-α and -γ are also affected by that drug.[324]

H. Hemin

Hemin was reported to differentially inhibit three molecular forms of neuroblastoma cell α-polymerase.[118] Byrnes et al.[325] reported that the activity of rabbit bone marrow α-polymerase was inhibited by hemin and that the inhibition was competitive with respect to the template, noncompetitive with substrates, and reversed by globin. Hemin was proposed to induce dissociation of α-polymerase from template DNA.

I. Plant Cell Agglutinins

Sensitivity of human neuroblastoma cell DNA polymerase-α to concanavalin A and to ricin has been reported by Basu and colleagues.[118,119] The results of these studies are discussed in Section III of this chapter.

J. Modified DNA Templates

A large number of template modifications inhibit or block DNA synthesis by DNA polymerases. For example, DNA polymerase-α as well as other classes of animal cell DNA polymerases, are inhibited by DNA cross-linked by mitomycin C,[326] X-ray irradiated DNA[327], and polyvinyladenosine-hybridized DNA.[328] Most of the template modifications do not exhibit specificity with respect to different classes of animal cell DNA polymerases. Frequently, the transient inhibition of chain elongation is correlated with enhanced misincorporation of noncomplementary nucleotides (cf. Chapter 8 for full discussion of the effect of template modification on fidelity of polymerases). The only reported selective inhibitory effect of any of these modified templates is of X-ray irradiated DNA, which inhibits α-polymerase activity 200-fold less efficiently than β-polymerase.[327] However, X-rays induce multiple changes in DNA involving various adducts, and thus the selective inhibition of α-polymerase is difficult to interpret.

K. 2′,3′-Dideoxythymide-5′-triphosphate

An analog of dTTP, 2′,3′-dideoxythymidine-5′-triphosphate (d_2TTP), is inhibitory to all DNA polymerases when added at an adequate excess over the normal monodeoxy-substrate. This phenomenon led to the development of the widely used Sanger method of sequencing of DNA. However, the different classes of animal cell DNA polymerases differ widely in their susceptibility to inhibition by d_2TTP and other dideoxy-derivatives of nucleoside triphosphates (see Reference 37 for a review). DNA polymerases-α from all studied sources are relatively resistant to inhibition by deoxynucleoside triphosphates, and high excess of the analog are required for full inhibition of these enzymes.[329] By contrast, DNA polymerases of the β and γ types are strongly inhibited at d_2TTP to dTTP ratios of 0.5 to 5.0.[37,329] Hence, d_2TTP is widely used to distinguish between α-polymerase and other types of animal cell DNA polymerases in crude extracts. The inhibitory activity and mechanism of action of d_2TTP on DNA polymerases β and γ are discussed in Chapter 4, Section V.D and Chapter 5, Section I.C.2, respectively.

L. Palmitoyl Coenzyme A

A recent report by Shimada et al.[330] describes preferential inhibition of α-polymerase from sea urchin by palmitoyl CoA. Polymerase-α from that organism was most sensitive to palmitoyl CoA inhibition (K_i 28 μM), γ-polymerase less susceptible ($K_i = 116$ μM), and polymerase-β was resistant. Both the long hydrocarbon side chain and the coenzyme moiety of palmitoyl CoA were required for inhibition of polymerase-α by palmitoyl CoA. Whether suppression of α-polymerase activity by palmitoyl Col also occurs for enzyme from other cells and tissues is not clear at present.

M. β-Lapachone

A naturally occurring compound isolated from a number of tropical trees, β-lapachone was found to inhibit calf thymus α-polymerase and viral reverse transcriptase but to have no effect on the activities of thymus-β-polymerase or *E. coli* DNA polymerase I.[331] β-Lapachone is quite a potent inhibitor, suppressing 50% of the activity of α-polymerase at 8 μm. Dithiothreitol was found to be required for its inhibitory action.[331]

N. Other Inhibitors

Several authors reported on diverse inhibitors of α-polymerase, which possessed different efficacies and mechanisms of action. All these inhibitors, however, also arrested the activities of DNA polymerases other than α type and thus do not appear to be selective for α-polymerase. Among this group of inhibitors, several derivatives of the antibiotic rifamycin were found to inhibit DNA polymerase-α and other polymerases.[332-334] These inhibitors were primarily developed in an attempt to search for selective inhibitory agents for retroviral reverse transcriptase. The fact that the various derivative of rifamycin exert inhibitory effects on disparate DNA polymerases attests to the likely possibility that these enzymes share similar basic mechanisms of catalysis. In analogy, captan, *N*-trichloromethylthio-4-cyclohexane 1,2-dicarboximide, has been shown to act as an inhibitor of α, β, and γ eukaryotic polymerases, prokaryotic DNA polymerases, and terminal deoxyribonucleotidyl polymerases.[335]

Of limited selectivity are the antileukemic drugs daunomycin and Adriamycin, which interact directly with animal cell DNA polymerases to inhibit competitively their interaction with the primer-template.[336,337] The K_i of both daunomycin and Adriamycin for α-polymerase was 2- to 15-fold lower than for β-polymerase.[337]

Anionic polysaccharides were found to be inhibitory to α-polymerases as well as to DNA polymerases-β, γ and to reverse transcriptase and terminal deoxynucleotidyl transferase. It was reported that heparin inhibited the activities of human cell DNA polymerases-α, -β, and -γ and that all three enzyme types bound to heparin-Sepharose,[342] It was suggested that heparin exerts its inhibitory activity by reversibly binding to the enzymes, thereby interfering with their interaction with the primer-template. Similarly, heparin at 1 μM was highly inhibitory to DNA polymerases-α, -β, and -γ from *X. laevis* ovaries.[11] Last, it was shown recently that several types of glycosaminoglycans are potent inhibitors of calf thymus DNA polymerase-α and that the degree of enzyme inhibition was proportional to the extent of sulfation of the polysaccharide and to the glycosaminoglycan to DNA ratio.[343]

VII. UPDATE

A. Structure of Polymerase-α

A DNA polymerase-α DNA primase complex has been purified extensively from CV-1 cultured cells of green monkey.[348] Denaturing electrophoresis of the 53,000-fold purified enzyme revealed that it consists of a cluster of large polypeptides with molecular sizes of 176, 158, 133, and 118 kdaltons and a group of at least four additional polypeptides of 62, 57, 53, and 30 kdaltons.[348] The general subunit composition of the CV-1 conventionally purified polymerase-α is similar to that of human KB cell DNA polymerase-α primase complex that was isolated by immunoaffinity chromatography.[109]

B. Auxiliary Activities Associated with DNA Polymerase-α

1. DNA Primase

a. Structural Aspects

Structural relationship between DNA primase and DNA polymerase I of yeast have been described in detail by several groups in early 1985. Unlike DNA primase of higher eukaryotes that, in most documented cases, is undissociable from polymerase-α, the major proportion of yeast primase activity is readily dissociated from DNA polymerase. DNA primase from *Saccharomyces* was purified to near homogeneity by Jazwinski and Edelman.[349] The isolated yeast primase was devoid of polymerase activity and corresponded to a single 65-kdalton polypeptide. A minor portion of the primase copurified, however, with DNA polymerase I — indicating functional association between these two activities.[349] Essentially similar results were documented by Wilson and Sugino[350] who obtained a major proportion of *Saccharomyces* DNA primase as a 60-kdalton monomeric polypeptide that was free of DNA polymerase activity. Yet, a minor amount of the yeast primase activity cofractionated with DNA polymerase I.[350] Using immunoaffinity purification procedures, Plevani et al.[351] isolated from yeast cell a DNA primase-DNA polymerase I complex which was dissociable into its two enzymic components and which could be reconstituted in vitro back into a complex form. Primase activity that was freed of polymerase I cofractionated with 58- and 48-kdalton polypeptides. Clearly, unlike most documented cases of tight and inseparable association between primase and polymerase-α from animal cells, yeast primase is dissociated with relative ease from polymerase I, but the two activities appear to form a functional complex in the cell.

An additional case of close association between mammalian DNA polymerase-α and DNA primase was reported for CV-1 cultured monkey cells.[348] In extracts of these cells, two subfractions of polymerase-α were obtained, one which was inseparable from primase activity and another that was devoid of primase. Note, however, that no polymerase-free primase activity was observed even in extensively purified fractions.[348]

b. Catalytic Aspects

Enzymic properties of DNA primase from yeast have been described by several investigators.[349,350,352] Yeast and animal cell primases are similar in respect to the size of their product RNA primer, their ability to incorporate both ribo- as well as deoxyribonucleotides into the primer and their resistance to the RNA polymerase inhibitor α-amanitin.[349,350] Badaracco and associates[352] demonstrated that RNA primers synthesized by the polymerase I-primase complex are shorter than the nine to ten nucleotide oligomers and multiples that the polymerase-free primase synthesizes.

Initiation, elongation, and termination of RNA-primed DNA synthesis in mammalian cells have been investigated by two groups. With CV-1 polymerase-α-primase complex, primers were found to always start with A or G whose ratio is determined by the relative concentration of ATP and GTP.[348] Further, nearest-neighbor analysis indicated a preference for purine nucleotides at the junction point between the RNA primer and the extended DNA chain. Increase in the relative concentration of a single ribonucleotide triphosphate slightly increased its representation at the RNA-DNA junction.[348] It appears that the base composition of the RNA primer as well as its size are modulated by the relative concentrations of the ribonucleotide triphosphate substrates that are supplied to the CV-1 primase-polymerase-α complex. Initiation and termination sites on the DNA template for primase-polymerase-α from Ehrlich ascites cells have been studied.[353] Highly purified primase-polymerase-α copied in vitro the 5081-nucleotide-long DNA of minute virus of mice in the presence of radiolabeled GTP. All the 3′-hydroxyl termini of the newly made DNA were located 2 to 14 nucleotides upstream to one of three hexanucleotide sequences $C_2A_2C_2$, C_2AC_3, or $C_2A_2T_2$. Digestion of the product DNA revealed that the remaining primer RNA began with GTP, was five to seven nucleotides long, and was located opposite C_4 or C_5 sequences that were situated three to nine bases upstream to the above-mentioned hexanucleotides: $C_2A_2C_2$, C_2AC_3, or $C_2A_2T_2$.[353] It has been suggested that hexamers of the general composition $C_2A_{1-2}(C_{2-3}/T_2)$ are involved in either the initiation and/or the termination of DNA synthesis.[352]

Immunoaffinity purified calf thymus DNA polymerase-α primase complex has been shown to catalyze synthesis of RNA primers on ϕX174 single-stranded DNA and to elongate them efficiently into full length double-stranded products.[358] Whereas primer and DNA synthesis occurred with both poly(dT) and poly(dC), the synthetic templates poly(dA) and poly(dI) did neither support initiation nor DNA synthesis. Sucrose density gradient centrifugation and nitrocellulose filter binding assays were used to demonstrate that the polymerase-α-primase complex binds to deoxypyrimidine polymers, but not to deoxypurine polymers. By use of appropriate mixed polymers, it has been shown that purine deoxynucleotides may serve as template for the polymerase only after a four- to eight-residues-long oligopyrimidine sequence has been copied by the primase.[358]

2. *DNA Binding Proteins*

Two single-strand-specific DNA binding proteins of 48 and 61 kdaltons have been isolated from calf thymus.[354] These proteins are highly susceptible to proteolytic degradation and their isoation in intact form depends on the rapidity of the purification process. These immunologically and functionally related polypeptides bind noncooperatively to single-strand DNA but not to duplex DNA. Both proteins stimulate the activity of polymerase-α 2- to 100-fold at different ratios of binding protein to activated DNA template.[354] With single-stranded phage fd DNA, it appears that the stimulatory proteins block nonproductive polymerase binding sites on DNA, thus increasing the probability of interaction between polymerase-α and the 3′-hydroxyl terminus of the primer.[354] It was speculated[354] that the two proteins may facilitate the action of C_1C_2-like primer recognition cofactors of polymerase.[52,53,70] It is not clear yet, however, whether the described bovine DNA binding proteins are required for synthesis of DNA in vivo.

Williams et al.[359] determined the complete amino acid sequence of the calf thymus single-stranded DNA binding protein UP1.[171] Partial sequencing of the helix destabilizing protein from mouse myeloma cell[183] revealed high degree of sequence homology with the bovine UP1. The amino acid sequences of the calf and mouse binding proteins appear to have no homology with any previously sequenced protein.[359] Hence, these proteins are probably representatives of a distinct class of nucleic acid binding proteins that may have a role in the metabolism of single-stranded RNA or DNA.

C. Chromosomal Localization of Human DNA Polymerase-α Gene

A pleasing outcome of the availability of monoclonal antibodies against human polymerase-α[54] (see Chapter 6, Section V) is the success in assigning the gene for human polymerase-α to chromosome X.[355,356] Wang et al.[355] detected the expression of human α-polymerase in human-rodent somatic cell hybrids by use of nonneutralizing monoclonal antibodies against polymerase-α whose specificity is restricted to the human enzyme. A panel of hybrid clones, which contained specific sorted human chromosome and were propagated in selective media, were screened for presence of human polymerase-α. Appearance of human polymerase-α was correlated with presence of human chromosome X in the hybrid cell. Further, examination of hybrid clones that contained different but overlapping regions of chromosome X indicated that a locus necessary and sufficient for expression of polymerase-α is located at a junction of $X_p21.3$ and $X_p22.1$ on the short arm of chromosome X.[354] This locus is close to the proposed Duchenne muscular dystrophy and Ornithine transcarbamilase loci, and it lies outside the noninactivated region of the X-chromosome.[355] The assignment of the gene for DNA polymerase-α to the X-chromosome has been confirmed. Hanaoka and associates[356] hybridized mouse FM3A mutant cell strain that contains heat-labile polymerase-α[357] with human diploid fibroblasts or lymphocytes. Expression of human polymerase-α in cell hybrids was verified either by application of elevated temperature that restricted the growth of mutant rodent cells but not of cells containing normal human enzyme or by use of selective monoclonal antibodies against the human polymerase. Correlation between temperature resistance and HAT resistance as well as direct karyotyping led to the assignment of the human α-polymerase gene to human chromosome X.[355]

REFERENCES

1. Bollum, F. J., Calf thymus polymerase, *J. Biol. Chem.*, 235, 2399, 1960.
2. Brakel, C. L. and Blumenthal, A. B., Multiple forms of *Drosophila* embryo polymerase: evidence for proteolytic conversion, *Biochemistry*, 16, 3137, 1977.
3. Brakel, C. L. and Blumenthal, A. B., Three forms of DNA polymerase from *Drosophila melanogaster* embryos. Purification and properties, *Eur. J. Biochemistry*, 88, 351, 1978.
4. Banks, G. R., Boezi, J. A., and Lohman, I. R., A high molecular weight DNA polymerase from *Drosophila melanogaster* embryos. Purification, structure and partial characterization, *J. Biol. Chem.*, 254, 9886, 1979.
5. Loeb, L. A., Purification and properties of deoxyribonucleic acid polymerase from sea urchin embryos, *J. Biol. Chem.*, 244, 1672, 1969.
6. Morris, P. W. and Rutter, W. J., Nucleic acid polymerizing enzymes in developing *Strongylocentrotus franciscanus* eggs, *Biochemistry*, 15, 3106, 1976.
7. De Petrocelli, D., Paris, E., Filosa, S., and Capasso, A., Separation and partial characterization of DNA polymerases in sea urchin *Paracentrotus lividus* eggs, *Biochem. Biophys. Res. Commun.*, 68, 954, 1976.
8. Racine, F. M. and Morris, P. W., DNA polymerases α and β in the California urchin, *Nucl. Acids Res.*, 5, 3945, 1978.
9. Tato, F., Gandini, D. A., and Tocchini-Valentini, G. P., Major DNA polymerases common to different *Xenopus laevis* cell types, *Proc. Natl. Acad. Sci. U.S.A.*, 71, 3706, 1974.
10. Zimmerman, W. and Weissbach, A., Participation of deoxyribonucleic acid polymerase-α in amplification of ribosomal deoxyribonucleic acid in *Xenopus laevis*, *Mol. Cell Biol.*, 1, 680, 1981.
11. Nelson, E. M., Stowers, D. J., Bayne, M. L., and Benbow, R. M., Classification of DNA polymerase activities from ovaries of the frog *Xenopus laevis*, *Dev. Biol.*, 96, 11, 1983.
12. Brun, G., Rougeon, F., Lauber, M., and Chapville, F., Purification and properties of DNA polymerases from chick embryo, *Eur. J. Biochem.*, 41, 241, 1974.
13. Fry, M. and Weisman-Shomer, P., Altered nuclear deoxyribonucleic acid α-polymerase in senescent cultured chick embryo fibroblasts, *Biochemistry*, 15, 4319, 1976.
14. Matsukage, A., Yamaguchi, M., Tanabe, K., Taguchi, Y. N., Nishizawa, M., and Takahashi, T., DNA chain elongation mechanism of DNA polymerase α, β, and γ in *New Approaches in Eukaryotic DNA Replication*, de Recondo, A. M., Ed., Plenum Press, New York, 1983, 81.
15. Yoneda, M. and Bollum, F. J., Deoxyribonucleotidyl-polymerizing enzymes of calf thymus gland. I. Large scale purification of terminal and replicative deoxynucleotidyl transferases, *J. Biol. Chem.*, 240, 3385, 1965.
16. Weissbach, A., Schlabach, A., Fridlender, B., and Bolden, A., DNA polymerases from human cells, *Nature (London) New. Biol.*, 231, 167, 1971.
17. Holmes, A. M., Hesslewood, I. P., and Johnson, I. R., The occurrence of multiple activities in the high molecular weight DNA polymerase fraction of mammalian tissue. A preliminary study of some of their properties, *Eur. J. Biochem.*, 43, 487, 1974.
18. Craig, R. K. and Keir, H. M., Deoxyribonucleic acid polymerases of BHK/21-C13 cells. Partial purification and characterization of the enzymes, *Biochem. J.*, 145, 215, 1975.
19. Matsukage, A., Sivarajan, M., and Wilson, S. H., Studies on DNA α-polymerase of mouse myeloma: partial purification and comparison of three molecular forms of the enzyme, *Biochemistry*, 15, 5305, 1976.
20. Yoshida, S., Kondo, T., and Ando, T., Multiple molecular species of cytoplasmic DNA polymerase from calf thymus, *Biochim. Biophys. Acta*, 353, 463, 1974.
21. Banks, G. R., Holloman, W. K., Kairis, M. V., Spanos, A., and Yarranton, G. T., A DNA polymerase from *Ustilago maydis*. I. Purification and properties of the polymerase activity, *Eur. J. Biochem.*, 62, 131, 1976.
22. McLennan, A. G. and Keir, H. M., Deoxyribonucleic acid polymerase of *Euglena gracilis*. Primer-template utilization of and enzyme activities associated with the two deoxyribonucleic acid polymerases of high molecular weight, *Biochem. J.*, 151, 239, 1975.
23. Wintersberger, E., Deoxyribonucleic acid polymerases from yeast. Further purification and characterization of DNA-dependent polymerases A and B, *Eur. J. Biochem.*, 50, 41, 1974.
24. Wintersberger, E., DNA-dependent DNA polymerases from eukaryotes, *Trends Biochem. Sci.*, 2, 58, 1977.
25. Chang, L. M. S., DNA polymerases from baker's yeast, *J. Biol. Chem.*, 252, 1873, 1977.
26. Ganz, P. R., and Pearlman, R. E., Purification from *Tetrahymena thermophila* of DNA polymerasesd and a protein which modifies its activity, *Eur. J. Biochem.*, 113, 159, 1980.
27. Creran, M. and Pearlman, R. E., Deoxyribonucleic acid polymerase from *Tetrahymena pyriformis*. Purification and properties of the major activity in exponentially growing cells, *J. Biol. Chem.*, 249, 3123, 1974.

28. Sakai, A. and Watanabe, Y., DNA polymerases of *Tetrahymena pyriformis*. I. Characterization of two *N*-ethylmaleimide-sensitive DNA polymerases from exponentially growing cells, *J. Biochem. (Tokyo)*, 91, 845, 1982.
29. Loomis, L. W., Rossomando, E. F., and Chang, L. M. S., DNA polymerase of *Dictyostelium discoideum, Biochim. Biophys. Acta*, 425, 469, 1976.
30. McLennan, A. G. and Keir, H. M., The deoxyribonucleic acid polymerases of non-vertebrate eukaryotes, *Biochem. Soc. Symp.*, 42, 55, 1977.
31. Weissbach, A., Baltimore, D., Bollum, F. J., Gallo, R. C., and Korn, D., Nomenclature of eukaryotic DNA polymerases, *Science*, 190, 401, 1975.
32. Weissbach, A., Eukaryotic DNA polymerases, *Ann. Rev. Biochem.*, 46, 25, 1977.
33. Falaschi, A. and Spadari, S., The three DNA polymerases of animal cells: properties and functions, in *DNA Synthesis: Present and Future*, Molineux, I. and Kohyanma, M., Eds., Plenum Press, New York, 1978, 487.
34. Huberman, J. A., New views on the biochemistry of eukaryotic DNA replication revealed by aphidicolin, an unusual inhibitor of DNA polymerase α, *Cell*, 23, 647, 1981.
35. Waqar, M. A., Evans, M. J., and Huberman, J. A., Effect of 2′, 3′-dideoxythymidine-5′-triphosphate on HeLa cell *in vitro* DNA synthesis: evidence that DNA polymerase α is the only polymerase required for DNA replication, *Nucl. Acids Res.*, 5, 1933, 1978.
36. Matsukage, A., Ono, K., Ohashi, A., Takahashi, T., Nakayama, C., and Saneyoshi, M., Inhibitory effect of 1-β-D-arabinofuranosylcytosine 5′-triphosphate on DNA polymerases from murine cells and oncornavirus, *Cancer Res.*, 38, 3076, 1978.
37. Fry, M., Eukaryotic DNA polymerases, in *Enzymes of Nucleic Acid Synthesis and Modification*, Jacob, S. T., Ed., CRC Press, Boca Raton, Fla., 1983, 39.
38. Holmes, A. M., Hesslewood, I. P., Wickremasinghe, R. F., and Johnston, I. R., The heterogeneity of deoxyribonucleic acid polymerase-α, *Biochem. Soc. Symp.*, 42, 17, 1977.
39. Hübscher, U., Spanos, A., Albert, V., Grummt, F., and Banks, G. R., Evidence that a high molecular weight replicative DNA polymerase is conserved during evolution, *Proc. Natl. Acad. Sci. U.S.A.*, 78, 6771, 1981.
40. Sauer, D. and Lehman, I. R., Immunological comparison of purified DNA polymerase-α from embryos of *Drosophila melanogaster* with forms of the enzyme present *in vivo, J. Biol. Chem.*, 257, 12394, 1982.
41. Sugino, A. and Nakayamara, K., DNA polymerase-α mutants from *Drosophila melanogaster* cell line, *Proc. Natl. Acad. Sci. U.S.A.*, 77, 7049, 1980.
42. Liu, P. K., Chang, C. C., Trosko, J. E., Dube, D. K., Martin, G. M., and Loeb, L. A., Mammalian mutator mutant with an aphidicolin-resistant DNA polymerase-α, *Proc. Natl. Acad. Sci. U.S.A.*, 80, 797, 1983.
43. Korn, D., Fisher, P. A., Battey, J., and Wang, T. S.-F., Structural and enzymological properties of human DNA polymerases α and β, *Cold Spring Harbor Symp. Quant. Biol.*, 43, 613, 1979.
44. Korn, D., Fisher, P. A., and Wang, T. S.-F., Enzymological characterization of human DNA polymerases-α and β, in *New Approaches in Eukaryotic DNA Replication*, de Recondo, A. M., Ed., Plenum Press, New York, 1983, 17.
45. Yagura, T., Kozu, T., and Seno, T., Mouse DNA polymerase accompanied by a novel RNA polymerase activity: purification and partial characterization, *J. Biochem. (Tokyo)*, 91, 607, 1982.
46. Conaway, R. C. and Lehman, I. R., A DNA primase activity associated with DNA polymerase-α from *Drosophila melanogaster* embryos, *Proc. Natl. Acad. Sci. U.S.A.*, 79, 2523, 1982.
47. Shioda, M., Nelson, E. M., Bayne, M. L., and Benbow, R. M., DNA primase activity associated with DNA polymerase-α from *Xenopus laevis* ovaries, *Proc. Natl. Acad. Sci. U.S.A.*, 79, 7209, 1982.
48. Riedel, H.-D., Konig, H., Stahl, H., and Knippers, R., Circular single stranded phage M13-DNA as a template for DNA synthesis in protein extracts from *Xenopus laevis* eggs: evidence for a eukaryotic DNA priming activity, *Nucl. Acids Res.*, 10, 5621, 1982.
49. Yoshida, S., Suzuki, R., Masaki, S., and Koiwai, O., RNA primase associated with 10S DNA polymerase from calf thymus: rapid isolation using monoclonal antibody and tryptic peptide mapping analysis, *Biochim. Biophys. Acta*, 741, 348, 1983.
50. Burke, J. F., Plummer, J., Huberman, J. A., and Evans, M. J., Restriction fragment primed ϕX174 single-stranded DNA as template for DNA polymerases α and β. Detection and partial purification of a DNA polymerase α stimulating factor, *Biochim. Biophys. Acta*, 609, 205, 1980.
51. Faust, E. A. and Rankin, C. D., *In vitro* conversion of MVM parvovirus single-stranded DNA to the replicative form by DNA polymerase α from Ehrlich ascites tumor cells, *Nucl. Acids Res.*, 10, 4181, 1982.
52. Pritchard, C. G. and DePamphilis, M. L., Preparation of DNA polymerase $\alpha \cdot C_1C_2$, *J. Biol. Chem.*, 258, 9801, 1983.
53. Pritchard, C. G., Weaver, D. T., Baril, E. F., and DePamphilis, M. L., DNA polymerase α cofactors C_1C_2 function as primer recognition proteins, *J. Biol. Chem.*, 258, 9810, 1983.

54. Tanaka, S., Hu, S.-Z., Wang, T. S.-F., and Korn, D., Preparation and preliminary characterization of monoclonal antibodies against human polymerase α, *J. Biol. Chem.*, 257, 8386, 1982.
55. Matsukage, A., Yamaguchi, M., Tanabe, K., Nishizawa, M., Takahashi, T., Seto, M., and Takahashi, T., Establishment of hybridoma clones which produce anti-chick embryo DNA polymerase α monoclonal antibodies *Gann*, 73, 850, 1982.
56. Masaki, S., Shiku, H., Kaneda, T., Koiwai, O., and Yoshida, S., Production and characterization of monoclonal antibody against 10S DNA polymerase α from calf thymus, *Nucl. Acids Res.*, 10, 4703, 1982.
57. Wang, T. S.-F., Hu, S.-Z., and Korn, D., DNA primase from KB cells. Characterization of a primase activity tightly associated with immunoaffinity-purified polymerase-α, *J. Biol. Chem.*, 259, 1854, 1984.
58. Wahl, A. F., Kowalski, S. P., Harwell, L. W., Lord, E. M., and Bambara, R. A., Immunoaffinity purification and properties of a high molecular weight calf thymus DNA α polymerase, *Biochemistry*, 23, 1895, 1984.
59. Deterea-Wadleigh, S., Karawya, E., and Wilson, S. H., Synthesis of DNA polymerase by *in vitro* translation of calf RNA, *Biochem. Biophys. Res. Commun.*, 122, 420, 1984.
60. Karawya, E., Swack, J., Albert, W., Fedorko, J., Minna, J. D., and Wilson, S. H., Identification of a higher M_r α-polymerase catalytic polypeptide in monkey cells by monoclonal antibody, *Proc. Natl. Acad. Sci. U.S.A.*, 81, 7777, 1984.
61. Liu, P. K. and Loeb, L. A., Transfection of DNA polymerase-α gene, *Science*, 226, 833, 1984.
62. Fisher, P. A. and Korn, D., DNA polymerase α. Purification and structural characterization of a near homogeneous enzyme from human KD cells, *J. Biol. Chem.*, 252, 6528, 1977.
63. Grummt, F., Waltl, G., Jantzen, H.-M., Hamprecht, K., Hübscher, U., and Kuenzle, C. C., Diadenosine $5',5'''$-p^1,p^4-tetraphosphate, a ligand of the 57-kilodalton subunit of DNA polymerase-α, *Proc. Natl. Acad. Sci. U.S.A.*, 76, 6081, 1979.
64. Grosse, F. and Krauss, G., Purification of a 9S DNA polymerase-α species from calf thymus, *Biochemistry*, 20, 5470, 1981.
65. Méchali, H. and de Recondo, A.-M., Structural analysis of eukaryotic DNA polymerase-α, in *New Approaches in Eukaryotic DNA Replication*, de Recondo, A.-M., Ed., Plenum Press, New York, 1983, 57.
66. Goscin, L. P. and Byrnes, J. J., Isolation of the catalytic core of DNA polymerase-α from rabbit bone marrow, *Nucl. Acids Res.*, 10, 6023, 1982.
67. Yamaguchi, M., Tanabe, K., Takahashi, T., and Matsukage, A., Chick embryo DNA polymerase α. Polypeptide components and their microheterogeneity, *J. Biol. Chem.*, 257, 4484, 1982.
68. Faust, E. A., Gloor, G., Macintyre, M.-F., and Nagy, R., ATP(CTP)-dependent conversion of MVM parvovirus single-stranded DNA to its replicative form by a purified 10S species of mouse DNA polymerase α, *Biochim. Biophys. Acta*, 781, 216, 1984.
69. Chen, Y.-C., Bohn, E. W., Planck, S. R., and Wilson, S. H., Mouse DNA polymerase α. Subunit structure and identification of a species with associated endonuclease, *J. Biol. Chem.*, 254, 11678, 1979.
70. Lamothe, P., Baril, B., Chi, A., Lee, L., and Baril, E., Accessory proteins for DNA polymerase α activity with single-strand DNA templates, *Proc. Natl. Acad. Sci. U.S.A.*, 78, 4723, 1981.
71. Hübscher, U., DNA polymerases in prokaryotes and eukaryotes: mode of action and biological implications, *Experientia*, 39, 1, 1983.
72. Salisbury, J. G., O'Connor, P. J., and Saffhill, R., Molecular size and fidelity of DNA polymerase α from the regenerating liver of the rat, *Biochim. Biophys. Acta*, 517, 181, 1978.
73. Méchali, M., Abadidebat, J., and de Recondo, A. M., Eukaryotic DNA polymerase α. Structural analysis of the enzyme from regenerating rat liver, *J. Biol. Chem.*, 255, 2114, 1980.
74. Holmes, A. M. and Johnston, I. R., Molecular asymmetry of rat liver cytoplasmic DNA polymerase, *FEBS Lett.*, 29, 1, 1973.
75. Masaki, S., Tanabe, K., and Yoshida, S., Large polypeptides of 10S DNA polymerase α from calf thymus: rapid isolation using monoclonal antibody and tryptic peptide mapping analysis, *Nucl. Acids. Res.*, 12, 4455, 1984.
76. Spanos, A., Sedgwick, S. G., Yarranton, G. T., Hübscher, U., and Banks, G. R., Detection of the catalytic activities of DNA polymerases and their associated exonucleases following SDS-polyacrylamide gel electrophoresis, *Nucl. Acids Res.*, 9, 1825, 1981.
77. Karawya, E., Swack, J. A., and Wilson, S. H., Improved conditions for activity gel analysis of DNA polymerase catalytic polypeptides, *Anal. Biochem.*, 135, 318, 1983.
78. Scovassi, A. I., Torsello, S., Plevani, Y., Badaracco, G. F., and Bertazzoni, U., Active polypeptide fragment common to prokaryotic and mitochondrial DNA polymerases, *EMBO J.*, 1, 1161, 1982.
79. Holmes, A. M., Hesslewood, I. P., and Johnston, I. R., *In vitro* conversion of a calf thymus DNA polymerase to a 7.3S species, *Nature (London)*, 255, 420, 1975.

80. Holmes, A. M., Hesslewood, I. P., and Johnston, I. R., Evidence that DNA polymerase α of calf thymus contains a subunit of molecular weight 155,000, *Eur. J. Biochem.,* 62, 229, 1976.
81. Wickremasinghe, R. G., Hesslewood, I. P., and Johnston, I. R., Differential *N*-ethylmaleimide inhibition of two enzymes of the DNA α-polymerase fraction from calf thymus, *FEBS Lett.,* 78, 139, 1977.
82. Hesslewood, I. P., Holmes, A. M., Wakeling, W. F., and Johnston, I. R., Studies on the purification and properties of a 6.8S DNA polymerase activity found in calf thymus DNA polymerase α fraction, *Eur. J. Biochem.,* 84, 123, 1978.
83. McKune, K. and Holmes, A. M., Further studies on partially purified calf thymus DNA polymerase-α, *Nucl. Acids Res.,* 6, 3341, 1979.
84. McKune, K. and Holmes, A. M., Studies on the processivity of highly purified calf thymus 8S and 7.3S DNA polymerase α, *Biochem. Biophys. Res. Commun.,* 90, 864, 1979.
85. Hockensmith, J. W. and Bambara, R. A., Kinetic characteristics which distinguish two forms of calf thymus polymerase α, *Biochemistry,* 29, 227, 1981.
86. Wright, G. E., Baril, E. F., and Brown, N. C., Butylanilinouracil: a selective inhibitor of HeLa cell DNA polymerase α, *Nucl. Acids. Res.,* 8, 99, 1980.
87. Homparler, R. L., Rossi, M., and Labitan, A., Partial purification and properties of two forms of deoxyribonucleic acid polymerase from calf thymus, *J. Biol. Chem.,* 248, 285, 1973.
88. Tsuruo, T., Hirayama, K., Satoh, H., and Ukita, T., A new form of high molecular weight DNA polymerase in the nuclei of rat ascites hepatoma cells, *J. Biochem. (Tokyo),* 78, 401, 1975.
89. Tanabe, K. and Takahashi, T., Studies on the molecular species of DNA polymerase extracted from rat ascites hepatoma cells, *J. Biochem. (Tokyo),* 79, 85, 1976.
90. Enomoto, T., Tanuma, S.-I., and Yamada, M.-A., Purification and characterization of two forms of DNA polymerase α from HeLa cell nuclei, *Biochemistry,* 22, 1134, 1983.
91. Brosius, S., Grosse, F., and Krauss, G., Subspecies of DNA polymerase α from calf thymus with different fidelity in copying synthetic template-primers, *Nucl. Acids Res.,* 11, 193, 1983.
92. Kozu, T., Yagura, T., and Seno, T., A novel subspecies of mouse DNA polymerase α: variations of the activity during postnatal tissue development and its subcellular location, *Cell. Struct. Funct.,* 7, 9, 1982.
93. Filipula, D., Fisher, P. A., and Korn, D., DNA polymerase α. Common polypeptide core structure of three enzyme forms from human KB cells, *J. Biol. Chem.,* 257, 2029, 1982.
94. Grosse, F. and Krauss, G., Purification and partial characterization of a DNA polymerase α species from calf thymus, *Nucl. Acids Res.,* 8, 5703, 1980.
95. Reckman, B., Grosse, F., and Krauss, G., The elongation of mismatched primers by DNA polymerase α from calf thymus, *Nucl. Acids Res.,* 11, 7251, 1983.
96. Masaki, S. and Yoshida, S., 10S DNA polymerase from calf thymus which copies both poly(rA)·oligo(dT) and activated DNA, *Biochim. Biophys. Acta,* 521, 74, 1978.
97. Yoshida, S., Masaki, S., and Koiwai, O., Further characterization of a poly(rA·oligo(dT) dependent activity of multiple DNA polymerase-α from calf thymus, *Biochim. Biophys. Acta,* 654, 194, 1981.
98. Masaki, S., Koiwai, O., and Yoshida, S., 10S DNA polymerase α of calf thymus shows a microheterogeneity in its large polypeptide component, *J. Biol. Chem.,* 257, 7172, 1982.
99. Albert, W., Grummt, F., Hübscher, U., and Wilson, S., Structural homology among calf thymus α-polymerase polypeptides, *Nucl. Acids Res.,* 10, 935, 1982.
100. Hübscher, U., Gerschwiler, P., and McMaster, G. K., A mammalian DNA polymerase α holoenzyme functioning on defined *in vivo*-like templates, *EMBO J.,* 1, 1513, 1982.
101. Hübscher, U. and Ottinger, H.-P., Mammalian DNA polymerase α holoenzyme functioning on defined *in vivo*-like templates, in *Mechanisms of DNA Replication and Recombination,* Cozzarelli, N., Ed., Alan R. Liss, New York, 1983, 517.
102. Fichot, O., Pascal, M., Méchali, M., and de Recondo, A. M., DNA polymerase α from regenerating rat liver. Catalytic properties of the highly purified enzyme, *Biochim. Biophys. Acta,* 561, 29, 1979.
103. Matsukage, A., Bohn, E. W., and Wilson, S. H., On the DNA polymerase III of mouse myeloma: partial purification and characterization, 14, 1006, 1975.
104. Villani, G., Sauer, B., and Lehman, I. R., DNA polymerase α from *Drosophila melanogaster* embryos. Subunit structure, *J. Biol. Chem.,* 255, 9479, 1980.
105. Villani, G., Fay, P. G., Bambara, R. A., and Lehman, I. R., Elongation of RNA-primed DNA templates by DNA polymerase α from *Drosophila melanogaster* embryos, *J. Biol. Chem.,* 256, 8202, 1981.
106. Kaguni, L. S., Rossignol, J.-M., Connaway, R. C., and Lehman, J. R., Isolation of an intact DNA polymerase-primase from embryos of *Drosophila melanogaster, Proc. Natl. Acad. Sci. U.S.A.,* 80, 2221, 1983.
107. Kaguni, L. S., Rossignol, J.-M., Conaway, R. C., Banks, G. R., and Lehman, I. R., Association of DNA primase with the β/γ subunits of DNA polymerase α from *Drosophila melanogaster* embryos, *J. Biol. Chem.,* 258, 9037, 1983.

108. Kaguni, L. S., DiFrancesco, R. A., and Lehman, I. R., The DNA polymerase-primase from *Drosophila melanogaster* embryos. Rate and fidelity of polymerization on single stranded DNA templates, *J. Biol. Chem.*, 259, 9314, 1984.
109. Wang, T.S.-F., Hu, S.-Z., and Korn, D., DNA primase from KB cells. Characterization of a primase activity tightly associated with immunoaffinity purified DNA polymerase-α, *J. Biol. Chem.*, 259, 1854, 1984.
110. Karawya, E. M. and Wilson, S. H., Studies on catalytic subunits of mouse myeloma α-polymerase, *J. Biol. Chem.*, 257, 13129, 1982.
111. Badaracco, G., Capucci, L., Plevani, P., and Chang, L. M. S., Polypeptide structure of DNA polymerase I from *Saccharomyces cerevisiae, J. Biol. Chem.*, 258, 10720, 1983.
112. Arendes, J., Kim, K., and Sugino, A., Yeast 2-μm plasmid DNA replication *in vitro*. Purification of the CDC gene product by complementation assay, *Proc. Natl. Acad. Sci. U.S.A.*, 80, 673, 1983.
113. Tsuchiya, E., Kimura, K., Miyakawa, T., and Sakuzo, F., Characteristic alteration in the nuclear DNA polymerase activity during the cell division cycle of *Saccharomyces cerevisiae, Nucl. Acids Res.*, 12, 3143, 1984.
114. Cobianchi, F., Riva, S., Mastromei, G., Spadari, S., Pedrali-Noy, G., and Falaschi, A., Enhancement of the rate of DNA polymerase α activity on duplex DNA by a DNA-binding protein and a DNA-dependent ATPase of mammalian cells, *Cold Spring Harbor Symp. Quant. Biol.*, 43, 639, 1978.
115. Boxer, L. M. and Korn, D., Structural and enzymological characterization of a deoxyribonucleic acid dependent adenosine triphosphatase from KB nuclei, *Biochemistry*, 19, 2623, 1980.
116. De Jong, P. J., Tommassen, J. P. M., Van der Vliet, P. C., and Jansz, H. S., Purification and characterization of DNA-dependent ATP phosphorylases from KB cells, *Eur. J. Biochem.*, 117, 179, 1981.
117. Méchali, M. and de Recondo, A. M., Eukaryotic DNA polymerase-α: tryptic map analysis and comparison between the rat liver and *Drosophila* enzymes, *Biochem. Int.*, 4, 465, 1982.
118. Bhattacharya, P., Simet, I., and Basu, S., Inhibition of human neuroblastoma DNA polymerase activities by plant lectins and toxins, *Proc. Natl. Acad. Sci. U.S.A.*, 76, 2218, 1979.
119. Bhattacharya, P., Simet, I., and Basu, S., Differential inhibition of multiple forms of DNA polymerase-α from IMR-32 human neuroblastoma cells, *Proc. Natl. Acad. Sci. U.S.A.*, 78, 2683, 1981.
120. Bhattacharya, P. and Basu, S., Probable involvement of a glycoconjugate in IMR-32 DNA synthesis: decrease of DNA polymerase-α activity after tunicamycin treatment, *Proc. Natl. Acad. Sci. U.S.A.*, 79, 1488, 1982.
121. Basu, S., Bhattacharya, P., and Basu, M., The role of eucaryotic cell surface glycoconjugates in DNA replication, *J. Biosci.*, 5 (Suppl. 1), 157, 1983.
122. Kornberg, A., *DNA Replication*, W. H. Freeman, San Francisco, 1980.
123. Kornberg, A., *DNA Replication. A Supplement*, W. H. Freeman, San Francisco, 1982.
124. Banks, G. R. and Yarranton, G. T., A DNA polymerase from *Ustilago maydis* II. Properties of the associated deoxyribonuclease activity, *Eur. J. Biochem.*, 62, 143, 1976.
125. Wintersberger, E., Yeast DNA polymerases: antigenic relationship, use of DNA primer and associated exonuclease activity, *Eur. J. Biochem.*, 84, 167, 1978.
126. Byrnes, J. J., Structural and functional properties of DNA polymerase delta from rabbit bone marrow, *Mol. Cell. Biochem.*, 62, 13, 1984.
127. DePamphilis, N. E. and Wasserman, P. M., Replication of eukaryotic chromosomes: a close-up of the replication fork, *Ann. Rev. Biochem.*, 49, 627, 1980.
128. Rowen, L. and Kornberg, A., Primase, the dnaG portion of *Esherichia coli.* An enzyme which starts DNA chains, *J. Biol. Chem.*, 253, 758, 1978.
129. Rowen, L. and Kornberg, A., A ribo-deoxyribonucleotide primer synthesized by primase, *J. Biol. Chem.*, 253, 770, 1978.
130. Benz, E. W., Jr., Reinberg, D., Vicuna, R., and Hurwitz, J., Initiation of DNA replication by the dnaG protein, *J. Biol. Chem.*, 255, 1096, 1980.
131. Scherzinger, E., Lanka, E., Morelli, G., Seiffert, D., and Yuki, A., Bacteriophage T7-induced DNA priming proteins. A novel enzyme involved in DNA replication, *Eur. J. Biochem.*, 72, 543, 1979.
132. Romano, L. J. and Richardson, C. C., Requirements for synthesis of ribonucleic acid primers during lagging strand synthesis by the DNA polymerase and gene 4 protein of bacteriophage T7, *J. Biol. Chem.*, 254, 10476, 1979.
133. Romano, L. J. and Richardson, C. C., Characterization of the ribonucleic acid primers and the deoxyribonucleic acid product synthesized by the DNA polymerase and gene 4 proteins of bacteriophage T7, *J. Biol. Chem.*, 254, 10483, 1979.
134. Nossal, N. G., RNA priming of DNA replication by bacteriophage T4 proteins, *J. Biol. Chem.*, 255, 2176, 1980.

135. Liu, C.-C., Burke, R. L., Hibner, V., Barry, J., and Alberts, B. M., Probing DNA replication mechanisms with the T4 bacteriophage *in vitro* system, *Cold Spring Harbor Symp. Quant. Biol.*, 43, 469, 1979.
136. Liu, C.-C. and Alberts, B. M., Characterization of RNA primer synthesis in the T4 bacteriophage *in vitro* DNA replication system, *J. Biol. Chem.*, 256, 2821, 1981.
137. Reichard, P., Eliasson, R., and Soderman, G., Initiator RNA in dicontinuous polyoma DNA synthesis, *Proc. Natl. Acad. Sci. U.S.A.*, 71, 4901, 1974.
138. Hunter, T. and Francke, B., *in vitro* polyoma DNA synthesis: involvement of RNA in discontinuous chain growth, *J. Mol. Biol.*, 83, 123, 1974.
139. Pigiet, V., Eliasson, R., and Reichard, P., Replication of polyoma DNA in isolated nuclei. III. The nucleotide sequence at the RNA-DNA junction of nascent strands, *J. Mol. Biol.*, 84, 197,1974.
140. Wagar, M. A. and Huberman, J. A., Covalent attachment of RNA to nascent DNA in mammalian cells, *Cell*, 6, 551, 1975.
141. Tseng, B. Y. and Goulian, M., Initiator RNA of discontinuous DNA synthesis in human lymphocytes, *Cell*, 12, 483, 1977.
142. Eliasson, R., and Reichard, P., Replication of polyoma DNA in isolated nuclei. VII. Initiator RNA synthesis during nucleotide depletion, *J. Mol. Biol.*, 129, 393, 1979.
143. Tseng, B. Y., Erickson, J. M., and Goulian, M., Initiator RNA of nascent DNA from animal cells, *J. Mol. Biol.*, 129, 531, 1979.
144. Tseng, B. Y. and Goulian, M., Initiator RNA synthesis upon ribonucleotide depletion. Evidence for substitutions, *J. Biol. Chem.*, 255, 2062, 1980.
145. Kaufman, G., Characterization of initiator RNA from replicating simian virus DNA synthesized in isolated nuclei, *J. Mol. Biol.*, 147, 25, 1981.
146. Chang, L. M. S. and Bollum, F. J., A chemical model for transcriptional initiation of DNA replication, *Biochem. Biophys. Res. Commun.*, 46, 1354, 1972.
147. Philippe, M., Sheinin, R., and de Recondo, A. M., Association between primase and DNA polymerase α in murine cells, in, *Proteins Involved in DNA Replication*, Hubscher, U. and Spadari, S., Eds., Plenum Press, New York, 1985, 295.
148. de Recondo, A. M., Lepesant, J. A., Richot, O., Grasset, L., Rossingnol, J. M., and Cazillis, M., Synthetic template specificity of a deoxyribonucleic acid polymerase from regenerating rat liver, *J. Biol. Chem.*, 248, 131, 1973.
149. Kaftory, A. and Fry, M., Highly efficient copying of single stranded DNA by eukaryotic cell chromatin, *Nucl. Acids Res.*, 5, 2679, 1978.
150. Hübscher, U., The mammalian primase is part of a high molecular weight DNA polymerase α polypeptide, *EMBO. J.*, 2, 133, 1983.
151. Hübscher, U., DNA polymerse holoenzymes, *Trend. Biochem. Sci.*, 9, 390, 1984.
152. Yagura, T., Kozu, T., and Seno, T., Mouse DNA replicase. DNA polymerase associated with a novel RNA polymerase activity to synthesize initiator RNA of strict size, *J. Biol. Chem.*, 257, 11121, 1982.
153. Yagura, T., Kozu, T., Seno, T., Saneyoshi, M., Hiraga, S., and Nagano, N., A novel form of DNA polymerase α associated with DNA primase activity of vertebrates, *J. Biol. Chem.*, 258, 13070, 1983.
154. Tseng, B. Y. and Ahlem, C. N., A DNA primase from mouse cells. Purification and partial characterization, *J. Biol. Chem.*, 258, 9845, 1983.
155. Kozu, T., Yagura, T., and Seno, T., *De novo* DNA synthesis by a novel mouse DNA polymerase associated with primase activity, *Nature (London)*, 298, 180, 1982.
156. Yagura, T., Kozu, T., and Seno, T., Mechanism of stimulation by a specific protein factor of *de novo* DNA synthesis by mouse DNA replicase with fd phage single stranded circular DNA, *Nucl. Acids Res.*, 11, 6369, 1983.
157. Nishizawa, M., Tanabe, K., Matsukage, A., Nakamura, M., and Yagura, T., Detection of DNA replicase activity in rat thymoma cells, *J. Biochem. (Tokyo)*, 93, 287, 1983.
158. Yagura, T., Tanaka, S., Kozu, T., Seno, T., and Korn, D., Tight association of DNA primase with a subspecies of mouse DNA polymerase-α, *J. Biol. Chem.*, 258, 6698, 1983.
159. Yagura, T., Kozu, T., Seno, T., Nishijima, Y., Ohtsuka, E., and Ikehara, M., Inhibition of a novel subspecies of DNA polymerase-α by 2′ deoxy-2′azidoadenosine 5′-triphosphate, *Biochem. Biophys. Res. Commun.*, 104, 1189, 1982.
160. Konig, H. Riedel, H. D., and Knippers, R., Reactions *in vitro* of the DNA polymerase primase from *Xenopus laevis* eggs. A role for ATP in chain elongation, *Eur. J. Biochem.*, 135, 435, 1985.
161. Méchali, M. and Harland, R. M., DNA synthesis in a cell free system from *Xenopus* eggs: priming and elongation on a single stranded DNA *in vitro*, *Cell*, 30, 93, 1982.
162. Grosse, F. and Krauss, G., The primase activity of DNA polymerase α from calf thymus, *J. Biol. Chem.*, 260, 1881, 1985.
163. Conaway, R. C. and Lehman, I. R., A DNA primase activity associated with DNA polymerase α from *Drosophila melanogaster* embryos, *Proc. Natl. Acad. Sci. U.S.A.*, 79, 2523, 1982.

164. Conaway, R. C. and Lehman, I. R., Synthesis by the DNA primase of *Drosophila melanogaster* of a primer with a unique chain length, *Proc. Natl. Acad. Sci. U.S.A.*, 79, 4585, 1982.
165. Tseng, B. Y. and Ahlem, C. N., DNA primase activity from human lymphocytes, *J. Biol. Chem.*, 257, 7280, 1982.
166. Gronostajski, R. M., Field, J., and Hurwitz, J., Purification of a primase activity associated with DNA polymerase-α from HeLa cells, *J. Biol. Chem.*, 259, 9479, 1984.
167. Hu, S.-Z., Wang, T. S.-F., and Korn, D., DNA primase from KB cells. Evidence for a novel model of primate catalysis by a highly purified primase/polymerase-α complex, *J. Biol. Chem.*, 259, 2602, 1984.
168. Kaufman, G. and Falk, H. H., An oligoribonucleotide polymerase from SV40-infected cells with properties of a primase, *Nucl. Acids Res.*, 7, 2309, 1982.
169. Tseng, B. Y. and Ahlem, C. N., DNA primase from mouse cells, in *Mechanisms of DNA Replication and Recombination*, Cozzarelli, N. Ed., Alan R. Liss, New York, 1983, 511.
170. Tseng, B. Y. and Ahlem, C. N., Mouse primase initiation sites in the origin region of simian virus 40, *Proc. Natl. Acad. Sci. U.S.A.*, 81, 2342, 1984.
171. Herrick, G. and Alberts, B., Purification and physical characterization of nucleic acid helix-unwinding proteins from calf thymus, *J. Biol. Chem.*, 251, 2124, 1976.
172. Herrick, G. and Alberts, B., Nucleic acid helix-coil transitions mediated by helix unwinding proteins from calf thymus, *J. Biol. Chem.*, 251, 2133, 1976.
173. Herrick, G., Delius, H., and Alberts, B., Single stranded DNA structure and DNA polymerase activity in the presence of nucleic acid helix-unwinding proteins from calf thymus, *J. Biol. Chem.*, 251, 2142, 1976.
174. Henner, D. and Furth, J. J., *De novo* synthesis of a polymer of deoxyadenylate and deoxythymidilate by calf thymus DNA polymerase α, *Proc. Natl. Acad. Sci. U.S.A.*, 72, 3944, 1975.
175. Henner, D. and Furth, J. J., Primed and unprimed synthesis of poly[d(AT)] by calf thymus DNA polymerase α, *J. Biol. Chem.*, 252, 1932, 1977.
176. Valentini, O., Biamonti, G., Mastromei, G., and Riva, S., Structural and functional heterogeneity of single-stranded DNA binding proteins from calf thymus, *Biochim. Biophys. Acta*, 782, 147, 1984.
177. Duguet, M., Soussi, T., Rossignol, J.-M., Méchali, M., and de Recondo, A. M., Stimulation of rat liver α- and β-type DNA polymerases by an homologous DNA unwinding protein, *FEBS Lett.*, 79, 160, 1977.
178. Duguet, M. and de Recondo, A. M., A deoxyribonucleic acid unwinding protein isolated from regenerating liver, *J. Biol. Chem.*, 253, 1660, 1978.
179. Bonne, C., Duguet, M., and de Recondo, A. M., Rat liver single-strand DNA binding proteins: physiological variations, *FEBS Lett.*, 106, 292, 1979.
180. Bonne, C., Duguet, M., and de Recondo, A. M., Single-strand DNA binding protein from rat liver: interaction with supercoiled DNA, *Nucl. Acids Res.*, 8, 4955, 1980.
181. Duguet, M., Bonne, C., and de Recondo, A. M., Single strand deoxyribonucleic acid binding protein from rat liver changes the helical structure of deoxyribonucleic acid, *Biochemistry*, 20, 3598, 1981.
182. Bonne, C., Santiere, P., Duguet, M., and de Recondo, A. M., Identification of a single-stranded DNA binding protein from rat liver with high mobility group protein I, *J. Biol. Chem.*, 257, 2722, 1982.
183. Planck, S. R. and Wilson, S. H., Studies on the structure of mouse helix-destabilizing protein-1. DNA binding and controlled proteolysis with trypsin, *J. Biol. Chem.*, 255, 11547, 1980.
184. Detera, S. D., Becerra, S. P., Swack, J. A., and Wilson, S. H., Studies on the mechanism of DNA polymerase-α. Nascent chain elongation, steady state kinetics and initiation phase of DNA synthesis, *J. Biol. Chem.*, 256, 6933, 1981.
185. Otto, B., Baynes, M., and Knippers, R., A single-strand-specific DNA-binding protein from mouse cells that stimulates DNA polymerase, *Eur. J. Biochem.*, 73, 17, 1977.
186. Richter, A., Knippers, R., and Otto, B., Interaction of a mammalian single strand specific DNA binding protein with DNA polymerase α, *FEBS Lett.*, 91, 293, 1978.
187. Danse, J. M., Egly, J. M., and Kempf, J., *In vitro* activation of DNA polymerase α by a protein kinase in chick embryo, *FEBS Lett.*, 124, 84, 1981.
188. Laks, M. S. and Jungmann, R. A., Prereplicative modulation of nuclear protein kinases in the regenerating rat liver, *Biochem. Biophys. Res. Commun.*, 96, 697, 1980.
189. Fry, M., Renart, Z., Weisman-Shomer, P., Lapidot, J., and Sharf, R., Factor D — a template-selective stimulatory protein of DNA polymerases: unchanged level in chromatin during liver regeneration, submitted for publication, 1985.
190. Fry, M., Lapidot, J., and Weisman-Shomer, P., A DNA polymerase α stimulatory protein from mouse liver: partial purification and study of its properties and interaction with polymerase α, *Biochemistry*, in press, 1985.
191. Fry, M., Weisman-Shomer, P., Lapidot, J., and Sharf, R., Stringent template selective stimulation of diverse DNA polymerases by murine DNA binding protein factor D, submitted, 1985.

192. Riva, S., Clivio, A., Valentini, O., and Cobianchi, F., DNA binding proteins from calf thymus with an enhanced ability to stimulate DNA polymerase α *in vitro, Biochem. Biophys. Res. Commun.,* 95, 34, 1980.

193. Benbow, R. M., Breaux, C. B., Joenje, H., Krauss, M. R., Lennox, R. W., Nelson, E. M., Wang, N. S., and White, S. R., Eukaryotic DNA replication: the first steps toward a multienzyme system from *Xenopus laevis, Cold Spring Harbor Symp. Quant. Biol.,* 43, 597, 1978.

194. Murakami, K. and Mano, Y., Stimulation of sea urchin DNA polymerase by protein factors, *Biochem. Biophys. Res. Commun.,* 55, 1125, 1973.

195. Murakami-Murofushi, K. and Mano, Y., Stimulation of sea urchin DNA polymerase by protein factors. II. Formation of active complex in DNA polymerase reaction, *Biochim. Biophys. Acta,* 475, 254, 1977.

196. Shimada, H., Cellular factor stimulating DNA synthesis in nuclei isolated from sea urchin embryos, *Dev. Biol.,* 97, 454, 1983.

197. Novak, B. and Baril, E. F., HeLa DNA polymerase activity *in vitro:* specific stimulation by a non-enzymic protein factor, *Nucl. Acids Res.,* 5, 221, 1978.

198. Kalf, G. F., Metrione, R. M., Kerlavage, B. L., and Koszalka, T. R., Variation of DNA polymerase activities of rat giant trophoblast cells in mid gestation, *Dev. Biol.,* 84, 351, 1981.

199. Kalf, G. F., Metrione, R. M., and Koszalka, T. R., A protein stimulatory factor for DNA polymerase α in rat giant trophoblast cells, *Biochem. Biophys. Res. Commun.,* 100, 566, 1981.

200. Steinberg, J. A., Otten, M., and Grindey, G. B., Isolation of a DNA polymerase α-associated regulatory protein from calf thymus, *Cancer Res.,* 39, 4330, 1979.

201. Méchali, M. and de Recondo, A. M., Detection of a DNA binding factor associated with mammalian DNA polymerase-α, *Biochem. Biophys. Res. Commun.,* 82, 255, 1978.

202. Lee, S.-C. and Lucas, Z. J., Regulatory factors produced by lymphocytes. II. Inhibition of cellular DNA synthesis associated with a factor inhibiting DNA polymerase α activity, *J. Immunol.,* 118, 88, 1977.

203. Kaftory, A., Weisman-Shomer, P., and Fry, M., Nuclear protein from cultured hepatoma cells preferentially inhibits copying of denatured DNA by isolated chromatin, *FEBS Lett.,* 99, 265, 1979.

204. Miyata, S. and Kihara, H., Properties of DNA-binding protein in a culture medium of thymus cells, *J. Biochem. (Tokyo),* 92, 1539, 1982.

205. Zamecnik, P. C., Stephenson, M. L., Janeway, C. M., and Randerath, K., Enzymatic synthesis of diadenosine tetraphosphate and diadenosine triphosphate with a purified lysyl-sRNA synthetase, *Biochem. Biophys. Res. Commun.,* 24, 91, 1966.

206. Randerath, K., Janeway, C. M., Stephenson, M. L., and Zemecnik, P. C., Isolation and characterization of dinucleotide tetra- and tri-phosphate in the presence of lysyl-sRNA synthetase, *Biochem. Biophys. Res. Commun.,* 24, 98, 1966.

207. Rapaport, E. and Zamecnik, P. C., Presence of diadenosine 5′,5‴-P^1,P^4-tetraphosphate (Ap_4A) in mammalian cells in levels varying widely with proliferative activity of the tissue: a possible positive "pleiotypic activator", *Proc. Natl. Acad. Sci. U.S.A.,* 73, 3984, 1976.

208. Weinman-Dorsch, C., Hedl, A., Grummt, I., Albert, W., Ferdinand, F.-J., Friis, R. R., Pierron, G. Moll, W., and Grummt, F., Drastic rise of intracellular adenosine (5′)tetraphospho (5′)adenosine correlates with onset of DNA synthesis in eukaryotic cells, *Eur. J. Biochem.,* 138, 179, 1984.

209. Probst, H., Hamprecht, K., Gekeler, V., Replication initiation and intracellular levals of ATP, ADP, AMP and diadenosine 5′,5‴-P^1,P^4-tetraphosphate in Ehrlich ascites cells cultured aerobically and anaerobically, *Biochem. Biophys. Res. Commun.,* 110, 688, 1983.

210. Grummt, F., Diadenosine 5′,5‴-P^1,P^4-tetraphosphate triggers initiation of *in vitro* DNA replication in baby hamster cells, *Proc. Natl. Acad. Sci. U.S.A.,* 75, 371, 1978.

211. Grummt, F., Diadenosine tetraphosphate triggers *in vitro* DNA replication, *Cold Spring Harbor Symp. Quant. Biol.,* 43, 649, 1978.

212. Rapaport, E., Zamecnik, P. C., and Baril, E. F., HeLa cell DNA polymerase-α is tightly associated with tryptophanyl-tRNA synthetase and diadenosine 5′,5‴-P^1P^4-tetraphosphate binding activities, *Proc Natl. Acad. Sci. U.S.A.,* 78, 838, 1981.

213. Rapaport, E., Zamecnik, P. C., and Baril, E. F., Association of diadenosine 5′,5‴-P^1,P^4-tetraphosphate binding protein with HeLa cell DNA polymerase-α *J. Biol. Chem.,* 256, 12148, 1981.

214. Baril, E., Bonin, P., Burstein, D., Mara, K., and Zamecnik, P., Resolution of the diadenosine 5′,5‴-P^1,P^4-tetraphosphate binding subunit from a multiprotein form of HeLa cell DNA polymerase-α *Proc. Natl. Acad. Sci. U.S.A.,* 80, 4931, 1983.

215. Rapaport, E., and Feldman, L., Adenosine(5′)tetraphosphate (5′)adenosine-binding protein of calf thymus, *Eur. J. Biochem.,* 138, 111, 1984.

216. Castroviejo, M., Fournier, M., Gatius, M., Gander, J. C., LaBouesse, B., and Litvak, S., Tryptophanyl t-RNA synthesis is found closely associated with and stimulates DNA polymerase-α-like activity from wheat embryos, *Biochem. Biophys. Res. Commun.,* 107, 294, 1982.

217. Zamecnik, P. C., Rapaport, E., and Baril, E. F., Priming of DNA synthesis by 5′, 5‴-P¹,P⁴-tetraphosphate with a double stranded octadecamer as a template and DNA polymerase-α, *Proc. Natl. Acad. Sci. U.S.A.*, 79, 1791, 1982.
218. Ono, K., Iwata, Y., Nakamura, H., and Matsukage, A., Selective inhibition of terminal deoxynucleotidyl transferase by diadenosine 5′, 5‴-P¹,P⁴-tetraphosphate, *Biochim. Biophys. Res. Commun.*, 95, 34, 1980.
219. Lawton, K. G., Wierowski, J. V., Schechter, S., Hilf, R., and Bambara, R. A., Analysis of the mechanism of ATP stimulation from calf thymus DNA α-polymerase, *Biochemistry*, 23, 4294, 1984.
220. Gross, F. and Krauss, G., Replication of M13mp7 single stranded DNA *in vitro* by the 9-S DNA polymerase-α from calf thymus, *Eur. J. Biochem.*, 141, 109, 1984.
221. Fisher, P. A., Wang, T. S., and Korn, D., Enzymological characterization of DNA polymerase-α. I. Basic catalytic properties, processivity and gap utilization of the homogeneous enzyme from human KB cells, *J. Biol. Chem.*, 254, 6128, 1979.
222. Wang, E. C., Henner, D., and Furth, J. J., Duplication of single strand DNA catalyzed by calf thymus DNA polymerase-α *Nucl. Acids Res.*, 3, 129, 1976.
223. Fisher, P. A., and Korn, D., Enzymological characterization of KB cell DNA polymerase-α II. Specificity of the protein nucleic acid interaction, *J. Biol. Chem.*, 254, 11033, 1979.
224. Fisher, P. A. and Korn, D., Enzymological characterization of KB cell DNA polymerase-α III. The polymerization reaction with single-stranded DNA, *J. Biol. Chem.*, 254, 11040, 1979.
225. Fisher, P. A. and Korn, D., Properties of the primer-binding site and the role of magnesium ion in primer-template recognition by KB cell DNA polymerase-α, *Biochemistry*, 20, 4560, 1981.
226. Fisher, P. A., and Korn, D., Ordered sequential mechanism of substrate recognition and binding by KB cell DNA polymerase-α, *Biochemistry*, 20, 4560, 1981.
227. Fisher, P. A., Chen, J. T., and Korn, D., Enzymological characterization of KB cell DNA polymerase-α. Regulation of template binding by nucleic acid base composition, *J. Biol. Chem.*, 256, 133, 1981.
228. Wilson, S. H., Matsukage, A., Bohn, E. W., Chen, Y. C., and Sivarjan, M., Polynucleotide recognition by DNA polymerase-α, *Nucl. Acids Res.*, 4, 3981, 1977.
229. de Recondo, A. M., Rossignol, J. M., Mĕchali, M., and Duguet, M., Properties and interactions of DNA polymerase-α, DNA polymerase-β and a DNA binding protein of regenerating rat liver, in *DNA synthesis Present and Future*, Molineux, I. and Kohiyama, M., Eds., Plenum Press, New York, 1978, 559.
230. Keller, W., RNA primed DNA synthesis *in vitro*, *Proc. Natl. Acad. Sci. U.S.A.*, 69, 1560, 1972.
231. Spadari, S. and Weissbach, A., RNA-primed DNA synthesis: specific catalysis by HeLa cell DNA polymerase-α *Proc. Natl. Acad. U.S.A.*, 72, 503, 1975.
232. Tsuruo, T., Hiriyama, K., and Utika, T., Three DNA polymerases of rat ascites hepatoma cells: properties of the enzymes and effect of RNA synthesis on the reactions, *Biochim. Biophys. Acta*, 383, 274, 1975.
233. Brun, G. and Weissbach, A., Initiation of HeLa cell DNA synthesis in a subcellular system, *Proc. Natl. Acad. Sci. U.S.A.*, 75, 5931, 1978.
234. Ikeda, J. E., Longiaru, M., Horwitz, M. S., and Hurwitz, J., Elongation of primed DNA templates by eukaryotic DNA polymerase, *Proc. Natl. Acad. Sci. U.S.A.* 77, 5827, 1980.
235. Dube, D. K., Kunkel, T. A., Seal, G., and Loeb, L. A., Distinctive properties of mammalian DNA polymerase, *Biochim. Biophys. Acta*, 561, 369, 1979.
236. Yoshida, S. and Masaki, S., Utilization *in vitro* of deoxyuridine triphosphate in DNA synthesis by DNA polymerase α and β from calf thymus, *Biochim. Biophys. Acta*, 561, 396, 1979.
237. Wist, E., Partial purification of deoxyuridine triphosphate nucleotide hydrolase and its effect on DNA synthesis in isolated HeLA nuclei, *Biochim. Biophys. Acta*, 565, 98, 1979.
238. Tanabe, K., Taguchi, Y. N., Matsukage, A., and Takahashi, T., Steady-state kinetics of mouse DNA polymerase-α, *J. Biochem. (Tokyo)*, 88, 35, 1980.
239. Chang, L. M. S., The distributive nature of enzymatic DNA synthesis, *J. Mol. Biol.*, 93, 219, 1975.
240. Bambara, R. A., Uremura, D., and Choi, T., On the processive mechanism of *Escherichia coli* DNA polymerase I. Quantitative assessment of processivity, *J. Biol. Chem.*, 253, 413, 1978.
241. Das, S. K. and Fujimara, R. K., Processiveness of DNA polymerases: comparative study using a simple procedure, *J. Biol. Chem.*, 254, 1227, 1979.
242. Grossberger, D. and Clough, W., Characterization of purified Epstein-Barr virus induced deoxyribonucleic acid polymerase: nucleotide turnover, processiveness and phosphoacetic acid sensitivity, *Biochemistry*, 20, 4049, 1981.
243. Wierowski, J. V., Lawton, K. G., Hockensmith, J. W., and Bambara, R. A., Stimulation of calf thymus DNA α-polymerase by ATP, *J. Biol. Chem.*, 258, 6250, 1983.
244. Riedel, H. D., Konig, W., and Knippers, R., ATP and the processivity of *Xenopus laevis* DNA polymerase-α, *Biochim. Biophys. Acta*, 783, 158, 1984.

245. Weaver, D. T. and DePamphilis, M. L., Specific sequences in native DNA that arrest synthesis by DNA polymerase-α, *J. Biol. Chem.*, 257, 2075, 1982.
246. Kaguni, L. S. and Clayton, D. A., Template-directed pausing *in vitro* DNA synthesis by DNA polymerase-α from *Drosophila melanogaster embryos, Proc. Natl. Acad. Sci. U.S.A.*, 99, 983, 1982.
247. Reckman, B., Grosse, F., Urbanke, C., Frank, R., Blocker, M., and Krauss, G., Analysis of secondary structures in M13mp8 (+) single stranded DNA by the pausing of polymerase-α, submitted for publication, 1984.
248. DePamphilis, M. L., Anderson, S., Cusick, M., Hay, R., Herman, T., Krokan, M., Shelton, B., Tack, L., Tappel, D., Weaver, D., and Wasserman, P. M., The interdependence of DNA replication, DNA sequence and chromatin structure in simian virus 40 chromosomes, in *Mechanistic Studies of DNA Replication and Genetic Recombination*, Alberts, B., Ed., Academic Press, New York, 1980, 55.
249. Matsukage, A., Nishizawa, M., Takahashi, T., and Hozumi, T., *in vitro* synthesis of short DNA pieces by DNA polymerase-α from mouse myeloma, *J. Biochem. (Tokyo)*, 88, 1869, 1980.
250. Wang, T. S. F. and Korn, D., Reactivity of KB cell deoxyribonucleic acid polymerases-α and -β with nicked and gapped deoxyribonucleic acid, *Biochemistry*, 19, 1782, 1980.
251. Bollum, F. J., Mammalian DNA polymerases, *Prog. Nucl. Acids Res. Mol. Biol.*, 15, 109, 1975.
252. Sarngadharan, M. G., Robert-Guroff, M., and Gallo, R. C., DNA polymerases of normal and neoplastic mammalian cells, *Biochim. Biophys. Acta*, 516, 419, 1978.
253. Chan, J. Y. H. and Becker, F. F., Differential inhibition of rat liver DNA polymerase *in vitro* by direct-acting carcinogens and the protective effect of thiol reducing agents, *Biochem. J.*, 193, 985, 1981.
254. Spadari, S., Sala, F., and Pedrali-Noy, C., Aphidicolin: a specific inhibitor of nuclear DNA replication in eukaryotes, *Trends Biochem. Sci.*, 7, 29, 1982.
255. Brundret, K. M., Dalziel, B., Hesp, B., Jarvis, J. A. J., and Neidl, S., X-ray crystallographic determination of the structure of the antibiotic aphidicolin. A tetracyclic diterpenoid containing a new ring system, *J. Chem. Soc. Chem. Commun.*, 18, 1027, 1972.
256. Bucknal, R. A., Moores, H., Simms, R., and Hesp, B., Antiviral effects of aphidicolin, a new antibiotic produced by *Cephalosporium aphidicola, Antimicrob. Agents Chemother.*, 4, 294, 1973.
257. Ikegami, S., Toguchi, T., Ohashi, M., Oguro, M., Nagano, H., and Mano, Y., Aphidicolin prevents mitotic cell division by interfering with the activities of DNA polymerase-α, *Nature (London)*, 275, 458, 1978.
258. Longiaru, M., Ikeda, J. E., Jarkovsky, Z., Horwitz, S. B., and Horwitz, M. S., The effect of aphidicolin on adenovirus DNA synthesis, *Nucl. Acids Res.*, 6, 3369, 1979.
259. Ichikawa, A., Negishi, M., Tomita, K., and Ikegami, S., Aphidicolin: a specific inhibitor of DNA synthesis in mastocytoma F-815 cells, *Jpn. J. Pharmacol.*, 30, 301, 1980.
260. Genskens, M., Hardt, N., Pedrali-Noy, G., and Spadari, S., An autoradiographic demonstration of nuclear DNA replication by DNA polymerase-α and of mitochondrial DNA synthesis by DNA polymerase-γ, *Nucl. Acids Res.*, 9, 1599, 1981.
261. Pedrali-Noy, G., Belvedere, M., Crepaldi, T., Focher, F., and Spadari, S., Inhibition of DNA replication and growth of several human and murine neoplastic cells by aphidicolin without detectable effect upon synthesis of immunoglobulins and HLA antigens, *Cancer Res.*, 42, 3810, 1982.
262. Yagura, T., Kozu, T., and Seno, T., Arrest of chain growth of replicon-sized intermediates by aphidicolin during rat fibroblast cell chromosome replication, *Eur. J. Biochem.*, 123, 15, 1982.
263. Spadari, S., Sala, F., and Pedrali-Noy, G., Aphidicolin and eukaryotic DNA synthesis, in *Proteins involved in DNA replication*, Hubscher, U. and Spadari, S., Eds. Plenum Press, New York; and *Adv. Exp. Med. Biol.*, 179, 169, 1984.
264. Ohashi, M., Taguchi, T., and Ikegami, S., Aphidicolin: a specific inhibitor of DNA polymerases in the cytosol of rat liver, *Biochem. Biophys. Res. Commun.*, 82, 1084, 1978.
265. Pedrali-Noy, G. and Spadari, S., Effect of aphidicolin on viral and human DNA polymerases, *Biochem. Biophys. Res. Commun.*, 88, 1194, 1979.
266. DiCicoccio, R. A., Chadha, K., and Srivastava, B. I. S., Inhibition of *Herpes simplex* virus-induced DNA polymerase, cellular DNA polymerase-α and virus production by aphidicolin, *Biochim. Biophys. Acta*, 609, 224, 1980.
267. Pedrali-Noy, G. and Spadari, S., Mechanism of inhibition of *Herpes simplex* virus and vaccinia virus DNA polymerases by aphidicolin, a highly specific inhibitor of DNA replication in eukaryotes, *J. Virol.*, 36, 457, 1980.
268. Krokan, H., Schaffer, P., and DePamphilis, M. L., Involvement of eukaryotic deoxyribonucleic acid polymerase-α and -β in the replication of cellular and viral deoxyribonucleic acid, *Biochemistry*, 18, 4431, 1979.
269. Dinter-Gottlieb, G. and Kauffman, G., Aphidicolin arrest irreversibly impairs simian virus 40 chromosomes, *J. Biol. Chem.*, 258, 3809, 1983.

270. Sala, F., Parisi, B., Burroni, D., Amileni, A. R., Pedrali-Noy, G., and Spadari, S., Specific and reversible inhibition by aphidicolin of the α-like DNA polymerase of plant cells, *FEBS Lett.*, 117, 93, 1980.
271. Sala, F., Galli, M. G., Levi, M., Burroni, D., Parisi, B., Pedrali-Noy, G., and Spadari, S., Functional roles of the plant α-like and γ-like DNA polymerases, *FEBS Lett.*, 124, 112, 1981.
272. Sala, F., Galli, M. C., Pedrali-Noy, G., and Spadari, S., Effect of aphidicolin on the replication of plant DNA, in *Molecular Form and Function of Plant Genomes,* van Vloten-Doting, L., Ed., Plenum Press, New York, 1985, in press.
273. Plevani, P., Badaracco, G., Ginelli, E., and Sora, S., Effect and mechanism of action on aphidicolin on yeast deoxyribonucleic acid polymrases, *Antimicrob. Agents Chemother.*, 18, 50, 1980.
274. Haraguchi, T., Oguro, M., and Nagano, H., Specific inhibitors of eukaryotic DNA synthesis and DNA polymerase-α, 3-deoxyaphidicolin and aphidicolin-a7-monoacetate, *Nucl. Acids Res.*, 11, 1197, 1983.
275. Oguro, M., Suzuki-Hori, C., Nagano, H., Mano, Y., and Ikegami, S., The mode of inhibitory action by aphidicolin on eukaryotic DNA polymerase-α, *Eur. J. Biochem.*, 97, 603, 1979.
276. Ono, K., Iwata, Y., and Nakane, H., Regulatory effect of aphidicolin on the activity of DNA polymerase-α from mouse myeloma, *Biomed. Pharmacother.*, 37, 27, 1983.
277. Krokan, H., Wist, E., and Krokan, K. H., Aphidicolin inhibits DNA synthesis by DNA polymerase-α and isolated nuclei by a similar mechanism, *Nucl. Acids Res.*, 9, 4709, 1981.
278. Holmes, A. M., Studies on the inhibition of highly purified calf thymus 8 S and 7.3 S DNA polymerase-α by aphidicolon, *Nucl. Acids Res.*, 9, 161, 1981.
279. Oguro, M., Shioda, M., Nagano, H., Mano, Y., Hanaoka, F., and Yamada, M., The mode of action of aphidicolin on DNA synthesis in isolated nuclei, *Biochim. Biophys. Res. Commun.*, 92, 13, 1980.
280. Ayusawa, D., Iwata, K., Ikegami, S., and Seno, T., Reversal of aphidicolin-directed inhibition of DNA synthesis *in vivo* and *in vitro* by deoxyribonucleosides or their 5′-triphosphates, *Cell Struct. Funct.*, 5, 147, 1980.
281. Oguro, M. and Nagano, H., The inhibitory mode of aphidicolin on DNA synthesis in NaCl-treated HeLa cell nuclei, *J. Biochem, (Tokyo)*, 91, 2091, 1982.
282. Ayusawa, D., Iwata, K., Kozu, T., Ikegami, S., and Seno, T., Increase in dATP pool in aphidicolin-resistant mutants of mouse FM3A cells, *Biochem. Biophys. Res. Commun.*, 91, 946, 1979.
283. Ayusawa, D., Iwata, K., and Seno, T., Isolation of mouse FM3A cells mutants with thermolabile thymidylate synthetase by resistance to aphidicolin, *Biochim. Biophys. Res. Commun.*, 96, 1654, 1980.
284. Ayusawa, P., Iwata, K., and Seno, K., Alteration of ribonucleotide reductase in aphidicolin-resistant mutants of mouse FM3A cells with associated resistance to arabinosyladenine and arabinosylcytosine, *Somat. Cell Genet.*, 7, 27, 1981.
285. Sabourin, C. L. K., Bates, P. F., Glatzer, L., Chang, C. C., Trosko, J. E., and Boezi, J. A., Selection of aphidicolin resistant CHO cells with altered levels of ribonucleoside reductase, *Somat. Cell Genet.*, 7, 255, 1981.
286. Nishimura, M., Yoshuda, H., Ikegami, S., Ohashi, M., and Yamada, M., Aphidicolin-resistant mutant of which DNA polymerase-α is induced by this drug, *Biochem. Biophys. Res. Commun.*, 91, 939, 1979.
287. Liu, P. K., Chang, C.-C., and Trosko, J. E., Association of mutator activity with UV sensitivity in an aphidicolin-resistant mutant of Chinese hamster V79 cells, *Mutat. Res.*, 106, 317, 1982.
288. Liu, P. K., Trosko, J. E., and Chang, C.-C., Hypermutability of a UV-sensitive aphidicolin-resistant mutant Chinese hamster fibroblasts, *Mutat. Res.*, 106, 333, 1982.
289. Liu, P. K., Chang, C.-C., and Trosko, J. E., Evidence for mutagenic repair in V79 Cell mutant with aphidicolin-resistant DNA polymerase-α, *Somat. Cell Mol. Genet.*, 10, 235, 1984.
290. Tanabe, K., Karawya, E., Fewell, J. W., Kuff, E. L., and Wilson, S. H., DNA polymerase and simian virus 40 infection of resting monkey cells: induction of aphidicolin resistant α polymerase, *Nucl. Acids Res.*, 11, 8253, 1983.
291. Scovassi, A. I., Cavalloro, R., and Bertazzoni, U., A DNA polymerase-α partially resistant to aphidicolin in cells and embryos of MED-fly *(Ceratitis capitate Wied), Biochem. Inter.*, 7, 481, 1983.
292. Wright, G. E., Baril, E. F., and Brown, N. C., Butylanilinouracil: a selective inhibitor of HeLa cell DNA synthesis and HeLa cell DNA polymerase-α, *Nucl. Acids Res.*, 8, 99, 1980.
293. Wright, G. E., Baril, E. F., Brown, V. M., and Brown, N. C., Design and characterization of N^2-arylaminopurines which selectively inhibit replicative DNA synthesis and replication-specific DNA polymerase: guanine derivatives active on mammalian DNA polymerase-α and bacterial DNA polymerase III, *Nucl. Acids Res.*, 10, 4431, 1982.
294. Rochowska, M., Siedlecki, J., Skurzak, H., Wright, G., and Zmudzka, B., Inhibition of calf thymus DNA polymerase α and of normal and cancer cell growth by butylanilinouracil and butylphenylguanine, *Biochim. Biophys. Acta*, 699, 67, 1982.

295. Khan, N., Wright, G. E., Dudycz, L. W., and Brown, N. C., Butylphenyl dGTP: a selective and potent inhibitor of mammalian DNA polymerase alpha, *Nucl. Acids Res.*, 12, 3695, 1984.
296. Fridland, A., Inhibition of deoxyribonucleic acid chain initiation: a new mode of action for 1-β-D-arabinofuranosylcytsine in human lymphoblasts, *Biochemistry*, 16, 5308, 1977.
297. Wist, E., Krokan, H., and Prydz, H., Effect of 1-β-D-arabinofuranosylcytosine triphosphate on DNA synthesis in isolated HeLa nuclei, *Biochemistry*, 15, 3647, 1976.
298. Yoshida, S., Yamada, M., and Masaki, S., Inhibition of DNA polymerase-α and -β of calf thymus by 1-β-D-arabinofuranosylcytosine 5′-triphosphate, *Biochim. Biophys. Acta*, 477, 144, 1977.
299. DiCicoccio, R. A. and Srivastava, S., Kinetics of inhibition of deosynucleotide-polymerizing enzyme activities from normal and leukemic cells by 9-β-D-arabinofuranosyladenine 5′-triphosphate and 1-β-D-arabinofuranosylcytosine 5′-triphosphate, *Eur. J. Biochem.*, 79, 411, 1977.
300. Schrecker, A. W., Smith, R. G., and Gallo, R. C., Comparative inhibition of purified DNA polymerases from murine leukemia virus and human lymphocytes by 1-β-D-arabinofuranosylcytosine 5′-triphosphate, *Cancer Res.*, 34, 286, 1974.
301. Lynch, W. E., Short, J., and Lieberman, I., The 7.1 S nuclear DNA polymerase and DNA replication in intact liver, *Cancer Res.*, 36, 901, 1976.
302. DePhilip, R. M., Lynch, W. E., and Lieberman, I., Nuclear polymerases of human carcinomas, *Cancer Res.*, 37, 707, 1977.
303. Wist, E., The role of DNA polymerases α, β, and γ in nuclear DNA synthesis, *Biochim. Biophys. Acta*, 562, 62, 1979.
304. Tanaka, M. and Yoshida, S., Altered sensitivity to 1-β-D-arabinofuranosylcytosine 5′-triphosphate of DNA polymerase-α from leukemic blasts of acute lymphoblastic leukemia, *Cancer Res.*, 42, 649, 1982.
305. Muller, W. E. G., Rhode, H. J., Beyer, R., Maidhof, A., Lachmann, M., Taschner, H., and Zahn, R. K., Mode of action of 9-β-arabinofuranosyladenine on the synthesis of DNA, RNA and protein *in vivo* and *in vitro*, *Cancer Res.*, 35, 2160, 1975.
306. Ono, K., Ohashi, A., Yamamoto, A., Matsukage, A., Takahashi, T., Saneyoshi, M., and Ueda, T., Inhibitory effects of 9-β-D-arabinofuranosylguanine 5′-triphosphate and 9-β-D-arabinofuranosyladenine 5′-triphosphate on DNA polymerases from murine cells and oncornavirus, *Cancer Res.*, 39, 4673, 1979.
307. Okura, A. and Shioda, S., Differential inhibition of DNA polymerases of calf thymus by 9-β-D-arabinofuranosyladenine 5′-triphosphate, *J. Biochem. (Tokyo)*, 84, 727, 1978.
308. Ono, K., Ohashi, A., Ogasawara, M., Matsukage, A., Takahashi, T., Nakayamara, C., and Saneyoshi, M., Inhibition of deoxyribonucleic acid polymerase from murine cells and oncornavirus by 5-alkylated derivative of 1-β-D-arbinofuranosyluracil 5′-triphosphate: substituent effects on inhibitory action, *Biochemistry*, 20, 5088, 1981.
309. Ono, K., Ogasawara, M., Ohashi, A., Matsukage, A., Takahashi, T., Nakayamara, C., and Saneyoshi, M., Inhibitory effects of various 5-halogenated derivatives of 1-β-D-arbinofuranosyluracil 5′-triphosphate on DNA polymerases from murine cells and oncovirus: substituent effect on inhibitory action, *Biochemistry*, 21, 1019, 1982.
310. Ono, K., Ogasawara, M., and Nakane, H., Inhibition of terminal deoxyribonucleotidyl transferase by various 5-alkylated derivatives of 1-β-D-arabinofuranosyluracil 5′-triphosphate: substituent effects on inhibitory action, *Biomed. Pharmacother.*, 37, 288, 1983.
311. Huang, E. S., Human cytomegalovirus. IV. Specific inhibition of virus-induced DNA polymerase activity and viral DNA replication by phosphoacetic acid, *J. Virol.*, 16, 1560, 1975.
312. Bolden, A., Aucker, J., and Weissbach, A., Synthesis of *Herpes simplex* virus, vaccinia virus and adenovirus in isolated HeLa cell nuclei, *J. Virol.*, 16, 1584, 1975.
313. Miller, R. L. and Rapp, F., Distinguishing cytomegalovirus, Mycoplasma and cellular DNA polymerases, *J. Virol.*, 20, 564, 1976.
314. Leinbach, S. S., Reno, J. M., Lee, L. F., Isbell, A. F., and Boezi, J. A., Mechanism of phosphonacetate inhibition of herpesvirus-induced DNA polymerase, *Biochemistry*, 15, 426, 1976.
315. Allandeen, U. S. and Bertino, J. R., Inhibition of activities of DNA polymerases α, β, and γ and reverse transcriptase of L1210 cells by phosphonacetic acid, *Biochim. Biophys. Acta*, 520, 490, 1978.
316. Allen, G. P., O'Callaghan, D. J., and Randall, C. C., Purification and characterization of equine herpesvirus-induced DNA polymerase, *Virology*, 76, 395, 1977.
317. Chang, L. M. S. and Bollum, F. J., A comparison of associated enzyme activities in various deoxyribonucleic acid polymerases, *J. Biol. Chem.*, 248, 3398, 1973.
318. Berger, N. A., Sikorski, G. W., and Adams, J. W., Phosphonacetate inhibition of deoxyribonucleic acid synthesis in intact and permeable eukaryotic cells, *Biochem. Pharmacol.*, 28, 2496, 1979.
319. Furman, P. A., St.Clair, M. H., Fyfe, J. A., Rideout, J. L., Keller, P. M., and Elion, G. B., Inhibition of *Herpes simplex* virus-induced DNA polymerase activity and viral DNA replication by 9-(2-hydroxyethoxymethyl)guanine and its triphosphate, *J. Virol.*, 32, 72, 1979.

320. Derse, D., Cheng, Y.-C., Furman, F. A., St.Clair, M. H., and Elion, G. B., Inhibition of purified human and *Herpes simplex* virus-induced DNA polymerases by 9-(2-hydroxyethoxymethyl)guanine triphosphate. Effect on primer template function, *J. Biol. Chem.*, 256, 11447, 1981.
321. Ashton, W. T., Karkas, J. D., Field, A. K., and Tolman, R. L., Activation by thymidine kinase and potent antiherpetic activity of 2-nor-2′-deoxyguansine (2′NDG), *Biochem. Biophys. Res. Commun.*, 108, 1716, 1982.
322. Field, A. K., Davis, M. E., DeWitt, C., Perry, H. C., Liou, R., Germershansen, J., Karkas, J. D., Ashton, W. T., Johnston, D. B. R., and Tolman, R. L., 9-{[2-hydroxy-1-(hydroxymethyl)ethoxy]methyl}guanine: a selective inhibitor of herpes group virus replication, *Proc. Natl. Acad. Sci. U. S. A.*, 80, 4139, 1983.
323. Germershansen, J., Bostedor, R., Field, A. K., Perry, N., Liou, R., Bull, H., Tolman, R. L., and Karkas, J. D., A comparison of the antiviral 2′-nor-deoxyguanosine and acyclovir: upake and phosphorylation in tissue culture and kinetics of *in vitro* inhibition of viral and cellular DNA polymerase by their respective triphosphates, *Biochem. Biophys. Res. Commun.*, 116, 360, 1983.
324. Edenberg, H. J., Novobiocin inhibition of simian virus 40 DNA replication, *Nature (London)*, 286, 529, 1980.
325. Byrnes, J. J., Downey, K. M., Esserman, L., and So, A. G., Mechanism of inhibition of erythroid cytoplasmic DNA polymerase, *Biochemistry*, 14, 796, 1975.
326. Tanaka, M. and Yoshida, S., Inhibition of deoxyribonucleic acid polymerase-α and -β and terminal deoxyribonucleotidyl transferase from calf thymus by mitomycin C-deoxyribonucleic acid, *Biochem. Pharmacol.*, 30, 299, 1981.
327. Nakamura, H., Morita, T., and Yoshida, S., Differential inhibition of mammalian DNA polymerases by X-irradiated DNA, *Int. J. Radiat. Biol.*, 38, 449, 1980.
328. Pitha, J. and Wilson, S. H., Template specific inhibition of mammalian DNA polymerase, *Nucl. Acids Res.*, 3, 825, 1976.
329. Waqar, A. M., Evans, M. J., Manly, K. F., Hughes, R. G., and Huberman, J. A., Effects of 2′,3′-dideoxynucleosides on mammalian cells and viruses, *J. Cell Physiol.*, 121, 402, 1984.
330. Shimada, H., Haraguchi, T., Nagano, H., Fujiwara, A., and Yasuman, I., Inhibition of DNA polymerases of sea urchin by palmitoyl coenzyme A, *Biochem. Biophys. Res. Commun.*, 110, 902, 1983.
331. Schuerch, A. R. and Wherli, W., β Lapachone, an inhibitor of oncornavirus reverse transcriptase and eukaryotic DNA polymerase-α. Inhibitory effect, thiol dependency and specificity, *Eur. J. Biochem.*, 84, 197, 1978.
332. DiCicoccio, R. A. and Srivastava, B. I. S., Structure-activity relationship and specificity of inhibition of DNA polymerases from normal and leukemic cells of man and from simian virus by rifamycin derivatives, *J. Natl. Cancer Inst.*, 61, 1187, 1978.
333. Wu, R. S., DeFilipps, M. K. W., and Quinn, F. R., Quantitative structure-activity correlations of rifamycins as inhibitors of viral RNA-directed DNA polymerases and mammalian α and β DNA polymerases, *J. Med. Chem.*, 23, 256, 1980.
334. Yagura, T., Kozu, T., and Seno, T., Inhibition of DNA polymerases α and β by rifamycin derivatives AF/013, *J. Biochem. (Tokyo)*, 90, 1397, 1981.
335. Dillwith, J. W. and Lewis, R. A., Inhibition of DNA polymerase by captan, *Biochim. Biophys. Acta*, 696, 245, 1982.
336. Tanaka, M. and Yoshida, S., Mechanism of inhibition of calf thymus DNA polymerases α and β by daunomycin and adriamycin, *J. Biochem (Tokyo)*, 87, 911, 1980.
337. Tanaka, M. and Yoshida, S., Mechanism of action of antileukemic agents inhibition of DNA polymeases, *Acta Haematol. Jpn.*, 43, 995, 1980.
338. Bose, K., Karran, P., and Strauss, B., Repair of depurinated DNA *in vitro* by enzymes purified from human lymphoblasts, *Proc. Natl. Acad. Sci. U.S.A.*, 75, 794, 1978.
339. Mosbaugh, D. W. and Linn, S., Excision repair and DNA synthesis with a combination of HeLa DNA polymease-β and DNase V, *J. Biol. Chem.*, 258, 108, 1983.
340. Mosbaugh, D. W. and Linn, S., Gap filling DNA synthesis by HeLa cell DNA polymerase-α in an in *vitro* base excision DNA repair scheme, *J. Biol. Chem.*, 259, 10247, 1984.
341. Chang, L. M. S., Rafter, E., Augl, C., and Bollum, F. J., Purification of a DNA polymerase-DNA primase complex from calf thymus glands, *J. Biol. Chem.*, 259, 14679, 1984.
342. DiCicoccio, R. A. and Srivastava, B. I. S., Inhibition of deoxynucleotide-polymerizing enzyme activities of human cells and of simian sarcoma virus by heparin, *Cancer Res.*, 38, 2401, 1978.
343. Furukawa, K. and Bhavandan, V. P., Influence of anionic polysaccharides on DNA synthesis in isolated nuclei and by DNA polymerase-α: correlation of observed effects with properties of the polysaccharides. *Biochim. Biophys. Acta*, 740, 466, 1983.
344. Fansler, B. and Loeb, L. A., Sea urchin nuclear DNA polymerase IV. Reversible association of DNA polymerase with nuclei during the cell cycle, *Exp. Cell Res.*, 75, 433, 1972.
345. Benbow, R. M., Stowers, D. J., Hiriyanna, K. T., Dunne, V. C., and Ford, E. C., DNA polymerase-α_2 from ovaries of the frog *Xenopus laevis*, submitted for publication, 1985.

346. Mikhailov, V. S. and Androsova, I. M., Effect of spermine on interaction of DNA polymerase-α from the loach *(Misgurnus fossilis)* eggs with DNA, *Biochim. Biophys. Acta,* 783, 6, 1984.
347. Shioda, M. and Benbow, R. M., in preparation.
348. Yamaguchi, M., Hendrickson, E. A., and DePamphilis, M. L., DNA primase-DNA polymerase α from simian cells. Modulation of RNA primer synthesis by ribonucleoside triphosphates, *J. Biol. Chem.,* 260, 6254, 1985.
349. Jazwinski, S. M. and Edelman, G. M., A DNA primase from yeast; purification and partial characterization, *J. Biol. Chem.,* 260, 4995, 1985.
350. Wilson, F. E. and Sugino, A., Purification of a DNA primase from the yeast *Saccharomyces cerevisiae.* Primase can be separated from DNA polymerase I, *J. Biol. Chem.,* 260, 8173, 1985.
351. Plevani, P., Foiani, M., Valsanini, P., Badaracco, G., Cheriathundam, E., and Chang, L. M. S., Polypeptide structure of DNA primase from a yeast DNA polymerase-primase complex, *J. Biol. Chem.,* 260, 7102, 1985.
352. Badaracco, G., Bianchi, M., Valsanini, P., Magni, G., and Plevani, P., Initiation, elongation and pausing of *in vitro* DNA synthesis catalyzed by immunopurified yeast DNA primase: DNA polymerase complex, *EMBO J.,* 4, 1313, 1985.
353. Faust, E. A., Nagy, R., and Davey, S. K., Mouse DNA polymerase α-primase terminates and reinitiates DNA synthesis 2—14 nucleotides upstream of $C_2A_{1-2}(C_{2-3}/T_2)$ sequences on a minute virus of mice DNA template, *Proc. Natl. Acad. Sci. U.S.A.,* 82, 4023, 1985.
354. Sapp, M., Konig, H., Riedel, H. D., Richter, A., and Knippers, R., A newly detected class of mammalian single strand-specific DNA binding proteins. Effect on DNA polymerase α-catalyzed DNA synthesis, *J. Biol. Chem.,* 260, 1550, 1985.
355. Wang, T. S.-F., Pearson, B. E., Suomalainen, H. A., Mohandas, T., Shapiro, L. J., Schroder, J., and Korn, D., Assignment of the gene for human DNA polymerase-α to the X chromosome, *Proc. Natl. Acad. Sci. U.S.A.,* 82, 5270, 1985.
356. Hanaoka, F., Tandai, M., Miyasawa, H., Murakami, Y., Hori, T., and Yamada, M., Assignment of human polymerase gene to the X chromosome, Proc. 8th Int. Human Gene Mapping Workshop, Abstr. 147, Helsinki, August 1985.
357. Murakami, Y., Yasuda, H., Miyazawa, H., Hanaoka, F., and Yamada, M.-A., Characterization of a temperature sensitive mutant of mouse FM3A cells defective in DNA replication, *Proc. Natl. Acad. Sci. U.S.A.,* 32, 1761, 1985.
358. Holmes, A. M., Cheriathundam, E., Bollum, F. J., and Chang, L. M. S., Initiation of DNA synthesis by the calf thymus DNA polymerase-primase complex, *J. Biol. Chem.,* 260, 10840, 1985.
359. Williams, K. R., Stone, K. L., LoPresti, M. B., Merrill, B. L., and Planck, S. R., Amino acid sequence of the UP1 calf thymus helix-destabilizing protein and its homology to an analogous protein from mouse myeloma, *Proc. Natl. Acad. Sci. U.S.A.,* 32, 5666, 1985.

Chapter 4

DNA POLYMERASE-β

I. INTRODUCTORY REMARKS

Once multiple DNA polymerases were documented in bacteria, a similar diversity was sought and found in animal cells. It became apparent that in addition to the high-molecular-weight DNA polymerases-α, a low-molecular-weight species is also present in extracts from animal cells. DNA polymerases of the β class were initially identified and distinguished from α type polymerase as low-molecular-size DNA polymerases in HeLa cells,[1] in rat liver cells[2], and in calf thymus.[3] Subsequently, low-size nuclear DNA polymerases of the β category were described in numerous organisms, among them sea urchins,[4-6] adult *Drosophila melanogaster* flies,[7] *Xenopus laevis* frogs,[8,9] chick embryo,[10,11] guinea pig liver,[12] rat Novikoff hepatoma cells,[13] rabbit bone marrow,[14] mouse myeloma cells,[15] and several types of human cells.[16]

Phylogenetic surveys of polymerase-β demonstrated its prevalence from metazoan to mammals.[17-19] Polymerases of the β type appear to be undetectable in protists, in higher plants and fungi, and in embryonic insect cells.[17-20] However, the existence of β-like polymerase in certain classes of eukaryotes is still debatable. Although polymerase-β was undetected in most species of free-living parasitic protozoans,[17] there is evidence for its presence in certain unicellular species.[21] Also, polymerase-β activity was not traced in embryonic tissues of insects,[17-19] but activity that possessed characteristics of polymerase-β was described in adult *D. melanogaster* flies[7] and in cockroach testes.[19] Last, although it was concluded that polymerase-β is absent in cells of protists, fungi, and higher plants,[20] some authors reported on DNA polymerases which had some features of polymerase-β in *Dictyostelium discoideum*[22] and in *Neurospora crassa*.[23]

Since in some of these studies the identification of DNA polymerase-β was based on mainly the size and resistance to reagents that interact with the thiol group, the data on some species will have to be reconfirmed with more specific techniques, including employment of monospecific antibodies. Although DNA polymerases of the β-type were identified more than a decade after the appearance of the first report on an animal cell polymerase-α, progress in enzyme identification, purification to homogeneity and characterization was much more rapid for β-polymerases than for α-type enzymes.

DNA polymerases of the β category appear as a minor fraction of the total polymerase activity in dividing animal cells, and their absolute intracellular activity remains practically constant during cell cycling.[24] In contrast to the complex structure of α polymerases, DNA polymerases of the β class consist of a single polypeptide chain of size ranging between 32 and 50 kdaltons in different species of eukaryotes (see Section II below). Polymerases of the β type are more resistant than polymerase of the α and γ classes to reagents that react with thiol groups, and are insensitive to aphidicolin, a specific inhibitor of polymerase-α, but are selectively sensitive to inhibition by deoxynucleoside triphosphates (see Section V.C below). The distinct size difference between α and β polymerases, as well as their different responses to selective inhibitors and to monospecific antibodies, were highly instrumental in establishing polymerase-β as a separate class of animal DNA cell polymerase (see Chapter 2). The relative lack of structural complexity of polymerase-β as well as its relative stability, enabled several investigators to purify to homogeneity β-polymerases from various cell types. Furthermore, monospecific antibodies were raised against the homogeneous enzyme and were used to study its evolution and intracellular localization. Monospecific immunoglobulins were also instrumental in procedures for the purification of polymerase-β specific messenger RNA (see Section V.C below). In this chapter we survey the main findings on the structure of polymerase-β, its association with auxiliary proteins, its catalytic properties, and its response to selective inhibitors. The evolution, biology, and in vivo roles of polymerase-β are reviewed in Chapters 6, 7, and 9.

II. STRUCTURE OF POLYMERASE-β

Physical separation of DNA polymerase-β from polymerases-α and -γ is usually attained by velocity gradient sedimentation or by column chromatography on DEAE-cellulose and phosphocellulose. Polymerase-β is isolated by virtue of its uniquely low sedimentation constant (∼3.5 S) and by its basic isoelectric point, which precludes its absorption to DEAE-cellulose at a low ionic strength. Using various combinations of conventional protein fractionation procedures, different authors purified polymerase-β from cells of chick embryo,[11] mouse myeloma,[25,26] Novikoff hepatoma,[13] guinea pig liver,[12] calf thymus, [27,28] rabbit and pig liver and rat hepatoma,[29] and human KB cultured cells.[30] So far, all homogeneous preparations of polymerase-β consisted of a single polypeptide, usually of 40 kdaltons,[11,27,29,30] although sizes of 32 kdaltons were defined for guinea pigs[12] and Novikoff hepatoma[13] β enzymes, and a 50-kdalton polymerase-β was purified from sea urchin embryo.[4] That β-polymerases display a conserved size of <50 kdaltons was shown by Chang[17] for the native enzyme from a wide range of eukaryotic organisms. Conservation of a single 50-kdalton polypeptide in β-polymerases was susequently shown by activity gel electrophoretic analysis of extracts from different animal cells.[19] This method has been shown to be particularly useful in analyzing DNA polymerase-β, since it is composed of a single relatively short polypeptide chain that readily renatures after SDS gel electrophoresis.[31,99]

Tanabe et al.[29] purified β-polymerases from several mammalian cell types, demonstrating that in all of the species examined, β-polymerases were comprised of a single 40-kdalton polypeptide whose tryptic map patterns were similar for different organisms. Comparing tryptic peptide maps of homogeneous chick and rat polymerase-β, Matsukage and colleagues[11] identified 14 identical fragments of a total of 24 peptides, which were obtained from each of the two enzymes. The high degree of structural conservation of polymerase-β was confirmed through an immunological approach employed by Chang et al.[18,28] Monospecific antibodies raised in the rabbit against polymerase-β from calf thymus cross-reacted with polymerase-β from a variety of vertebrate species, but did not react with other classes of eukaryotic or prokaryotic DNA polymerases nor with terminal deoxyribonucleotidyl transferases.[18,28] Immunochemical probing of polymerase-β recently yielded information on the structure of the newly made enzyme and led to the isolation of the polymerase-β specific mRNA, which could be translated in vitro into enzymatically active 40-kdalton polymerase.[32,33]

Yamaguchi et al.[32] raised in the rabbit monospecific antibody to chick polymerase-β, which precipitated a 40-kdalton polypeptide from both [^{35}S]-methionine-labeled crude cell extracts and from partially purified polymerase-β preparation. Furthermore, only a 40-kdalton polypeptide could be discerned in the immunoprecipitate both after a short [^{35}S]-methionine pulse and after a prolonged chase period. It was concluded, therefore, that the 40-kdalton polypeptide, which constitutes the native polymerase-β molecule, is likely to be the primary translation product of polymerase-β mRNA.[32] In a subsequent study, Yamaguchi et al.[33] further showed that immunoprecipitated 40-kdalton polypeptide, which displayed DNA polymerase activity upon electrophoresis on activity gels, was the product of the in vitro translation of chick cell mRNA. These studies imply that posttranslational modification of DNA polymerase-β is unlikely to be required for activation of the enzyme. Moreover, an 1800-nucleotide long mRNA for polymerase-β was purified and shown to constitute 0.001% of the total poly(A)-rich cellular RNA molecule population.[33] In addition to establishing that the size of the primary product of mRNA translation is equal to that of the single 40-kdalton polypeptide of the mature polymerase-β molecule, these studies open avenues to cloning and altering polymerase-β through the use of recombinant DNA techniques.

Although the size, composition, and much of the primary structure of β-polymerases appear to be highly uniform in animal cells, the native enzyme often exhibits higher than expected apparent size. First, in many cases, but not all,[7,11] native β-polymerases enter into high-molecular-weight aggregates.[34-37] It was reported that a nonenzymatic protein, which was separated from KB cell DNA polymerase-β by gel filtration, promoted aggregation of the enzyme when the two proteins were reconstituted together.[35] Furthermore, some evidence indicates that aggregates of polymerase-β may contain, in addition to multiple molecules of the enzyme, other accessory proteins.[36-38] Early observation by de Recondo and group[36] indicated that polymerase-β from regenerating rat liver partly appears in high molecular size forms. These forms were not self-aggregates of the β enzyme, but rather contained additional proteins that are specific to the regenerating liver.

In a series of studies, Meyer and associates[38-40] demonstrated close association in aggregates formed between DNA polymerase-β and a deoxyribonuclease that stimulates it. After carefully adjusting the ionic strength of Novikoff hepatoma extracts, Mosbaugh et al.[38] observed appearance of polymerase-β in four forms, which sedimented in 7.3 S, 5.8 S, 4.1 S, and 3.3 S. The different isoelectric points of the 7.3 S and 4.1 S forms suggested that the former contains proteins other than polymerase-β in addition to the enzyme. Furthermore, during purification, the apparent size of the larger forms was reduced to 4.2 S, and at the final stage of DNA cellulose chromatography, the 4.1-S species was converted into a 3.3-S form, releasing a 12-kdalton accessory protein.[38] The dissociated small protein was subsequently identified as a polymerase-β stimulatory DNase V (for a discussion of DNase V see Section III.A below).[39,40]

Similarly, Kunkel et al.[12] observed that during the final step of purification of guinea pig hepatic polymerase-β, a stimulatory protein was dissociated from the enzyme. Krauss and Linn[37] showed that placental human polymerase-β could be resolved into three subspecies by phosphocellulose chromatography and glycerol gradient velocity sedimentation. By increasing the ionic strength, an 8.6-S aggregated form was dissociated into a 4.6-S species. Most interestingly, disaggregation entailed a decline in the fidelity of the copying of deoxyribopolymers so that replication conducted by the 6.3-S and 4.6-S forms was less faithful than synthesis conducted by the 8.6-S aggregate.[37] The tantalizing possibility that the high-molecular-weight aggregate contained a fidelity enhancing cofactor was not pursued. However, these reports, together with additional observations on multiple enzymic forms in crude preparations of polymerase-β,[16,36,41] suggest that within the cell polymerase-β may be a component of a supramolecular complex, which contains factors that enhance polymerase activity, fidelity or other enzymic properties. Such a complex might be functional in repair of cellular DNA in analogy with a putative DNA replication complex (cf. Chapter 7).

III. PROTEINS AUXILIARY TO DNA POLYMERASE-β

Preparations of highly purified or apparently homogeneous polymerase-β from different cell types were found to be devoid of detectable $5' \rightarrow 3'$ or $3' \rightarrow 5'$ exonuclease activity.[8,10,12,14,30,42] Furthermore, purified β-polymerases were unable to conduct pyrophosphate exchange,[23,30,42] pyrophosphorolysis,[25,42] and hydrolysis of deoxyribonucleoside triphosphates[42] or of RNA.[8] Indeed the structural simplicity of polymerase-β which is comprised of only a single 40-kdalton polypeptide, minimizes the probability of the existence of polymerase-associated activities. As noted in the previous section, however, cruder preparations of polymerase-β appear to include auxiliary protein. in this section, accessory proteins which stimulate or modify the activity of polymerase-β are described.

A. DNase V

Meyer and colleagues[38,43] discerned in Novikoff hepatoma cell extracts several protein factors that stimulated the in vitro activity of mammalian DNA polymerases severalfold. Stimulatory factors from hepatoma cells were resolved by column chromatography into three molecular species which could also be detected in normal rat liver, though at a lower level.[43] Subsequent to these preliminary observations, the authors turned their attention to the particular stimulatory factor, which copurified with a 4.1-S form of hepatoma cell polymerase-β. DNA cellulose column chromatography of the 4.1-S polymerase-β led to the conversion of the enzymes to a pure 3.3-S form through the dissociation of a 12-kdalton protein designated hepatoma factor IV.[38] Reconstitution of the 3.3-S polymerase-β with factor IV led to stabilization of the enzyme and to several fold stimulation of its in vitro activity.[38] Analogous stimulatory factor was resolved from guinea pig hepatic polymerase-β during its purification.[12]

Studying in detail the properties of factor IV, Mosbaugh and Meyer[40] found that upon its addition to a polymerase-β directed DNA synthesizing reaction, the factor increased the initial velocity of the reaction and led to prolongation of the initial rate of DNA synthesis. Polyacrylamide gel electrophoresis analysis indicated that polymerase-β is complexed with the factor at a 1:1 stoichiometry. Observing that in the presence of factor IV, DNA synthesis was followed by the degradation of DNA, it was established that the factor is an exodeoxyribonuclease which hydrolyzes DNA in both the 3′ → 5′ and 5′ → 3′ directions.[39,40] The unique catalytic parameters of factor IV/DNase activity led to its present designation as DNase V. The tight association between polymerase-β and DNase V, and the specific stimulatory activity exerted by the exonuclease on the polymerase, led to the proposition that these two proteins constitute a part of a supramolecular DNA repair complex.[40] Mosbaugh and Linn[44] demostrated recently that concerted activity of HeLa cell polymerase-β and DNase V could perform long-patch DNA synthesis from a single nucleotide gap in DNA. Furthermore, incisions made in DNA by *E. coli* endonuclease III were activated by DNase V, which removed the 3′-deoxyribose termini, thus supporting polymerase-β directed DNA synthesis.[44] HeLa cell DNAse V also enhanced the rate and extent of DNA synthesis by polymerase-β from incisions made in depurinated DNA by apurinic/apyrimidinic endonuclease. In that case, DNase V removed the baseless sugar phosphate from the 5′ terminus, allowing nick translation to occur.[44] These results confirm and expand the data obtained with Novikoff hepatoma polymerase-β and DNase V, and indicate that the two enzymes may combine to participate in short- and long-patch DNA repair, as well as in in vivo repair of apurinic damage to DNA (see Section IV.3 below, and Chapter 7, Sections V.C and VI).

B. ATPase

Thomas and Meyer[45] identified in extracts of Novikoff hepatoma five chromatographically distinct DNA-dependent ATPase activities. One of these enzymes, ATPase III, was purified to apparent homogeneity and shown to be a dimer of 65-kdalton subunits.[45] ATPase III stimulated DNA polymerase-β, but not polymerase-α on duplex DNA. It was indicated that it was unlikely that ATPase III acts as a DNA topoisomerase or as helicase.[45]

Boxer and Korn[46] purified a 75-kdalton DNA-dependent ATPase from nuclei of human KB cells, which stimulated homologous DNA polymerase-β threefold, but also stimulated polymerase-α fivefold. Also, in contrast to the Novikoff hepatoma ATPase III, which could exert its stimulatory effect with either ATP or dATP, the human cell ATPase required ATP exclusively.[46] That ATPase stimulatory effect on DNA polymerase-β may not always be specific is indicated by the finding of a calf thymus ATPase that stimulated homologous polymerase-α, and whose effect was also enhanced by a DNA binding protein.[47] Hence, the possible in vivo functional association between polymerase-β and stimulatory ATPase is only speculative at present. One suggested role for ATPase III, functioning in concert with polymerase-β, is the catalysis of a limited displacement of DNA at the 5′ end of an apurinic nick in advance of DNA synthesis by polymerase-β.[45]

C. DNA Binding Proteins

Blue and Weissbach[48] isolated from HeLa cells nuclei an 85-kdalton protein that interacted with DNA polymerase-β to increase its synthetic activity. The factor bound to both double-stranded and single-stranded DNA was heat stable, and conferred heat stability on polymerase-β. No nuclease activity or stimulation of DNA polymerase-α or -γ could be detected for the HeLa cell protein factor.[46] A single-strand DNA binding protein that possessed helix-destabilizing activity was purified from regenerating rat liver and designated HD25 by Duguet and de Recondo.[49] The HD25 protein stimulated DNA polymerase-β only on short-gapped DNA having primer-to-template ratio >0.5.[50,51] However, HD25 had a pronounced stimulatory activity on polymerase-α, which is attributed to its helix-destabilizing activity (cf. Chapter 3, Section IV.C.1).

A single-strand DNA binding protein that exists as a globular monomer of 48 kdaltons and which was named accordingly SSB-48, has been isolated from Novikoff hepatoma.[101] This protein has been shown to stimulate significantly the activity of homologous DNA polymerase-β but to affect only slightly the rate of DNA synthesis catalyzed by polymerase-α. Enhancement of polymerase-β activity is highest with low enzyme concentration and with large gap sizes that exceed 250 bases.[101] No evidence has been found for direct binding between polymerase-β and SSB-48, whereas the latter binds weakly but cooperatively to DNA and possesses helix unwinding activity. It has been proposed, therefore, that SSB-48 places the template strand in a conformation that facilitates rapid DNA polymerization by polymerase-β.[101]

IV. CATALYTIC PROPERTIES OF DNA POLYMERASE-β

Shortly after the initial identification of DNA polymerase-β in animal cells, Chang[41] and Wang et al.[30] defined the basic catalytic features of DNA polymerase-β from calf thymus and human KB cells, respectively. Reactivity of the purified enzymes depended on the presence of template, primer, deoxyribonucleotidyl triphosphate substrates, and a divalent metal activator.[30,41]In an early study[41], it was indicated that homopolymer replication by polymerase-β, which was purified from calf thymus, appeared to involve a single binding site for dNTP substrates and one binding site for the primer. Even though polymerases α and β differ in their structure, fine catalytic properties, and in vivo functions, it is likely that they share similar catalytic mechanisms.

A. Interaction of DNA Polymerase-β with Primer-Template DNA

Under standard assay conditions with Mg^{2+} as metal activator, the template that is maximally copied in vitro by DNA polymerase-β is optimally gapped ("activated") DNA.[7,52] However, a higher rate of utilization of oligo(dT)·poly(dA) was reported for sea urchin polymerase-β.[4] Although earlier reports documented a failure of polymerase-β to copy the ribo strand of oligo(dT)·poly(rA),[52] it was soon shown repeatedly that under appropriate reaction conditions, polymerase-β from various cell types copies efficiently a polyriboadenelic acid template.[4,7,8,10,35,53] The efficacy of in vitro copying of oligo(dT)·poly(rA) by rat ascites hepatoma polymerase-β was radically affected by the primer-to-template ratio, attaining maximum activity at a 2:1 ratio.[54] The Km for 3′-hydroxyl termini of the primer was 11.1 μM for polymerase-β, compared with Km = 0.04 μM for polymerase-γ, which copies poly(rA) at a maximum rate.[54] Competing DNA templates and *N*-ethylmaleimide both affected the reactivity of polymerase-β with oligo(dT)·poly(rA) differently from its activity with activated DNA template.[55] Kinetic analysis suggested that a multimeric, perhaps dimeric, form of DNA polymerase-β was responsible for the copying of oligo(dT)·poly(rA).[55] Polyriboadenylic acid, however, is the only homoribopolymer that polymerase-β utilizes as a template, and attempts to make polymerase-β copy primed natural RNA did not succeed.[53]

Wang and Korn[56] conducted a careful analysis of template preferences of human KB cell polymerase-β. Competition experiments between gapped DNA primer-template and other molecules indicated that utilization of the former by β-polymerases was not affected by covalently closed intact duplex DNA, or by linear restriction fragments of DNA that contained blunt-ended 3′-hydroxyl and 5′-PO_4 termini. Hence, polymerase-β does not display detectable kinetic affinity for intact circular or linear duplex DNA molecules.[56] In sharp contrast to KB cell polymerase-β which exhibits strong affinity for single-stranded DNA,[57] polymerase-β activity on gapped DNA was only slightly inhibited by single-stranded natural heteropolymeric DNA or by synthetic homopolymers, and the inhibition was fully noncompetitive with the template DNA.[56] The noncompetitive linear pattern of inhibition of polymerase-β activity with gapped DNA by single-stranded DNA was maintained under a variety of incubation conditions with either Mn^{2+} or Mg^{2+} present. Thus a clear difference in the affinities of polymerase-α and polymerase-β for single-stranded DNA was established.[56] Furthermore, competition experiments conducted with duplex DNA, which contained nicks with either 3′-hydroxyl or 3′-phosphate termini, demonstrated that in contrast to polymerase-α, which did not bind to such DNA,[57,58] polymerase-β displayed strong avidity to both types of nicked DNA.[56] This result is supported by observations on the efficient utilization of nicked DNA by DNA polymerase-β (see Section IV.B).

B. Primer-Template Binding and Divalent Metal Ion Requirement

Kinetic analysis of the interaction of polymerase-α with the primer template yielded clear distinction between the initial template binding event and the subsequent enzyme primer interaction (see Chapter 3, Section V.B and V.E). Parallel studies with polymerase-β indicated that the partial reactions of template binding and primer binding could not be resolved kinetically for KB cell DNA polymerase-β[56]. It was proposed, therefore, that in contrast to polymerase-α, primer and template binding by polymerase-β may occur by a concerted event. For the analysis of the minimum effective template length for polymerase-β binding and activity, restriction fragments of PBR322 DNA with staggered ends of defined length were used as template. It was found that the choice of divalent metal cation had a radical effect on the minimum length of template for binding. Hence, with Mg^{2+} as metal activator, the minimum length of template required was equal or greater than five nucleotides, while with Mn^{2+} a single nucleotide template sufficed.[56]

Some controversy surrounds the relative efficacies of Mn^{2+} and Mg^{2+} as metal activators for DNA polymerase-β. Korn and associates[66] reported that maximum activity of KB cell polymerase-β was attained with Mg^{2+} as cation activator when template DNA and deoxyribonucleoside triphosphate were saturating. However, at very low concentrations of both template and substrates, Mn^{2+} became the preferred activator and polymerase activity was diminished in the presence of Mg^{2+}.[66] In contrast, Tanabe et al.[25] reported that Mg^{2+} persisted as the preferred metal activator for murine DNA polymerase-β at all levels of DNA template and substrates. Studies on enzyme activity with Mn^{2+} are difficult to interpret in the absence of measurements of free metal ion due to the steepness of the activating curve. Furthermore, comparison of metal activators is unlikely to be relevant to in vivo DNA synthesis since Mg^{2+} is the predominantly available metal activator within cells, and it is possible that it alone functions in total DNA polymerization.

C. Template DNA Gap Size Preference of DNA Polymerase-β

Considerable attention has been dedicated to the question of the ability of polymerase-β to recognize and fill gaps of defined different sizes in duplex DNA. This parameter serves to distinguish polymerase-β from DNA polymerases of other classes and may also illuminate some aspects of the in vivo functional properties of polymerase-β (see Chapter 7, Section V.A). Wang and Korn[59] compared the ability of KB cell polymerases-α and -β to utilize duplex DNA templates which contained a limited number of nicks (i.e., cleaved single phosphodiester bonds) or gaps of defined average size. First, it was found that in contrast to polymerase-α which was unreactive with nicks, polymerase-β performed a limited strand displacement synthesis on nicked DNA, incorporating roughly 15 nucleotides at each nick.[59] This result extends an independent demonstration of a high affinity of polymerase-β for nicked DNA.[56] On examining the utilization of gaps of defined size by KB cell polymerase-β, it was found that the enzyme exhibited maximal activity on short gaps of approximately ten nucleotides in length and was able to fill them to completion. In contrast, polymerase-α preferred 30 to 60 nucleotide-long gaps and was unable to fill them to completion.[56]

Mosbaugh and Linn[44] confirmed these observations by demonstrating that HeLa cell polymerase-β was capable of performing limited strand displacement synthesis from 3′-hydroxyl termini made in duplex DNA and cut by HeLa apurinic/apyrimidinic endonuclease at the 5′ side of an apurinic site. The HeLa cell polymerase-β could not, however, utilize nicks with 3′-deoxyribose termini created by *E. coli* endonuclease III at the 3′ side of apurinic sites. However, when the baseless sugar phosphate was removed from the 3′ terminus and a single nucleotide gap was created, polymerase-β catalyzed one nucleotide repair synthesis, inserting a single complementary nucleotide into gaps that lacked it.[44] When combined with DNase V, polymerase-β carried out nick-translation synthesis from single nucleotide gaps, converting them into large-patch repaired regions.[44] In a recent work, Evans and Linn[60] demonstrated that HeLa cell polymerase-β was capable of filling to completion short gaps created in SV40 minichromosomes after pyrimidine dimer excision by a combination of T4 UV endonuclease, HeLa apurinic/apyrimidinic endonuclease II, and HeLa DNase V. In contrast to the limited polymerase-β directed strand displacement, which was observed at nicks with naked DNA, no such displacement was taking place in SV40 chromatin, and repair entailed just the insertion of the single missing nucleotide.[60] These latter studies may better mimic the mechanisms of in vivo DNA repair synthesis (cf. Chapter 7, Section VI).

D. The Ordered Initial Interaction of Polymerase-β with Template-Primer and Nucleotide Substrates

Tanabe et al.[25] employed a steady-state kinetic analysis approach to investigate the order of interaction of polymerase-β with its interactants, primer-template, and deoxyribonucleoside triphosphate substrates. Using a nearly homogeneous preparation of mouse myeloma polymerase-β, these authors concluded that the pyrophosphate product inhibition pattern suggested an ordered bi-bi mechanism of polymerization, so that the enzyme interacts first with the template and second with nucleotide substrates.[25] That conclusion was confirmed by Wang and Korn,[56] who examined the mechanism of substrate addition to homogeneous KB cell polymerase-β, using d_2TTP as a dead end competitive inhibitor of the normal dTTP substrate. The analog was completely uncompetitive with respect to the DNA primer template, and it was thus concluded that polymerase-β interacts with DNA followed by substrate nucleotides. This ordered sequence appears to be the only pathway available for the enzyme, and thus its interaction with DNA and substrates is defined as rigidly ordered.[56]

E. Processivity of DNA Synthesis Catalyzed by Polymerase-β

Several authors have addressed the question of the processivity of DNA synthesis by β-polymerases that were isolated from a variety of cell types. Using the relatively insensitive technique of template competition, Chang[61] concluded that polymerase-β is completely nonprocessive, polymerizing a single deoxynucleoside monophosphate per template binding event. Bambara and colleagues[62] applied their method of assessing processivity of polymerase-β by comparing rates of DNA synthesis in reaction mixtures that contained full or partial complement of deoxynucleoside triphosphate substrates (see Chapter 3, Section V.F.1). A similar conclusion of total nonprocessivity of polymerase-β was reached by these researchers.[62] The processivity of mouse and chick polymerase-β was measured by following rates of product DNA elongation and effect of primer template concentration on the rate of the polymerization reaction. Both methods yielded similar results, indicating that the β-polymerases are totally distributive.[63,64]

Das and Fujimura,[65] however, using a method for the direct determination of number of residues polymerized by polymerase-β per single template binding event, concluded that nine to ten nucleotides were polymerized during one synthetic cycle. Using the same method for measurement of the processivity of KB cell polymerase-β, Wang and Korn[56] showed that on both nicked and gapped DNA, synthesis was essentially distributive in the presence of Mg^{2+} with an average one nucleotide incorporated per polymerization cycle. However, with Mn^{2+} as metal cation activator, the mechanism of polymerization became quasi-processive, with an average of four to six residues incorporated per template binding event.[56] Hence, it appears that under standard assay conditions with Mg^{2+} present, KB cell polymerase-β is totally distributive. However, with Mn^{2+} present the enzyme may become modestly processive. One must inquire whether there are cellular factors that increase the processivity of polymerase-β with Mg^{2+} so that the enzyme can fill at least short gaps in DNA without dissociating from the template.

V. INHIBITORS OF DNA POLYMERASE-β ACTIVITY

This section is concerned with inhibitors of polymerase-β, their selectivity, their potency, and their mechanisms of action. The use of inhibitors to elucidate roles of DNA polymerase-β in cellular DNA metabolism is discussed in Chapter 7.

A. Sulfhydryl Group Reactive Reagents

As discussed in Section VI.A of Chapter 3, DNA polymerases of the β class are usually resistant to the thiol group interactant *N*-ethylmaleimide, whereas both α-type and γ-type polymerases are readily inactivated by that agent. It was noted, however, that the activity of DNA polymerase-β from Novikoff hepatoma cells was inhibited about 40% by a high concentration (20 m*M* of *N*-ethylmaleimide.[13,67] DNA polymerase-β, like α and γ polymerases, is inactivated by *p*-chloromercuribenzoate (pCMB) and *p*-hydroxymercuribenzoate (pHMB).[8,67-70] However, polymerases of the β-type are more resistant to these agents than polymerase-α.[1] Inhibition of mouse cell polymerase-β by pHMB is due to the mercury atom, since $HgCl_2$ was found to be as inhibitory as pHMB, whereas hydroxybenzoate did not exert inhibitory effects on the enzyme. Also, inhibition of Novikoff hepatoma cell polymerase-β by pHMB could be reversed by mercaptoethanol, and moreover, the combination of that thiol-group reducing agent and pHMB stimulated the activity of the polymerase.[67] Similar observations were made earlier for mouse polymerase-β by Wilson and associates.[70]

B. Inorganic Phosphates

Chang and Bollum[42] were the first to note the sensitivity of polymerase-β activity to inhibition by inorganic phosphate (P_i). The inhibitory effect of P_i on polymerase-β was used to distinguish that enzyme from the more resistant polymerase-γ.[71] Inhibition of polymerase-β activity by P_i was found to be competitive with respect to the deoxyribonucleoside triphosphate substrates.[11] In a recent work, Modak and colleagues[72] investigated the mechanism of P_i-mediated inhibition of calf thymus polymerase-β. Whereas other authors[42,71] observed inhibition of polymerase-β activity by $\sim$50 mM P_i when Mg^{2+} acted as metal cation activator, Modak et al.[72] reported that <2 mM P_i was inhibitory to calf thymus polymerase-β only when Mn^{2+} was the divalent cation. At low phosphate concentration, Mg^{2+} did not support P_i mediated inhibition of polymerase-β, nor did it relieve it. Inhibition by P_i was strongly affected by the stoichiometry between P_i and Mn^{2+}, was competitive with respect to deoxyribonucleoside triphosphate concentration, and higher levels of Mn^{2+} were required for inhibition to occur when the concentration of the deoxyribonucleoside triphosphate substrates was raised. These indirect indications that Mn^{2+}, substrates, and P_i participate together in P_i-mediated inhibition of polymerase-β activity were supported by the direct chromatographic detection of a complex formed from among these three components.[72] It was proposed that the differential ability of various DNA polymerases to recognize the tripartite complex may serve as a basis for the selective inhibitory effects of inorganic phosphate on polymerase-β and retroviral DNA polymerase.[72]

C. 2′,3′-Dideoxynucleoside-5′-triphosphates

The 2′,3′-dideoxynucleoside-5-triphosphates are analogs of the 2′-deoxynucleoside 5′-triphosphate substrates for DNA polymerase which lack the 3′-oxygen in the deoxyribose moiety in addition to the 2′deoxy position. The analog 2′,3′-dideoxythymidine-5′-triphosphate (d_2TTP), as well as other dideoxy-analogs of the nucleoside triphosphates, inhibits the activity of animal cell DNA polymerases. However, the three classes of animal cell polymerases differ in their sensitivity to d_2TTP-mediated inhibition.

Edenberg et al.[73] found that DNA polymerase-α from CV-1 monkey cells was strikingly more resistant to d_2TTP than either β- or γ-polymerases from that cell type. Whereas polymerases-β and -γ were almost completely blocked at d_2TTP-to-dTTP ratios of 0.5 to 5.0, polymerase-α was resistant to 15-fold excess of analog over normal substrate.[73,75] Essentially similar results were obtained for HeLa cell DNA polymerases.[74] DNA polymerase-β, as well as polymerase-γ, were completely inhibited at a d_2TTP/dTTP ratio of 1.0, whereas polymerase-α activity remained unaffected at analog excesses of 10 to 50.[74] Human cell polymerase-β was 50% inhibited at a d_2TTP/dTTP ratio of 0.06, whereas polymerase-α required an excess of analog over substrate of 30 to be similarly inhibited.[74] The K_i values for d_2TTP were determined to be 6 μM and 29 μM for KB cell polymerases-β and -α, respectively.[76] Furthermore, the selective inhibitory action of d_2TTP on polymerase-β was enhanced by the reciprocal K_m values of these two enzymes for dTTP: β $K_m = 29$ μM and α $K_m = 4$ μM.[76]

Similar results on the selective inhibition of polymerases-β and -γ by low levels of d_2TTP were reported for enzymes from other cell types.[77-79] Three possible mechanisms for the inhibitory effect of d_2TTP on the activity of DNA polymerases were postulated.[74] First, d_2TTP may act as a DNA chain growth terminator if it is incorporated into DNA, since it contains no hydroxyl group at the 3′ position. Second, an alternate mechanism of inhibition is by direct competition between d_2TTP and dTTP for the same substrate site in the polymerase molecule. Third, the sensitive polymerase may be that one which becomes irreversibly bound to nascent DNA chains, which are terminated with d_2TTP. Ono et al.[78] demonstrated that mouse cell polymerase-α, which was relatively resistant to d_2TTP-mediated inhibition in the presence of Mg^{2+}, became highly sensitive to the analog in the presence of Mn^{2+} cations and at a high pH. In light of the enhanced sensitivity of polymerase-α to d_2TTP under these conditions, these authors favored the proposition that the analog acted as a direct competitor of dTTP for a binding site on the polymerase. Analysis of the conditions for inhibition by d_2TTP of a variety of prokaryotic, eukaryotic, and viral DNA polymerases yielded results in support of the competition mechanism of inhibition.[80] Since DNA polymerase-γ was the only type of animal cell polymerase that could incorporate d_2TTP into a growing DNA chain, it is likely that the extreme sensitivity of polymerase-γ to inhibition by d_2TTP is due to chain termination by the analog.[81]

However, the susceptibility of polymerase-β to d_2TTP should be due to a different mechanism, most possibly competition with dTTP for a substrate binding site. In a recent work, Huberman and colleagues[82] demonstrated that d_2GTP, d_2CTP, and d_2ATP selectively inhibit polymerases-β and -γ in the same fashion as d_2TTP. The four nonphosphorylated dideoxynucleosides, however, possess negligible effects on the growth of cultured 3T3 cells, probably because of their inefficient intracellular phosphorylation.[82] The higher sensitivity of DNA polymerase-β to d_2TTP-mediated inhibition has been used to distinguish between α-class and β-class polymerases in various crude replicative systems.[73,74,80,83] Recently, 2′,3′-deoxy-3′-aminonucleoside-5′-triphosphates, analogs of the normal substrates for DNA polymerase, were described.[84] These analogs, containing an amino group in place of a hydroxyl at the 3′position of the ribose moiety, are incorporated into the growing DNA chain and act as chain terminators, thus inhibiting DNA polymerase-β as well as polymerase-α.[84]

D. Other Inhibitors

Several unrelated compounds have been reported to inhibit the activity of DNA polymerase-β. Most of these different inhibitors do not appear to act selectively and exhibit similar potency toward several classes of DNA polymerases. Rifamycin derivatives act as nondiscriminatory inhibitors of α-, β-, and γ-polymerases, as well as of terminal deoxynucleotidyl transferase.[85-87] Several 5-alkylated derivatives of ara UTP were found to inhibit the activities of DNA polymerase-β as well as polymerases-α and -γ.[88] However, the potency of each derivative in inhibiting each enzyme type depended on the nature of the 5-alkyl side chain.[88] Similarly, 5-halogenated derivatives of ara UTP act as inhibitors of all classes of animal cell DNA polymerases, but their potency toward polymerase-β varies both with the halogen substituent and with the primer-template.[89] DNA, cross-linked by mitomycin C, was found to be inhibitory for both α- and β-polymerases.[90] X-ray-irradiated DNA was 200-fold more inhibitory for polymerase-β than for polymerase-α.[91]

Heparin was found to be an effective inhibitor of DNA polymerase-β as well as polymerases-α and -γ from human and *X. laevis* cells.[9,92]

E. Preferential Sensitivity of DNA Polymerase-β to Heat-Inactivation

Loeb and associates[93,94] reported that DNA polymerase-β from several cell types was more heat labile than polymerase-α, and that the two enzymes could thus be distinguished in crude cell extract by their differential heat sensitivity. This initial observation was confirmed by several investigators,[95-97] although one exception was noted.[98]

Preferential heat lability of polymerase-β was used for studies on its possible role in repair synthesis following damage by irradiation (cf. Chapter 7, Section IV.A).[95,98]

VI. UPDATE

A. Chromosomal Localization of Human DNA Polymerase-β Gene

The molecular cloning of c-DNA to mouse brain poly A^+ RNA in λgt11 phage has been instrumental in assigning the human gene for polymerase-β to chromosome 8.[100] A phage clone that contained a 1150 base pair c-DNA tract cross-reacted with polyclonal antibody against β-polymerase and was found to encode an open reading frame of a 40,900-dalton polypeptide. Partial amino acid sequencing of authentic polymerase- revealed identity with sequence predicted from the order of nucleotides in the c-DNA.[100] Radiolabeled cloned inserts were used for *in situ* hybridization with chromosomes from a number of human-mouse and human-hamster cell hybrids. The polymerase-β c-DNA was found to hybridize with a locus that cosegregated with human chromosome 8.[100]

REFERENCES

1. Weissbach, A., Schlabach, A., Fridlender, B., and Bolden, A., DNA polymerase from human cells, *Nature (London) New Biol.*, 231, 167, 1971.
2. Baril, E. F., Brown, O. E., Jenkins, M. D., and Laszlo, J., Deoxyribonucleic acid polymerases with rat liver ribosomes and smooth membranes. Purification and properties of the enzymes, *Biochemistry*, 10, 1981, 1971.
3. Chang, L. M. S. and Bollum, F. J., Low molecular weight deoxyribonucleic acid polymerase in mammalian cells, *J. Biol. Chem.*, 246, 5835, 1971.
4. Suzuki-Hori, C., Nagano, H., and Mano, Y., DNA polymerase β from the nuclear fraction of sea urchin embryo: characterization of the purified enzyme, *J. Biochem. (Tokyo)*, 82, 1613, 1977.
5. Hobart, P. M. and Infante, A. A., A low molecular weight DNA polymerase β in the sea urchin *Stronglyocentrotus prupruatus*, *J. Biol. Chem.*, 253, 8229, 1978.
6. Racin, F. M. and Morris, P. W., DNA polymerases α and β in the California urchin, *Nucl. Acids Res.*, 5, 3945, 1979.
7. Furia, M., Polito, L. C., Locorotondo, G., and Grippo, P., Identification of a DNA polymerase-β-like form in *Drosophila melanogaster* adult flies, *Nucl. Acids Res.*, 6, 3399, 1979.
8. Joenje, H. and Benbow, R. M., A low molecular weight DNA polymerase from ovaries of the frog *Xenopus laevis*. DNA polymerase β (ovarian), *J. Biol. Chem.*, 253, 2640, 1978.
9. Nelson, E. M., Stowers, D. J., Bayne, M. L., and Benbow, R. M., Classification of DNA polymerase activities from ovaries of the frog *Xenopus laevis*, *Dev. Biol.*, 96, 11, 1983.
10. Brun, G., Rougeon, F., Lauber, M., and Chapville, F., Purification and properties of DNA polymerases from chick embryo, *Eur. J. Biochem.*, 41, 241, 1974.
11. Yamaguchi, M., Tanabe, K., Taguchi, Y. N., Nishizawa, M., Takahashi, T., and Matsukage, A., Chick embryo DNA polymerase β. Purified enzyme consists of a single Mr = 40,000 polypeptide, *J. Biol Chem.*, 255, 9942, 1980.
12. Kunkel, T. A., Tcheng, J. E., and Meyer, R. R., Purification and properties of DNA polymerase β from guinea pig liver, *Biochim. Biophys. Acta*, 520, 302, 1978.
13. Stalker, D. M., Mosbaugh, D. W., and Meyer, R. R., Novikoff hepatoma deoxyribonucleic acid polymerase. Purification and properties of a homogenous β polymerase, *Biochemistry*, 15, 3114, 1976.
14. Chang, L. M. S. and Bollum, F. J., Low molecular weight deoxyribonucleic acid polymerase from rabbit bone marrow, *Biochemistry*, 11, 1264, 1972.
15. Matsukage, A., Bohn, E. W. and Wilson, S. H., Multiple forms of DNA polymerase in mouse myeloma, *Proc. Natl. Acad. Sci. U.S.A.*, 71, 578, 1974.
16. Srivastava, B. I. S., Deoxynucleotide-polymerizing enzymes in normal and malignant human cells, *Cancer Res.*, 34, 1015, 1974.
17. Chang, L. M. S., Phylogeny of DNA polymerase β, *Science*, 191, 1183, 1976.

18. Chang, L. M. S., Plevani, P., and Bollum, F. J., Evolutionary conservation of DNA polymerase β structure, *Proc. Natl. Acad. Sci. U.S.A.*, 79, 758, 1982.
19. Scovassi, A. E., Torsello, S., Plevani, P., Badaracco, G. F., and Bertazzoni, U., Active polypeptide fragments common to prokaryotic, eukaryotic, and polypeptide mitochondrial DNA polymerases, *EMBO J.*, 1, 1161, 1982.
20. McLennan, A. G. and Keir, H. M., The deoxyribonucleic acid polymerases of non-vertebrate eukaryotes, *Biochem. Soc. Symp.*, 42, 55, 1977.
21. Chang, L. M. S., Cheriathundam, E., Mahoney, E. M., and Cerami, A., DNA polymerase in parazytic protozoans differ from host enzymes, *Science*, 208, 510, 1980.
22. Baril, E. F., Scheiner, C., and Pederson, T., A β-like DNA polymerase activity in the slime mold *Dictyostelium discoideum*, *Proc. Natl. Acad. Sci. U.S.A.*, 77, 3317, 1980.
23. Stauder, G., Riesemann, H., Joester, W. M., and Joester, K. E., Purification and properties of a low molecular weight DNA polymerase from *Neurospora crassa*, *Biochim. Biophys. Acta*, 741, 308, 1983.
24. Fry, M., Eukaryotic DNA polymerases, in *Enzymes of Nucleic Acid Synthesis and Modification*, Jacob, S. T., Ed., CRC Press, Boca Raton, Fla., 1983, 39.
25. Tanabe, K., Bohn, E. W., and Wilson, S. H., Steady state kinetics of mouse DNA polymerase β, *Biochemistry*, 18, 3401, 1979.
26. Planck, S. R., Tanabe, K., and Wilson, S. H., Distinction between mouse DNA polymerase α and β by tryptic peptide mapping, *Nucl. Acids Res.*, 8, 2771, 1980.
27. Chang, L. M. S., Low molecular weight deoxyribonucleic acid polymerase from calf thymus chromatin. I. Preparation of homogenous enzyme, *J. Biol. Chem.*, 248, 3789, 1973.
28. Chang, L. M. S., and Bollum, F. J., Immunological reagents for comparisons of DNA polymerases — α and DNA polymerase β, *J. Biol. Chem.*, 256, 494, 1981.
29. Tanabe, K., Yamaguchi, M., Matsukage, A., and Takahasi, T., Structural homology of DNA polymerase β from various mammalian cells, *J. Biol. Chem.*, 256, 3098, 1981.
30. Wang, T. S.-F., Sedwick, W. D., and Korn, D., Nuclear deoxyribonucleic acid polymerase. Purification and properties of the homogenous enzyme from KB cells, *J. Biol. Chem.*, 249, 841, 1974.
31. Karawya, E., Swack, J. A., and Wilson, S. H., Improved conditions for activity gel analysis of DNA polymerase catalytic polypeptides, *Anal. Biochem.*, 135, 318, 1983.
32. Yamaguchi, M., Matsukage, A., Takahashi, T., and Toshitada, T., Biosynthesis of DNA polymerase β in chick embryonic cells, *J. Biol. Chem.*, 257, 3932, 1982.
33. Yamaguchi, M., Takahashi, T., Yasuda, K., Shimura, Y., and Matsukage, A., Characterization of messenger RNA for chick-embryo DNA polymerase β and its translation product *in vitro*, *Eur. J. Biochem.*, 133, 277, 1983.
34. Hecht, N. B., The relationship between two murine DNA-dependent DNA polymerases from cytosol and the low molecular weight DNA polymerase, *Biochem. Biophys. Acta*, 383, 388, 1975.
35. Wang, T. S.-F., Sedwick, W. D., and Korn, D., Nuclear deoxyribonucleic acid polymerase. Further observations on the structure and properties of the enzyme from human KB cells, *J. Biol. Chem.*, 250, 7040, 1975.
36. de Recondo, A. M. and Abadiedebat, J., Regenerating rat liver DNA polymerases: dissimilitude or relationship between nuclear and cytoplasmic enzymes, *Nucl. Acids Res.*, 3, 1823, 1976.
37. Krauss, S. W. and Linn, S., Fidelty of fractionated deoxyribonucleic acid polymerases from human placenta, *Biochemistry*, 19, 220, 1980.
38. Mosbaugh, D. W., Stalker, D. M., Probst, G. S., and Meyer, R. R., Novikoff hepatoma deoxyribonucleic acid polymerase. Identification of a stimulatory protein bound to the β polymerase, *Biochemistry*, 16, 1512, 1977.
39. Meyer, R. R. and Mosbaugh, D. W., Characterization of mammalian DNaseV, a double strand bidirectional exonuclease which interacts with DNA polymerase-β in the Novikoff hepatoma, in *Mechanistic Studies of DNA Replication and Genetic Recombination*, Alberts, B., Ed., Academic Press, New York, 1980, 639.
40. Mosbaugh, D. W. and Meyer, R. R., Interaction of mammalian deoxyribonuclease V, a double strand 3′ → 5′ and 5′ → 3′ exonuclease with deoxyribonucleic acid polymerase-β from Novikoff hepatoma, *J. Biol. Chem.*, 255, 10239, 1980.
41. Chang, L. M. S., Low molecular weight deoxyribonucleic acid polymerase from calf thymus chromatin. II. Initiation and fidelty of homopolyers replication, *J. Biol. Chem.*, 248, 6983, 1973.
42. Chang, L. M. S. and Bollum, F. J., A comparison of associated enzyme activities in various deoxyribonucleic acid polymerases, *J. Biol. Chem.*, 248, 3398, 1973.
43. Probst, G. S., Stalker, D. M., Mosbaugh, D. W., and Meyer, R. R., Stimulation of DNA polymerase by factors isolated from Novikoff hepatoma, *Proc. Natl. Acad. Sci. U.S.A.*, 72, 1171, 1975.
44. Mosbaugh, D. W. and Linn, S., Excision repair and DNA synthesis with a combination of HeLa DNA polymerase β and DNaseV, *J. Biol. Chem.*, 258, 108, 1983.

45. Thomas, D. C. and Meyer, R. R., Deoxyribonucleic acid dependent adenosinetriphosphatases from the Novikoff hepatoma. Characterization of a homogenous adenosinetriphosphatase that stimulates DNA polymerase β, *Biochemistry,* 21, 5060, 1982.
46. Boxer, L. M. and Korn, D., Structural and enzymological characterization of a deoxyribonucleic acid dependent adenosine triphosphatase from KB cell nuclei, *Biochemistry,* 19, 2623, 1980.
47. Cobianchi, F., Riva, S., Mastromei, G., Spadari, S., Pedrali-Noy, G., and Falaschi, A., Enchancement of the rate of DNA polymerase activities on duplex DNA by a DNA-binding protein and a DNA-dependent ATPase of mammalian cells, *Cold Spring Harbor Symp. Quant. Biol.,* 43, 639, 1979.
48. Blue, W. T. and Weissbach, A., Detection of a DNA polymerase β stimulating protein in HeLa cell nuclear extracts, *Biochem. Biophys. Res. Commun.,* 84, 603, 1978.
49. Duguet, M. and de Recondo, A. M., A deoxyribonucleic acid unwinding protein isolated from regenerating liver, *J. Biol. Chem.,* 253, 1660, 1978.
50. Duguet, M., Soussi, T., Rossignol, J.-M., Mechali, M., and de Recondo, A. M., Stimulation of rat liver α and β-type DNA polymerases by an homologous DNA unwinding protein, *FEBS Lett.,* 79, 160, 1977.
51. de Recondo, A. M., Rossignol, J.-M., Mechali, M., and Duguet, M., Properties and interactions of DNA polymerase α, DNA polymerase β and a DNA binding protein of regenerating rat liver, in *DNA Synthesis Present and Future,* Molineux, I. and Kohiyama, M., Eds., Plenum Press, New York, 1978, 559.
52. Schlabach, A., Fridlender, B., Bolden, A., and Weissbach, A., DNA dependent DNA polymerases from HeLa cell nuclei. II. Template and substrate utilization, *Biochem. Biophys. Res. Commun.,* 44, 879, 1971.
53. Chang, L. M. S., Replication of initiated polyriboadenylic acid by mammalian low molecular weight deoxyribonucleic acid polymerase, *J. Biol. Chem.,* 249, 7441, 1974.
54. Ono, K., Ohashi, A., Tanabe, K., Matsukage, A., Nishizawa, M., and Takahashi, T., Unique requirement for template-primers of DNA polymerase β from rat ascites hepatoma, *Nucl. Acids Res.,* 7, 715, 1979.
55. Yoshida, S., Yamada, M., and Masaki, S., Novel properties of DNA polymerase β with poly(rA)-oligo(dT) template-primer, *J. Biochem. (Tokyo),* 85, 1387, 1979.
56. Wang, T. S.-F. and Korn, D., Specificity of the catalytic interaction of human DNA polymerase β with nucleic acid substrates, *Biochemistry,* 21, 1597, 1982.
57. Fisher, P. A. and Korn, D., Enzymological characterization of KB cell DNA polymerase-α. II. Specificity of the protein-nucleic acid interaction, *J. Biol. Chem.,* 254, 11033, 1979.
58. Fisher, P. A. and Korn, D., Enzymological characterization of KB cell DNA polymerase-α. III. The polymerization reaction with single stranded DNA, *J. Biol. Chem.,* 254, 11040, 1979.
59. Wang, T. S.-F. and Korn, D., Reactivity of KB cell deoxyribonucleic acid polymerases α and β with nicked and gapped deoxyribonucleic acid, *Biochemistry,* 19, 1782, 1980.
60. Evans, D. H. and Linn, S., Excision repair of pyrimidine dimers from Simian virus 40 minichromosomes *in vitro, J. Biol. Chem.,* 259, 10252, 1984.
61. Chang, L. M. S., The distributive nature of enzymatic DNA synthesis, *J. Mol. Biol.,* 93, 219, 1975.
62. Bambara, R. A., Uyemura, D., and Choi, T., On the processive mechanism of *Escherichia coli* DNA polymerase I. Quantitative assessment of processivity, *J. Biol. Chem.,* 253, 413, 1978.
63. Matsukage, A., Nishizawa, M., and Takahashi, T., Difference in the mechanism of poly(dT) synthesis by DNA polymerases β and γ, *J. Biochem. (Tokyo),* 85, 1551, 1979.
64. Matsukage, A., Yamaguchi, M., Tanabe, K., Taguchi, Y. N., Nishizawa, M., and Takahashi, T., DNA chain elongation of DNA polymerases α, β, and γ, in *New Approaches to Eukaryotic DNA Replication,* de Recondo, A. M., Ed., Plenum Press, New York, 1983, 81.
65. Das, S. K. and Fujimura, R. K., Processiveness of DNA polymerases: a comparative study using a simple procedure, *J. Biol. Chem.,* 254, 1227, 1979.
66. Wang, T. S.-F., Eichler, D. C., and Korn, D., Effect of Mn^{2+} on the *in vitro* activity of human deoxyribonucleic acid polymerase β, *Biochemistry,* 16, 4927, 1977.
67. Mosbaugh, D. W., Kunkel, T. A., Stalker, D. M., Tcheng, J. E., and Meyer, R. R., Novikoff hepatoma deoxyribonucleic acid polymerase. Sensitivity of the β polymerase to sulfhydryl blocking agents, *Nucl. Acids Res.,* 3, 2341, 1976.
68. Zunino, F., Gambetta, R., Colombo, A., Luoni, G., and Zaccara, A., DNA polymerases of rat liver. Partial characterization and effect of various inhibitors, *Eur. J. Biochem.,* 60, 495, 1975.
69. Matsukage, A., Bohn, E. W., and Wilson, S. H., Differential sensitivity of low molecular weight DNA polymerase to sulfahydryl-blocking reagents, *Biochem. Biophys. Acta,* 383, 338, 1975.
70. Bohn, E. W., Matsukage, A., and Wilson, S. H., Stimulation of DNA polymerase activity by the combination of *p*-hydroxymercuribenzoate and dithiothreitol, *Biochem. Biophys. Res. Commun.,* 59, 243, 1974.

71. Knopf, K., Yamada, M., and Weissbach, A., HeLa cell DNA polymerase γ: further purification and properties of the enzyme, *Biochemistry,* 15, 4540, 1976.
72. Modak, M. J., Rao, K., and Marcus, S. L., The mechanism of inorganic phosphate-mediate inhibition of calf thymus DNA polymerase β and Raucher leukemia virus DNA polymerase, *Biochem. Biophys. Res. Commun.,* 107, 811, 1982.
73. Edenberg, H. J., Anderson, S., and DePamphilis, M. L., Involvement of DNA polymerase α in simian virus 40 DNA replication, *J. Biol. Chem.,* 253, 3237, 1978.
74. Waqar, M. A., Evans, M. J., and Huberman, J. A., Effect of 2′-3′-dideoxythymidine-5′-triphosphate on HeLa cell *in vitro* DNA synthesis: evidence that DNA polymerase α is the only polymerase required for DNA replication, *Nucl. Acids Res.,* 5, 1933, 1978.
75. Van der Vliet, P. C. and Kwant, M. M., Role of DNA polymerase γ in adenovirus DNA replication, *Nature (London),* 276, 532, 1978.
76. Fisher, P. A., Wang, T. S.-F., and Korn, D., Enzymological characterization of DNA polymerase α. Basic catalytic properties, processivity and gap utilization of the homogenous enzyme from human KB cells, *J. Biol. Chem.,* 254, 6128, 1979.
77. Abboud, M. M. and Horwitz, M. S., The DNA polymerases associated with the adenovirus type 2 replication complex: effect of 2′-3′-dideoxythymidine-5′-triphosphate on viral DNA synthesis, *Nucl. Acids Res.,* 6, 1025, 1979.
78. Ono, K., Ogasawara, M., and Matsukage, A., Inhibition of the activity of DNA polymerase α by 2′-3′-dideoxythymidine-5′-triphosphate, *Biochem. Biophys. Res. Commun.,* 88, 1255, 1979.
79. Allaudeen, H. S., Inhibition of deoxyribonucleic acid polymerases of human leukemic leukocytes by 2′-3′-dideoxythymidine triphosphate, *Biochem. Pharmacol.,* 29, 1149, 1980.
80. Krokan, H., Schaffer, P., and DePamphilis, M. L., Involvement of eukaryotic deoxyribonucleic acid polymerase α and γ in the replication of cellular and viral deoxyribonucleic acid, *Biochemistry,* 18, 4431, 1979.
81. Van der Vliet, P. C. and Kwant, M. M., Role of DNA polymerase γ in adenovirus DNA replication. Mechanism of inhibition of 2′-3′-dideoxythymidine-5′-triphosphate, *Biochemistry,* 20, 2628, 1981.
82. Waqar, M. A., Evans, M. J., Manly, K. F., Hughes, R. C., and Huberman, J. A., Effects of 2′,3′-dideoxynucleosides on mammalian cells and viruses, *J. Cell Physiol.,* 121, 402, 1984.
83. Fry, M., Shearman, C. W., Martin, G. M., and Loeb, L. A., On the accuracy of DNA synthesis directed by chromatin-associated DNA polymerase β, *Biochemistry,* 19, 5939, 1980.
84. Chigearadze, Z. G. and Beabealashvilli, R. Sh., 2′-3′ deoxy-3′ aminonucleoside 5′-triphosphates are the terminators of DNA synthesis catalyzed by DNA polymerases, *Nucl. Acids Res.,* 12, 1671, 1984.
85. Di Cicoccio, R. A. and Srivastava, B. I. S., Structure-activity relationship and specificity of inhibition of DNA polymerases from normal and leukemic cells of man and from simian virus by rifamycin derivatives, *J. Natl. Cancer Inst.,* 61, 1187, 1978.
86. Wu, R. S., De Filipps, M. K. W., and Quinn, F. R., Quantitative structure-activity correlations of rifamycins as inhibitors of viral-directed DNA polymerases and mammalian α and β DNA polymerases, *J. Med. Chem.,* 23, 256, 1980.
87. Yagura, R., Kozu, T., and Seno, T., Inhibition of DNA polymerases α and γ by rifamycin derivative AF/013, *J. Biochem. (Tokyo),* 90, 1394, 1981.
88. Ono, K., Ohashi, A., Ogasawara, M., Matsukage, A., Takahashi, T., Nakayama, C., and Saneyoshi, M., Inhibition of DNA polymerase from murine cells and oncovirus by 5′-alkylated derivatives of 1-β-D-arabinofuranosyluracil 5′-triphosphate: substituent effects on inhibitory action, *Biochemistry,* 20, 5088, 1981.
89. Ono, K., Ogasawara, M., Matsukage, A., Takahashi, T., Nakayama, C., and Saneyoshi, M., Inhibitory effects of various 5-halogenated derivatives of 1-β-D-arabinofuranosyluracil 5-triphosphate on DNA polymerases from murine cells and oncornavirus: substituent effects on inhibitory action, *Biochemistry,* 21, 1019, 1982.
90. Tanaka, M. and Yoshida, S., Inhibition of deoxyribonucleic acid polymerases α and β and terminal deoxynucleotidyl transferase from calf thymus by mitomycin C-deoxyribonucleic acid, *Biochem. Pharmacol.,* 30, 299, 1981.
91. Nakamura, H., Morita, T., and Yoshida, S., Differential inhibition of mammalian DNA polymerases by X-irradiated DNA, *Int. J. Radiat. Biol.,* 38, 449, 1980.
92. Di Cicoccio, R. A. and Srivastava, B. I. S., Inhibition of deoxynucletide-polymerizing enzyme activities of human cells and of simian sarcoma virus by heparin, *Cancer Res.,* 38, 2401, 1978.
93. Dube, D. K., Seal, G., and Loeb, L. A., Differential heat sensitivity of mammalian DNA polymerase, *Biochem. Biophys. Res. Commun.,* 76, 1675, 1979.
94. Dube, D. K., Kunkel, T. A., Seal, G., and Loeb, L. A., Distinctive properties of mammalian DNA polymerases, *Biochem. Biophys. Acta,* 561, 369, 1979.
95. Wawra, E. and Dlejs, I., Evidence for the function of DNA polymerase β in unscheduled DNA synthesis, *Nucl. Acid Res.,* 7, 1675, 1979.

96. Spiro, I. J., Denman, D. L., and Dewey, W. C., Effect of hyperthermia on CHO DNA polymerases α and β, *Radiat. Res.,* 89, 134, 1982.
97. Silber, J. R., Fry, M., Martin, G. M., and Loeb, L. A., Fidelity of DNA polymerase isolated from regenerating liver chromatin of aging *Mus musculus, J. Biol. Chem.,* 260, 1304, 1985.
98. Zunino, F., Gambetta, R., Colombo, A., Louni, G., and Zaccara, A., DNA polymerases of rat liver. Partial characterization and effect of various inhibitors, *Eur. J. Biochem.,* 60, 495, 1975.
99. Blank, A., personal communication.
100. McBride, O. W., Merry, D. E., Zmudzka, B. Z., and Wilson, S. H., Human DNA polymerase β is on chromosome 8, Proc. 8th Int. Human Gene Mapping Workshop, Abstr. 241, Helsinki, August, 1985.
101. Koerner, T. J. and Meyer, R. R., A novel single-stranded DNA binding protein from Novikoff hepatoma which stimulates DNA polymerase β. Purification and general characterization, *J. Biol. Chem.,* 258, 3126, 1983.

Chapter 5

DNA POLYMERASE-γ AND DNA POLYMERASE-δ

I. DNA POLYMERASE-γ

In a landmark discovery, Temin and Mizutani[1] and Baltimore[2] identified in 1970 activity of RNA-dependent DNA polymerase (reverse transcriptase) in retroviruses. Shortly thereafter, several investigators found that animal cells also contain polymerase activity that copies the ribo strand of the synthetic primer-templates oligo(dT)·poly(rA) or oligo(rU)·poly(rA). Gallo et al.[3] first demonstrated that human acute leukemia cells exhibit considerable activity capable of utilizing synthetic RNA templates. The question was posed whether this activity in malignant cells is of viral or cellular origin. Scolnick et al.[4] and Penner et al.[5] showed that analogous polymerase activity can be detected in normal human fibroblasts and lymphocytes, respectively. DNA polymerase(s) that utilized oligo(dT)·poly(rA) template and required 0.1 to 0.3 *M* salt and manganese ion activator for maximum activity were also shown to be present in normal chick embryo cells. These enzymes apparently differed from viral reverse transcriptase in that they failed to utilize natural RNA as template.[6,7]

In early 1972 it became evident that the newly described DNA polymerase-β is also able to copy oligo(dT)·poly(rA), and that maximum utilization of this template requires the presence of manganese ions and salt (see also the preceding chapter).[8] It was not clear, therefore, whether the synthetic RNA-dependent DNA synthesizing activity that was detected in various types of animal cells is due to polymerase of the β class, to a viral reverse transcriptase, or perhaps to a third category of polymerase. First indications that indeed eukaryotic cells possess a third polymerase were provided by Weissbach et al.[9] who fractionated by column chromatography HeLa cell extracts and separated cytoplasmic α-type, nuclear β-type, and a third cytoplasmic polymerase that utilized preferentially oligo(dT)·poly(rA). The three activities were distinguished by their intracellular location and chromatographic properties and under the assay conditions used, their preferential template-primer utilization.[9] Shortly thereafter, similar polymerase activity that utilized oligo(dT)·poly(rA) in preference to other primer-templates and required manganese and 0.1 to 0.3 *M* KCl for maximum activity was identified in a series of animal cell types.[10-12] Firmer criteria to establish that indeed a third class of DNA polymerase exists in animal cells were provided in 1974 by several authors. Wilson and associates[13] showed that mouse myeloma cells contain, in addition to α- and β-polymerases a 6- to 8-S membrane or mitochondria-bound polymerase that utilizes oligo(dT)·poly(rA) with high efficiency. Gallo and collaborators[14] demonstrated that the third type of human cell DNA polymerase can be distinguished immunologically from the other two cellular polymerases and from viral reverse transcriptase. Furthermore, it has been shown by several investigators that in contrast with retrovirus reverse transcriptase, the third cellular polymerase is able to copy only the ribo strand of oligo(dT)·poly(rA), but fails to effectively utilize oligo(dG)·poly(rC) or natural RNA as template.[9,14-17] Ensuing detailed studies established unequivocally that animal cells contain a third DNA polymerase with a unique size, antigenicity, and catalytic properties.[15-18] This third class of eukaryotic polymerase was designated in 1975 as DNA polymerase-γ.[19]

Following its initial detection in mammalian cells, DNA polymerase-γ was described in a wide variety of eukaryotic species including sea urchin,[20] *Xenopus laevis*,[21-23] and birds.[10,21,24] Although it was initially thought that mitochondria possess a fourth, distinct type of DNA polymerase,[19] it was shown by Weissbach and colleagues[25] and other investigators[26-28] that polymerase-γ is actually the mitochondrial polymerase. Yet a catalytically indistinguishable polymerase of the γ-type is also present in the cell nucleus of the majority of cells examined.[29-31] (cf. Chapter 6, Section III.A). The use of monospecific antibodies directed against polymerase-γ,[14,15,17,18] as well as the employment of selective inhibitors established that polymerase-γ is distinct from cellular α- and β-polymerases, and that it is likely to be responsible for the replication of mitochondrial DNA. However, its function in the cell nucleus is unknown at present.

In this chapter, the major structural and catalytic properties of DNA polymerase-γ are surveyed. The in vivo roles and biology of polymerase-γ are considered in Chapters 6 and 9, respectively.

A. Structure

Extensive purification of DNA polymerase-γ from extracts of dividing animal cells has been hindered by the difficulty of obtaining workable amounts of the enzyme, which constitutes only 1 to 10% of the total DNA polymerase activity as monitored with activated DNA.[21,34,35] Furthermore, polymerases of the γ class often display multiple molecular forms.[9,14,24,28,36] Column chromatography resolved multiple forms of DNA polymerase-γ from extracts of rat spleen,[32] calf thymus,[12] and HeLa,[9,28] and KB human cells.[36] Disparate molecular species of polymerase-γ could also be distinguished by the different efficacy with which they copied synthetic vs. natural DNA templates,[36] by their different response to salt[36] and to divalent metal ion activator,[32] and by their molecular size.[24,37] It is not clear, however, whether the multiplicity of forms seen in vitro is a reflection of in vivo variance within the polymerase-γ molecule population. Kozu et al.[32] reported that the appearance of one of the subspecies of rat spleen polymerase-γ depended on the mode of fractionation of the original crude extract. The possibility that microheterogeneity of polymerase-γ in that system is an artifact of the preparative procedure was raised.[32] Knopf et al.[38] reported that only one of two previously observed[16] forms of HeLa cell polymerase-γ was discerned when a modified purification scheme was employed. It is unclear, therefore, whether the microheterogeneity of polymerase-γ is due to limited proteolysis, self-aggregation, serendipitous association with other proteins, or to an authentic in vivo molecular variance of the polymerase.

The partial purification and purification to apparent homogeneity of γ-polymerases from various cell types was reported by several authors. Partial purification was achieved for DNA polymerase-γ in early studies. Spadari and Weissbach[16] isolated two chromatographically distinct forms of HeLa cell polymerase-γ, both exhibiting similar apparent sizes of 110 kdaltons. In a successive work, Knopf et al.[38] purified more extensively one of these forms by velocity sedimentation and estimated its size to be 160 kdaltons or 275 to 330 kdaltons by gel electrophoresis and gel filtration, respectively. Yoshida et al.[12] reported that purified polymerase-γ from calf thymus contained two polypeptides of 63 and 68 kdaltons. Caution is required with respect to earlier reports on the isolation of a 125 kdaltons polymerase-γ from nuclear membrane of adenovirus-infected KB cells,[36] since a distinct adenovirus-coded DNA polymerase of 140 kdaltons was recently discovered.[41-44]

Extensive purification of DNA polymerase of the γ class from mouse myeloma cells was reported by Wilson and associates.[17] The murine polymerase, purified 18,000-fold, exhibited apparent size of 315, 270, and 230 kdaltons as assessed by gel filtration, gel electrophoresis, and glycerol gradient centrifugation, respectively. Denaturing gel electrophoresis of the highly purified, but probably not homogeneous enzyme revealed polypeptides of 185, 135, 118, 66, 44, and 26 kdaltons. The purified myeloma γ enzyme was devoid of DNase, RNase, RNase H, and nucleotide diphosphokinase activities.[17] Human lymphoblast polymerase-γ was purified more than 9000-fold by Robert-Guroff et al.[15] A single form of the enzyme was observed, and the purified native polymerase exhibited a size of 120 kdalton. Although a 120-kdalton polypeptide was detected upon denaturing gel electrophoresis of the lymphoblastic enzyme, subunit structure was not proposed due to nonhomogeneity of the polymerase.[15]

Using embroynic chick as starting material, Matsukage et al.[24,37] purified polymerase-γ 1.5 × 10^6-fold to near homogeneity. Two forms of the chick polymerase-γ were separated by gel filtration, having apparent sizes of 280 and 180 kdalton.[24,37] Each of the forms was purified separately to near homogeneity and analyzed by denaturing gel electrophoresis for polypeptide composition. The 280-kdalton form contained almost exclusively a 47-kdalton polypeptide, and the amount of that polypeptide in fractions collected during purification correlated with polymerase activity. Both 47 and 135-kdalton polypeptides were found to be present in the purified 180-kdalton polymerase-γ. However, the specific activity of the native enzyme correlated well with the amount of the 47-kdalton polypeptide, but not with that of the 135-kdalton component.[24,37] The molecular size of one of the two native enzymes (180 kdaltons) and its S value (7.5 S) suggested that it is composed of four identical 47-kdalton subunits.[24,37] It is not clear, however, what is the composition of the 280-kdalton form, although a hexameric structure of this polymerase-γ form is a reasonable possibility. Matsukage et al.[39] also reported the extensive purification of MOPC104E mouse myeloma cell DNA polymerase-γ. The purified murine polymerase-γ yielded on denaturing gel electrophoresis a major polypeptide of 47 kdaltons and a minor component of 149 kdaltons. Analogous to the avian enzyme, it was suggested that the 180- to 280-kdalton murine polymerase-γ[17] is composed of four identical 47-kdalton polypeptide subunits.[39] Comparative tryptic peptide mapping of the purified avian and murine γ-polymerases revealed that 50% of the tryptic fragments were common to both enzymes.[37,39] Moreover, observed serological cross reaction between these two enzymes also indicated structural similarity between the two γ-polymerases.

It is clear that in the face of the considerable technical difficulties which hinder the preparation of workable amounts of highly purified polymerase-γ, data on its structure are still fragmentary. More work is needed to elucidate the molecular details of the composition of γ-polymerase from diverse sources. A possible way to achieve definitive analysis of this minor polymerase is by its overproduction through recombinant DNA techniques.

B. Catalytic Properties

By necessity, most studies on the catalytic properties of DNA polymerase-γ were performed with polymerases that were not homogeneous. Yet, clearly defined catalytic parameters distinguish polymerase-γ from DNA polymerases-α and -β, and from viral DNA polymerases.

1. Primer-Template Utilization

The primary distinguishing catalytic property of DNA polymerase-γ is its preference for oligo(dT)·poly(rA) template over synthetic deoxynucleotide homopolymers or activated DNA.[9,13,14,16,17,32,38] DNA polymerases of the γ type resemble retroviral RNA-dependent DNA polymerases in their preference for primed polyriboadenylic acid template. However, viral reverse transcriptase and polymerase-γ are antigenically distinguishable[14,15,18,44] and are physically separable in extracts of virus-infected cells.[14,45] Moreover, unlike viral reverse transcriptase, polymerase-γ cannot detectably copy in vitro natural RNA templates,[9,14-16] and does not copy at all, or utilizes poorly, the synthetic template oligo(dG)·poly(rC).[14-16,32,38,44] Last, poly(2-o-methylcytidylate) template primed with oligo(dG) is copied efficiently by retroviral DNA polymerase, but is not efficiently utilized by polymerase-γ.[46,47] It should be noted, however, that usually low activities of polymerase-γ were assayed for ability to copy defined templates. Therefore, the absolute inability of polymerase-γ to utilize certain templates cannot be definitely determined unless massive amounts of enzyme are assayed. Although DNA polymerase-β is also capable of utilizing oligo(dT)·poly(rA) primer-template, catalytic parameters of this reaction differ for enzymes of the β and γ classes. The preferred metal ion activator for polymerase-γ with oligo(dT)·poly(rA) is Mn^{2+} at 0.1 to 0.5 m*M*, and maximum enzymic activity is attained in the presence of 100 to 300 m*M* KCl and phosphate.[15,17,38,48] In contrast, reactivity of polymerase-β is abolished by phosphate (cf. Chapter 4, Section V.B). The Km values of γ-polymerases for the 3′-hydroxyl terminus of the oligo(dT) primer are 5- to 275-fold lower than those of polymerase-β.[37,49] Furthermore, the Km values of γ-polymerases for dTTP with oligo(dT)·poly(rA) template are 10- to 20-fold lower than those of β-polymerases.[37,48] It may be speculated that the high affinity of polymerase-γ to deoxynucleotide substrates reflects the low concentration of nucleotides available for DNA replication in mitochondria.

Although polymerases of the γ category are able to extend in vitro synthetic oligoribonucleotide primers,[15,37,46,50] HeLa cell polymerase-γ was reported to be unable to utilize natural RNA primers at a detectable level.[51]

2. Processivity and Long-Stretch DNA Synthesis

The most prominent catalytic property of polymerase-γ, which distinguishes it from the quasi-processive and distributive polymerases-α and -β, respectively, is its highly processive mode of DNA polymerization. First indication for the high processivity of polymerase-γ was obtained by Wilson and co-workers[17] with highly purified mouse myeloma enzyme. It was found that in the presence of a large excess of primer template, polymerase-γ synthesized only full-length copies of the template, thus suggesting processive mechanism of replication. In an extension of this work, Matsukage et al.[52] showed that poly(dT) copied in vitro by mouse cell polymerase-γ from excess oligo(dT)·poly(rA) template increased in length very rapidly. In contrast, poly(dT) polymerized by comparable activity of purified polymerase-β increased in size progressively at a slower rate. Kinetic analysis of those data suggested that polymerase-β simultaneously extended about 30 primer termini in a highly distributive fashion, thus completing synthesis after an extended period of incubation. In contrast, polymerase-γ elongated one primer at a time, bringing the polymerization to completion before a new primer was utilized.[52] Similar technique was employed to demonstrate highly processive synthesis by homogeneous chick cell DNA polymerase-γ.[37,53] The processivity vs. distributiveness of DNA polymerization by polymerases-γ and -β, respectively, were corroborated in experiments showing that increasing template concentration led to reduction of elongation rate and to an increase in product molecule number for polymerase-β. In contrast, product number or DNA elongation rate of polymerase-γ remained unaffected by template levels. These results were interpreted as an indication for the highly processive mode of polymerization by DNA polymerase-γ which, once it attaches to a template molecule, does not dissociate from it to interact with excess template molecules, as does the distributive polymerase-β.[37]

The highly processive DNA polymerization catalyzed by DNA polymerase-γ enables that enzyme to use natural self-primed single-stranded DNA and to synthesize a full unit length complementary product. Kollek and Goulian[54] demonstrated the ability of purified polymerase-γ to carry out the in vitro replication of parvovirus H-1 DNA, thus converting the single-stranded 3′ hairpin-primed DNA into a double-stranded replicative form. The product of the polymerase-γ-catalyzed reaction was a full-length, double-stranded replicative form, although several discrete sizes of partially double-stranded products were also discerned.[54] Robertson et al.[55] reported that DNA polymerase-γ purified from fetal bovine liver replicated in vitro parvoviral single-stranded DNA to a double-stranded, full-length DNA molecule with the product covalently linked to the 3′ hairpin primer. Kollek et al.[56] used nuclei isolated from parvovirus H-1-infected cells to characterize host polymerases, which may be involved in the in vivo replication of the viral genome. On the basis of aphidicolin and d_2TTP sensitivity of virus DNA replication, these authors suggested that polymerase-α may participate in the initial strand displacement stage of parvovirus DNA replication, whereas polymerase-γ converted and displaced single-stranded viral DNA into a double-stranded replicative form.[56] Results demonstrating the ability of polymerase-γ to replicate in vitro full-length molecules of parvovirus DNA suggest that the γ enzyme is capable of conducting the reaction alone.[55]

C. Inhibitors

1. Sulfhydryl Group Reactive Reagents

Unlike DNA polymerase-β and similar to polymerase-α, polymerases of the γ class are inactivated by the thiol group reactive agent *N*-ethylmaleimide.[9,17,18,20,23,32] In the presence of 0.5 to 1.0 m*M* of the inhibitor, 50 to 100% of the activity of polymerase γ is inhibited, whereas polymerase-β remains unaffected.[15,17]

2. 2′,3′-Dideoxythymidine-5-triphosphate

Of the three major classes of animal cell DNA polymerases, γ type polymerases are the most sensitive to inhibition by the dTTP analog 2′,3′-dideoxythymidine-5-triphosphate (d_2TTP). Inactivation of polymerase-γ by (d_2TTP) has been demonstrated for ovarian *X. laevis* enzyme,[23] for monkey CV-1 cell enzyme,[57] and for human HeLa cells,[58] to mention some selected examples. Van der Vliet and Kwant[59] demonstrated that 50% inactivation of polymerase-γ by d_2TTP required levels of analog which were 500- and 17-fold lower than for 50% inactivation of DNA polymerase-α and β respectively. In analogous experiments, it was demonstrated that at d_2TTP to dTTP ratio of 1.0, human lymphoblasts polymerase-α activity was 50% inhibited, and polymerase-γ was 70% unaffected, polymerase-β was uninhibited.[56] Moreover, d_2TTP was incorporated into newly made DNA only during polymerase-γ catalyzed DNA polymerization, but it was not incorporated into DNA when DNA synthesis was directed by polymerases-α and β.[60] Hence the potency of the analog toward polymerase-γ is explicable in light of its action as a chain terminator, as well as a substrate analog (cf. Chapter 4, Section V.C).

3. Intercalating Agents

Early observations by Meyer and Simpson[61] indicated that intercalating dyes are preferential inhibitors of the synthesis of mitochondrial DNA, and that the mitochondrial DNA polymerase is more sensitive to inhibition by ethidium bromide and acriflavin than nuclear polymerase. That early report was later verified and expanded by Tarrago-Litvak et al.,[62] who showed that the mitochondrial polymerase-γ was strongly inhibited by ethidium bromide, whereas α- and β-polymerases were resistant to that intercalator. Spectrophotometric analysis suggested that the intercalated form of ethidium bromide interacted with the active site of polymerase-γ.[62] The preferential intercalation of mitochondrial DNA compared with nuclear DNA is predictable by the circularity of the mitochondrial genome. The fact that polymerase-γ is uniquely sensitive to intercalators may imply that the polymerase has a site to accommodate circular DNA, which also accommodates intercalators.

4. Other Inhibitors

In contrast to polymerase-α which was found to be singularly sensitive to several selective inhibitors, so far there has been no inhibitor specific to polymerase-γ, nor has the use of inhibitors of polymerase-γ been instrumental in deciphering the in vivo role of DNA polymerase-γ (cf. Chapter 6). DNA polymerase-γ is resistant to the commonly used polymerase-α inhibitors aphidicolin, butylanilinouracil, and N^2-*p*-*n*(butylphenyl) guanine (cf. Chapter 3, Sections VI.B and VI.C). Polymerase-γ is also relatively resistant to arabinose analogs of deoxynucleotides (Chapter 3, Section VI.D). Polymerases of the γ types are inhibited, however, by phosphonacetic acid.[63,64] Rifamycin derivatives are inhibitors of γ polymerases, as they are of polymerases-β and -α.[65-67] Kinetic analysis suggested, however, that the mode of inhibition of polymerase-γ by rifamycin AF/013 differs from the mechanism of inhibition of polymerase-α.[67]

II. DNA POLYMERASE-δ

Highly purified or homogeneous DNA polymerases of the α, β, and γ categories are usually devoid of associated deoxyribonuclease activity. Although in some instances exonuclease activity was detected in impure preparations of polymerase-α, in the great majority of cases reported the exonuclease is removed with further purification, and purified polymerase-α is free of any detectable nuclease activity (see Chapter 3, Section IV.A). In analogy, DNA polymerase-β from Novikoff hepatoma initially copurifies with a specific exonuclease DNase V. However, in the final stages of purification, the nuclease is dissociated from the β-enzyme polypeptide (cf. Chapter 4, Section I). DNA polymerase-γ was also found to be devoid of deoxyribonuclease activity (this chapter, Section I.A.). It is possible that exonuclease, which associates with animal cell DNA polymerase(s) in vivo, has a physiological role within DNA replication and/or repair complexes (see Chapters 6 and 7). However, it appears that in contrast to bacterial and viral DNA polymerases, exonuclease does not constitute an integral part of polymerases-α, -β, and -γ polypeptides. One year after criteria were set in 1975 for the classification of eukaryotic DNA polymerases into the α, β, and γ categories,[19] a distinct DNA polymerase that was initimately associated with exodeoxyribonuclease was reported to be present in rabbit bone marrow by Byrnes et al.[68] In recent years, evidence was provided to show that this enzyme, although sharing some properties with polymerase-α, is singular in that the polymerase and exonuclease appear to reside in the same polypeptide, in analogy with some prokaryotic and animal virus DNA polymerases.[69] Association of a 3′→ 5′ exonuclease activity with DNA polymerase could confer proofreading capability on the polymerase and thus increase the fidelity of the DNA synthesis.[70]

The importance of establishing the existence of an exonuclease containing animal cell DNA polymerase and characterizing its properties is, therefore, apparent. At the time of this writing, only rabbit bone marrow and calf thymus polymerase-δ have been characterized at some depth, and little evidence exists on the presence and properties of δ-type polymerases in other tissues and cells. Most of the information surveyed in this section is limited, therefore, to studies of calf thymus and bone marrow polymerase-δ. The tantalizing question of the prevalance of δ-polymerases among other animal species and cell types remains open at present.

A. Structure

DNA polymerase-δ was purified extensively to apparent homogeneity from rabbit bone marrow.[69,71,72] Polymerase-δ constituted 10 to 25% of the total activity of polymerases-α and -δ in bone marrow extracts, and it was separated from two forms of polymerase-α by successive chromatographic steps. Whereas sedimentation coefficient values for α_1 and α_2 rabbit polymerases indicated molecular sizes of 215 and 100 kdaltons, respectively, polymerase-δ sedimented at 6.5 S having an estimated size of 122 kdaltons.[69,71] Apparently homogeneous rabbit bone marrow polymerase-δ exhibited a single 122 kdalton polypeptide upon sodium dodecyl sulfate polyacrylamide gel electrophoresis, suggesting that one polypeptide consisted of both DNA polymerase and 3′→ 5′ exodeoxyribonuclease activites.[69,72]

Downey and co-workers[73-75] also reported the purification of a δ-type DNA polymerase from calf thymus, an enzyme that exhibited both DNA polymerase and 3′→ 5′ exonuclease activity and which behaved as an asymmetric molecule, displaying different sizes upon velocity centrifugation (140 to 160 kdaltons) and gel filtration (200 kdaltons).[73,74] Denaturing electrophoresis revealed that the nearly homogeneous thymic polymerase-δ contained two major polypeptides of 49 and 60 kdaltons.[74] Being concerned about partial possible proteolysis of the polymerase-δ molecule during its isolation, these authors developed recently a new enzyme isolation scheme that minimizes proteolysis, and in particular removes calcium-dependent proteases from cell extracts at an early stage of the purification procedure.[75] In using this preparative procedure, calf thymus polymerase-δ was separated from polymerase-α by either hydrophobic column chromatography or by hydroxyapatite chromatography, and subsequently purified to apparent homogeneity. The homogeneous native enzyme had an apparent size of 173 kdaltons and upon denaturing gel electrophoresis, two polypeptides of 125 and 48 kdaltons were detected in the enzyme preparation.[75] It was suggested that limited proteolysis of the 125 kdalton polypeptide was responsible for the appearance of the 60 kdalton and perhaps of the 49 kdalton polypeptides in previous preparations of calf thymus polymerase-δ.[74] The reason for size and subunit discrepancy between the calf thymus and rabbit bone marrow polymerase-δ is not clear at present.

B. The Association Between Polymerase-δ and 3′→ 5′ Exonuclease

Clearly, the most intriguing property of polymerase-δ is its intimate association with a 3′→ 5′ exonuclease activity. The rabbit exonuclease, which removes nucleotides from the 3′ terminus of DNA, copurified to homogeneity with the DNA polymerase activity.[69,72] The tight association between polymerase and nuclease in the 122-kdalton δ polypeptide was demonstrated by their coordinate inactivation under a variety of conditions. Thus both exonuclease and polymerase activities were inactivated with indistinguishable kinetics by heat and by *N*-ethylmaleimide.[68,69] Furthermore, hemin and rifamycin AF/013, which were shown to inhibit in vitro synthesis of DNA by dissociating polymerase from template, inhibited both DNA polymerase and exonuclease to the same extent.[68,69] Rabbit DNA polymerase-δ exhibited sensitivity to aphidicolin, being 50% inhibited by 4 μg/mℓ aphidicolin. Interestingly, the associated 3′→ 5′ exonuclease activity was also inhibited by aphidicolin, showing loss of 50% activity at 10 μg/mℓ aphidicolin.[68] Also, calf thymus polymerase-δ, as well as the associated 3′→ 5′ exonuclease, were inhibited by aphidicolin on duplex poly d(A-T), with indistinguishable kinetics.[75] However, the exonuclease was not affected by the antibiotic on single-stranded poly(dT). The inhibitory effect of aphidicolin on the DNA polymerase-δ function with poly d(A-T) template and dTTP substrate is intriguing in light of similar highly specific inhibition exerted by that antibiotic on α-polymerases interacting with similar primer-templates and substrate (cf. Chapter 3, Section VI.B). Moreover, the blocking of the exonuclease activity by aphidicolin is the only example known to us of inhibition of a non-DNA polymerase enzymic activity by that antibiotic.

Although both rabbit polymerase-δ and 3′→ 5′ exonucleolytic activity appear to reside on a single 122 kdalton polypeptide, selective inhibition of the latter function has been achieved.[68,76] Mercaptopurine ribonucleoside 5′-monophosphates were reported to selectively inhibit the 3′→ 5′ nuclease activity, thus enhancing net DNA synthesis by polymerase-δ.[76] The selective inhibition of the exonuclease function decreased the accuracy of poly d(AT) copying by the polymerase-δ.[76] Recently Byrnes et al.[68] reported that 5′AMP and 5′GMP selectively inhibited the 3′→ 5′ exonuclease functions of polymerase-δ, and that the ensuing increase in DNA polymerization rate in the presence of the ribonucleoside monophosphates was a direct result of the inhibition of nuclease activity.

C. Open Questions Concerning DNA Polymerase-δ

Certainly the elucidation of the properties, mechanism of action, prevalence, and role in vivo of DNA polymerases of the δ type is of fundamental interest, since such a polymerase constitutes the only known example of an exonuclease tightly associated with an animal cell DNA polymerase. Some questions remain open, however, concerning polymerase-δ

Only two cell types, rabbit bone marrow and calf thymus, were found to contain δ-class polymerase. Wilson and colleagues[77] also reported on a subspecies of mouse cell DNA polymerase-α, which contained a bidirectional exonuclease activity. Association of exonuclease and that polymerase-α subspecies within a single polypeptide was not determined. It is not clear, therefore, whether that polymerase was a δ-type enzyme. In other instances, polymerase-α was found to copurify with exonuclease. Yet in most cases the nuclease was dissociated from the polymerase activity. The uniqueness of polymerase-δ lies, therefore, in its inseparability from exonuclease. The limited number of cases in which the presence of polymerase-δ was documented requires that the existence of such activity should be substantiated in other tissues and cells.

Another open question is the relationship of polymerase-δ to polymerase-α. Both enzymes are inhibited by *N*-ethylmaleimide, aphidicolin, hemin, and rifamycin AF/013.[68,69,71,75] However, in light of recent studies on the size and subunit composition of α-polymerases (cf. Chapter 3, Table 1), it appears that rabbit and calf thymus polymerase-δ differ structurally from each other as well as from polymerase-α[68,76] Some catalytic properties of calf thymus polymerase-δ also seem to distinguish it from homologous polymerase-α.[76] However, the multiplicity of forms of α-polymerases in that tissue, and their varied catalytic behavior, pose difficulties in the interpretation of these data. In summary, a more complete understanding of the place of polymerase-δ among other eukaryotic DNA polymerases, and its unique role in DNA metabolism, must await its isolation from different cell types and its rigorous physical and enzymological characterization. Thus monoclonal antibodies against polymerase-α should be applied to polymerase-δ to study structural relationship between the two enzymes. Further, analysis of the fidelity of DNA polymerase-δ should reveal whether the 3′→ 5′ exonuclease works in concert with the polymerase to increase its accuracy in DNA synthesis.

REFERENCES

1. Temin, H. M. and Mizutani, S., RNA-dependent DNA polymerase in virions of Rous sarcoma virus, *Nature (London)*, 226, 1211, 1970.
2. Baltimore, D., RNA-dependent DNA polymerase in virions of RNA tumor viruses, *Nature (London)*, 226, 1209, 1970.
3. Gallo, R. C., Yang, S. Y., and Ting, R. C., RNA dependent DNA polymerase of human acute leukemia cells, *Nature (London)*, 228, 927, 1971.
4. Scolnick, E. M., Aaronson, S. A., Todaro, G. J., and Parks, W. P., RNA dependent DNA polymerase activity in mammalian cells, *Nature (London)*, 229, 318, 1971.
5. Penner, P. E., Cohen, L. H., and Loeb, L. A., RNA-dependent DNA polymerase in human lymphocytes during gene activation by phytohaemagglutinin, *Nature (London) New Biol.*, 232, 58, 1971.

6. Maia, J. C. C., Rougeon, F., and Chapeville, F., Chick embryo poly (rA:dT)-dependent DNA polymerase, *FEBS Lett.*, 18, 130, 1971.
7. Stavrianopoulos, J. G., Karkas, J. D., and Chargaff, E., Nucleic acid polymerases of the developing chick embryo: a DNA polymerase preferring hybrid template, *Proc. Natl. Acad. Sci. U.S.A.*, 68, 2207, 1971.
8. Chang, L. M. S. and Bollum, F. J., Low molecular weight deoxyribonucleic acid polymerase from rabbit bone marrow, *Biochemistry*, 11, 1264, 1972.
9. Fridlender, B., Fry, M., Bolden, A., and Weissbach, A., A new synthetic RNA-dependent DNA polymerase from human tissue culture cells, *Proc. Natl. Acad. Sci. U.S.A.*, 69, 452, 1972.
10. Bolden, A., Fry, M., Muller, R., Citarella, R., and Weissbach, A., The presence of a polyriboadenylic acid-dependent DNA polymerase in eukaryotic cells, *Arch. Biochem. Biophys.*, 153, 26, 1972.
11. Suzuka, I. and Muto, M., Presence of poly(rA):poly(dT) dependent DNA polymerase in mouse myeloma cells, *Biochem. Biophys. Res. Commun.*, 46, 1874, 1972.
12. Yoshida, S., Ando, T., and Kondo, T., Synthetic RNA-dependent DNA polymerase from calf thymus, *Biochem. Biophys. Res. Commun.*, 60, 1193, 1974.
13. Matsukage, A., Bohn, E. W., and Wilson, S. H., Multiple forms of DNA polymerase in mouse myeloma, *Proc. Natl. Acad. Sci. U.S.A.*, 71, 578, 1974.
14. Lewis, B. J., Abrell, J. W., Smith, R. G., and Gallo, R. C., Human DNA polymerase III (R-DNA polymerase): distinction from DNA polymerase I and reverse transcriptase, *Science*, 183, 867, 1974.
15. Robert-Guroff, M., Schrecker, A. W., Brinkman, B. J., and Gallo, R. C., DNA polymerase γ of human lymphocytes, *Biochemistry*, 16, 2866, 1977.
16. Spadari, S. and Weissbach, A., HeLa cell R-deoxyribonucleic acid polymerase. Separation and characterization of two enzymatic activities, *J. Biol. Chem.*, 249, 5809, 1974.
17. Matsukage, A., Bohn, E. W., and Wilson, S. N., On the DNA polymerase III of mouse myeloma: partial purification and characterization, *Biochemistry*, 14, 1006, 1975.
18. Robert-Guroff, M. and Gallo, R. C., Serological analysis of cellular and viral DNA polymerase by an antiserum to DNA polymerase γ of human lymphoblasts, *Biochemistry*, 16, 2874, 1977.
19. Weissbach, A., Baltimore, D., Bollum, F. J., Gallo, R. C., and Korn, D., Nomenclature of eukaryotic DNA polymerases, *Science*, 190, 401, 1975.
20. Habara, A., Nagano, H., and Mano, Y., Identification of γ-like DNA polymerase from sea urchin embryos, *Biochem. Biophys. Acta*, 561, 17, 1979.
21. Bertazzoni, U. and Scovassi, A. E., Mitochondrial DNA polymerase of eukaryotic cell, in *The Organization and Expression of the Mitochondrial Genome*, Kroon, A. M. and Saccone, C., Eds., Elsevier/North Holland, 1980, 287.
22. Zimmerman, W. and Weissbach, A., Participation of deoxyribonucleic acid polymerase α in amplification of ribosomal deoxyribonucleic acid in *Xenopus laevis*, *Mol. Cell. Biol.*, 1, 680, 1981.
23. Nelson, E. M., Stowers, D. J., Bayne, M. L., and Benbow, R. M., Classification of DNA polymerase activities from ovaries of the frog *Xenopus laevis*, *Dev. Biol.*, 96, 11, 1983.
24. Yamaguchi, M., Matsukage, A., and Takahashi, T., Chick embryo DNA polymerase γ. Purification and structural analysis of nearly homogenous enzyme, *J. Biol. Chem.*, 255, 7002, 1980.
25. Bolden, A., Pedrali-Noy, G., and Weissbach, A., DNA polymerase of mitochondria is γ polymerase, *J. Biol. Chem.*, 252, 3351, 1977.
26. Tanaka, S. and Koike, K., DNA polymerase-γ is localized in mitochondria, *Biochem. Biophys. Res. Commun.*, 89, 791, 1978.
27. Radsak, K. and Shütz, E., Changes of mitochondrial DNA polymerase-γ activity in synchronized mouse cell cultures, *Eur. J. Biochem.*, 89, 3, 1978.
28. Adams, W. J. and Kalf, G. F., DNA polymerase isolated from mitochondrial chromosomes appears to be identical to DNA polymerase-γ, *Biochem. Biophys. Res. Commun.*, 95, 1875, 1980.
29. Genskens, M., Hardt, N., Pedrali-Noy, G., and Spadari, S., An autoradiographic demonstration of nuclear DNA replication by DNA polymerase α and of mitochondrial DNA synthesis by DNA polymerase γ, *Nucl. Acids Res.*, 9, 1599, 1981.
30. Bertazzoni, U., Scovassi, A. I., and Brun, G. M., Chick-embryo DNA polymerase γ. Identity of γ polymerases purified from nuclei and mitochondria, *Eur. J. Biochem.*, 81, 237, 1977.
31. Hübscher, U., Keunzle, C. C., and Spadari, S., Identity of DNA polymerase γ from synaptosomal mitochondria and rat brain nuclei, *Eur. J. Biochem.*, 81, 249, 1977.
32. Kozu, T., Kurihara, T., and Seno, T., Heterogenous forms of poly(rA)-oligo(dT)-directed DNA polymerase activity from rat spleen, *J. Biochem. (Tokyo)*, 89, 551, 1981.
33. Zimmerman, W., Chen, S. M., Bolden, A., and Weissbach, A., Mitochondrial DNA replication does not involve DNA polymerase α, *J. Biol. Chem.*, 255, 11847, 1980.
34. Bollum, F. J., Mammalian DNA polymerases, *Prog. Nucl. Acids Res. Mol. Biol.*, 15, 109, 1975.
35. Weissbach, A., Eukaryotic DNA polymerases, *Annu. Rev. Biochem.*, 46, 25, 1977.

36. Ito, K., Arens, M., and Green, M., Characterization of DNA polymerase associated with nuclear membrane fractions from uninfected and adenovirus 1-infected KB cells, *Biochem. Biophys. Acta,* 447, 340, 1976.
37. Matsukage, A., Yamaguchi, M., Tanabe, K., Taguchi, Y. N., Nishizawa, M., and Takahashi, T., DNA chain elongation mechanism of DNA polymerases α, β and γ, in *New Approaches in Eukaryotic DNA Replication,* de Recondo, A. M., Ed., Plenum Press, New York, 1983, 81.
38. Knopf, K.-W., Yamada, M., and Weissbach, A., HeLa cell DNA polymerase γ: further purification and properties of the enzyme, *Biochemistry,* 15, 4540, 1976.
39. Matsukage, A., Tanabe, K., Yamaguchi, M., Taguchi, Y. N., Nishizawa, M., Takahashi, T., and Takahashi, T., Identification of a polypeptide component of mouse myeloma DNA polymerase γ, *Biochem. Biophys. Acta,* 655, 269, 1981.
40. Lichy, J. H., Field H., Horwitz, M. S., and Hurwitz, J., Separation of the adenovirus terminal protein precursor from its associated DNA polymerase: role of both proteins in the initiation of adenovirus DNA replication, *Proc. Natl. Acad. Sci. U.S.A.,* 79, 5225, 1982.
41. Stillman, B. W., Tamanoi, F., and Mathews, M. B., Purification of an adenovirus-coded DNA polymerase that is required for initiation of DNA replication, *Cell,* 31, 613, 1982.
42. Lichy, J. H., Nagata, K., Friefeld, B. R., Enomoto, T., Field, J., Guggenheimer, R. A., Ikeda, J. E., Horwitz, M. S., and Hurwitz, J., Isolation of proteins involved in the replication of adenoviral DNA *in vitro, Cold Spring Harbor Symp. Quant. Biol.,* 47, 731, 1983.
43. Field, J., Gronostajski, R. M., and Hurwitz, J., Properties of the adenovirus DNA polymerase, *J. Biol. Chem.,* 259, 9487, 1984.
44. Gallagher, R. E., Todaro, G. J., Smith, R. G., Livingston, D. M., and Gallo, R. C., Relationship between RNA-directed DNA polymerase (reverse transcriptase) from human acute leukemic blood cells and primate type C-virus, *Proc. Natl. Acad. Sci. U.S.A.,* 71, 1309, 1974.
45. Lewis, B. J., Abrell, J. W., Smith, R. G., and Gallo, R. C., DNA polymerase in human lymphoblastoid cells infected with simian sarcoma virus, *Biochem. Biophys. Acta,* 349, 148, 1974.
46. Gerard, G. F., Rottman, F., and Green, M., Poly(2-*O*-methylcytidylate)-oligodeoxyguanylate as a template for the ribonucleic acid directed deoxyribonucleic acid polymerase in ribonucleic acid virus particles and a specific probe for the ribonucleic acid-directed enzyme in transformed murine cells, *Biochemistry,* 13, 1632, 1974.
47. Gerard, G. F., Poly(2-*O*-methylcytidylate)-oligodeoxyguanylate, a template-primer specific for reverse transcriptase is not utilized by HeLa cell γ polymerase, *Biochem. Biophys. Res. Commun.,* 63, 706, 1976.
48. Falaschi, A. and Spadari, S., The three DNA polymerases of animal cells: properties and functions, in *DNA Synthesis: Present and Future,* Molineux, I. and Kohiyama, M., Eds., Plenum Press, New York, 1978, 487.
49. Ono, K., Ohashi, A., Tanabe, K., Matsukage, A., Nishizawa, M., and Takahashi, T., Unique requirement for template-primers of DNA polymerase β from rat ascites hepatoma, *Nucl. Acids Res.,* 7, 715, 1979.
50. Wilson, S. H., Matsukage, A., Bohn, E. W., Chen, Y. C., and Sivarajan, M., Polynucleotide recognition by DNA α-polymerase, *Nucl. Acids Res.,* 4, 3981, 1977.
51. Spadari, S. and Weissbach, A., RNA-primed DNA synthesis: specific catalysis by HeLa cell DNA polymerase α, *Proc. Natl. Acad. Sci. U.S.A.,* 72, 503, 1975.
52. Matsukage, A., Nishizawa, M., and Takahashi, T., Difference in the mechanism of poly(dT) synthesis by DNA polymerases β and γ, *J. Biochem. Tokyo,* 85, 1551, 1979.
53. Yamaguchi, M., Matsukage, A., and Takahashi, T., Continuous synthesis of long DNA chains by chick embryo DNA polymerase γ, *Nature (London),* 285, 45, 1980.
54. Kollek, R. and Goulian, M., Synthesis of parvovirus H-1 replicative form from viral DNA by DNA polymerase γ, *Proc. Natl. Acad. Sci. U.S.A.,* 78, 6206, 1981.
55. Robertson, A. T., Bates, R. C., and Stout, R. E., Purified DNA polymerase γ replicates bovine parvovirus DNA to a unit-length product, *Biochem. Biophys. Res. Commun.,* 117, 580, 1983.
56. Kollek, R., Tseng, B. Y., and Goulian, M., DNA polymerase requirements for parvovirus H-1 DNA replication, *in vitro, J. Virol.,* 41, 982, 1982.
57. Edenberg, H. J., Anderson, S., and DePamphilis, M. L., Involvement of DNA polymerase α in simian virus 40 DNA replication, *J. Biol. Chem.,* 253, 3273, 1978.
58. Waquar, M. A., Evans, M. J., and Huberman, J. A., Effect of 2′-3′dideoxythymidine-5′-triphosphate on HeLa cell *in vitro* DNA synthesis: evidence that DNA polymerase α is the only polymerase required for DNA replication, *Nucl. Acids Res.,* 5, 1933, 1978.
59. Van der Vliet, P. C. and Kwant, M. M., Role of DNA polymerase γ in adenovirus DNA replication, *Nature (London),* 276, 532, 1978.
60. Van der Vliet, P. C. and Kwant, M. M., Role of DNA polymerase γ in adenovirus DNA replication. Mechanisms of inhibition by 2′-3′dideoxythymidine-5′-triphosphate, *Biochemistry,* 20, 2628, 1981.

61. Meyer, R. R. and Simpson, M. V., DNA biosynthesis in mitochondria: differential inhibition of mitochondrial and nuclear DNA polymerases by the mutagenic dyes ethidium bromide and acriflavin, *Biochem. Biophys. Res. Commun.*, 34, 238, 1969.
62. Tarrago-Litvak, L., Viratelle, O., Darriet, D., Dalibart, R., and Litvak, S., The inhibition of mitochondrial DNA polymerase γ from animal cells by intercalating agents, *Nucl. Acids. Res.*, 5, 2197, 1978.
63. Bolden, A., Aucker, J., and Weissbach, A., Synthesis of herpes simplex virus, vaccinia virus and adenovirus in isolated HeLa cell nuclei, *J. Virol.*, 16, 1584, 1975.
64. Allaudeen, H. S. and Bertino, J. R., Inhibition of activities of DNA polymerases α, β, γ and reverse transcriptase of L 1210 cells by phosphonacetic acid, *Biochem. Biophys. Acta*, 520, 490, 1978.
65. DiCicoccio, R. A. and Srivastava, B. I. S., Structure-activity relationship and specificity of inhibition of DNA polymerases from normal and leukemic cells of man and from simian virus by rifamycin derivatives, *J. Natl. Cancer Inst.*, 61, 1187, 1978.
66. Wu, R. S., De Filipps, M. K. W., and Quinn, F. R., Quantitative structure-activity correlations of rifamycins as inhibitors of viral RNA-directed DNA polymerases and mammalian α and β DNA polymerases, *J. Med. Chem.*, 23, 256, 1980.
67. Yagura, R., Kozu, T., and Seno, T., Inhibition of DNA polymerases α and γ by rifamycin derivative AF/013, *J. Biochem.*, 90, 1397, 1981.
68. Byrnes, J. J., Downey, K. M., Black, V. L., and So, A. G., A new mammalian DNA polymerase with 3′ to 5′ exonuclease activity: DNA polymerase δ, *Biochemistry*, 15, 2817, 1976.
69. Byrnes, J. J., Structural and functional properties of DNA polymerase delta from rabbit bone marrow, *Mol. Cell. Biochem.*, 62, 13, 1984.
70. Kornberg, A., *DNA replication*, W.H. Freeman, San Francisco, 1980 (supplement 1982).
71. Byrnes, J. J. and Black, V. L., Comparison of DNA polymerases α and δ from bone marrow, *Biochemistry*, 17, 4226, 1978.
72. Goscin, L. A. and Byrnes, J. J., DNA polymerase δ: one polypeptide, two activities, *Biochemistry*, 21, 2513, 1982.
73. Lee, M. Y. W. T., Tan, C.-K., So, A., and Downey, K. M., Purification of deoxyribonucleic acid polymerase δ from calf thymus: partial characterization of physical properties, *Biochemistry*, 19, 2096, 1980.
74. Lee, M. Y. W. T., Tan, C.-K, Downey, K. M., and So, A. G., Structural and functional properties of calf thymus DNA polymerase δ, *Prog. Nucl. Acids Res. Mol. Biol.*, 26, 82, 1981.
75. Lee, M. Y. W. T., Tan, C.-K, Downey, K. M., and So, A. G., Further studies on calf thymus DNA polymerase δ purified to homogeneity by a new procedure, *Biochemistry*, 23, 1906, 1984.
76. Byrnes, J. J., Downey, K. M., Que, B. G., Lee, M. Y. W. T., Black, V. L., and So, A. G., Selective inhibition of the 3′ to 5′ exonuclease activity associated with DNA polymerases: a mechanism of mutagenesis, *Biochemistry*, 16, 3740, 1977.
77. Chen, Y.-C., Bohn, E. W., Planck, S. R., and Wilson, S. H., Mouse DNA polymerase α. Subunit structure and identification of a species with associated exonuclease, *J. Biol. Chem.*, 254, 11678, 1979.

Chapter 6

ROLES OF ANIMAL CELL DNA POLYMERASES IN REPLICATION OF CELLULAR DNA

By copying, the ancient models should be perpetuated.

Hsieh Ho

I. INTRODUCTORY REMARKS

Replication of cellular DNA is one of the most complex and stringently regulated processes in living cells. Elucidation of the molecular details of the replication of small bacteriophage DNA indicated that the replicative process in prokaryotes and their viruses requires a concerted, tightly controlled action of a large number of replicative proteins.[1] Although the details of the replication of cellular DNA in eukaryotes are still largely obscure, there is no reason to assume that it is a less complex process than prokaryotic replication. At least as many, and quite possibly more, proteins and cofactors may be required.

In contrast to prokaryotes, animal cells independently replicate their nuclear and mitochondrial genomes. The replication of nuclear DNA in eukaryotic cells is sequestered to the synthetic period (S phase) of the cell cycle, and entry of cells into the replicative phase is a stringently controlled process.[2] Furthermore, the size of genomic DNA in animal cells, its association with numerous protein and RNA molecules, and its intricate spatial organization render it exceedingly more complex than viral and bacterial genomes.[3] The study of the regulated and accurate replication of genomic and mitochondrial DNA in animal cells is likely to involve the elucidation of the activities and mechanisms of presently unknown numbers of proteins and cofactors. Yet it is clear that a central component of any replicative machinery must be DNA polymerase(s) responsible for the accurate selection and efficient polymerization fo deoxyribonucleoside triphosphates.

Since it became evident that animal cells contain several distinct classes of DNA polymerases, researchers embarked on the task of assigning roles to the different types of polymerase in the various aspects of DNA metabolism. Different approaches were employed to elucidate the functions of DNA polymerases of the α, β, and γ categories in the replication of mitochondrial and genomic DNA. The major body of evidence on functions of DNA polymerases in DNA replication was collected from studies on correlations between levels of DNA polymerases assayed in vitro and replicative activity in vivo. These investigations included determinations of the intracellular localization of the various polymerases, yielding evidence that whereas all three types of DNA polymerase are present in the cell nucleus, only polymerase-γ exists in mitochondria. Hence, role for DNA polymerase-γ in the replication of mitochondrial DNA was suggested. In other studies, the intracellular levels of polymerases-α, -β, and -γ were measured in dividing and postmitotic cells and at different phases of the cell cycle. These experiments indicated that among the three polymerase types, only changes in polymerase-α activity were correlated with the replication of nuclear DNA. Studies of animal cell mutants that possess a modified polymerase-α also indicated that a central role in replication is played by polymerase-α.

The development of an arsenal of selective inhibitors to DNA polymerases allowed investigators to analyze their effects on DNA replication in vivo. Again, these studies indicated that polymerase-α is a central component of the replicative machinery in nuclei of animal cells. Use of selective inhibitors also suggests that polymerase-γ is responsible for the synthesis of mitochondrial DNA. In related investigations, monospecific antibodies against polymerase-α were employed to demonstrate its vital role in the replication of nuclear DNA. Last, correlations are drawn between the unique enzymological properties of DNA polymerase-α and catalytic properties required from a replicative enzyme.

In particular, the exclusive association of DNA polymerase-α with DNA primase and other ancillary proteins, which may be involved in replication of genomic DNA, strongly suggests that polymerase-α is the main replicative polymerase. Hence, physiological and enzymological studies, as well as the use of selective inhibitors of DNA polymerases, strongly support the perception that DNA polymerases-α and -γ play central roles in the replication of nuclear and mitochondrial DNA, respectively. Yet some cardinal questions regarding the participation of DNA polymerases in replicative processes in vivo have not been answered. It is argued that DNA polymerase-β plays a major role in DNA repair synthesis, but it is not yet clear what role, if any, it plays in DNA replication. Part of the total cellular activity of DNA polymerase-γ is also found in the cell nucleus. The role of that polymerase in DNA replication or other aspects of the metabolism of nuclear DNA has not been determined at the present time. Lessons from prokaryotic DNA metabolism taught that DNA polymerases of different types can be functionally interchangeable to a certain degree.[1] The possibility that non-α-polymerase can partly or completely be responsible for DNA replication in the nucleus is yet unexplored. Whatever DNA polymerase conducts replicative DNA synthesis, it has been proposed that replicative polymerase comprises a component of a supramolecular multienzyme replicative complex. The existence of such a complex, the mode of association of the replicative polymerase with its other constituents, as well as the mechanism of action of the replicative machinery, are still largely obscure.

The evidence for assignment of roles for DNA polymerases-α and -γ in the replication of nuclear and mitochondrial DNA, respectively, are reviewed in this chapter. Open questions regarding the participation of DNA polymerases in replicative processes in animal cells are also addressed.

II. REPLICATION OF NUCLEAR DNA

A. Participation of Polymerase-α in Nuclear DNA Replication

1. Intracellular Localization of DNA Polymerases

A correlation between the site of genomic DNA replication in vivo and localization of a replicative DNA polymerase would require that that enzyme be sequestered to the nucleus. Several lines of evidence indicated that polymerase-α is the major replicative polymerase in animal cells (see below). Yet the assignment of polymerase-α to the cell nucleus proved to be a most perplexing and hotly contested question. Whereas shortly after the discovery of polymerase-β, its nuclear localization was established, and DNA polymerase-γ was found to be associated with both nuclei and mitochondria, two opposing bodies of evidence indicated that polymerase-α is compartmentalized either within the nucleus or in the cytoplasm.

The recent development of a selection of monoclonal antibodies against polymerase-α has provided fresh evidence to further establish the nuclear localization of polymerase. By employing *in situ* labeling of polymerase-α with monospecific antibodies, it was possible to demonstrate its exclusive nuclear compartmentalization in a variety of cell types (see below). In this section, the evidence for the nuclear localization of DNA polymerases-α and -β, as well as part of the cellular activity of polymerase-γ, is surveyed. Further, recent evidence on the association of polymerase-α with the nuclear matrix, a subnuclear structure implicated in the spatial organization of DNA replication, is reviewed.

a. *Association of DNA Polymerases-α, -β, and -γ with the Cell Nucleus*

(1) DNA Polymerase-α

Early efforts to determine the subcellular localization of DNA polymerase-α usually involved fractionation of animal cells in aqueous media and determiniation of the in vitro activities of DNA polymerase-α in the separated cytoplasmic and nuclear fractions. The surprising finding was that polymerase-α was entirely or predominantly recovered in the cytoplasm.[4-10] The argument for cytoplasmic compartmentalization of polymerase was further supported by observations of cell types, which are naturally devoid of a nucleus and yet contain polymerase-α activity. Rabbit reticulocytes that have lost their nucleus possess detectable levels of an α-type polymerase.[11] Similarly, both α- and γ-polymerase activities are present in platelets that are formed by cytoplasmic division of megakaryotes in bone marrow and are thus devoid of a nucleus.[12]

Antithetical results were obtained in studies with cells that divide at an exceptionally rapid rate. Assuming a correlation between rate of replication in vivo and amount of a replicative enzyme, it should be expected that nuclei in which DNA synthesis is conducted at a particularly high rate will contain maximum amounts of DNA polymerase. Mazia and Hinegardner[13] established that early embryos of sea urchin that divide once every hour indeed contain exceptionally high levels of DNA polymerase activity. In a series of studies by Loeb and colleagues,[14-18] DNA polymerase-α was found to be localized in nuclei which were isolated from rapidly dividing sea urchin embryos. In fact, recovery of DNA polymerase-α in the nucleus was so pronounced that isolation of nuclei constituted an initial enrichment step in a purification scheme for sea urchin polymerase-α.[19]

From these studies, the concept emerged that polymerase-α was synthesized in the cytoplasm of unfertilized sea urchin egg, and it was translocated to the nucleus with each division cycle after fertilization. These studies were extended by Hobart and Infante,[20] who showed that virtually all the activity of DNA polymerases-α and -β was detected in the cytoplasm of unfertilized eggs of sea urchin, but during the blastula stage of embryonal development, polymerase-α became entirely nuclear. Similar relocalization of sea urchin polymerase-α during gastrulation was found, and further, prior to its migration to the nucleus, the α enzyme was associated with the rough endoplasmic reticulum.[21,22] It could be argued, however, that the nuclear localization of polymerase-α in early sea urchin embryos constitutes a singular requirement of embryogenesis, and that it is not a general phenomenon in somatic cells. Yet, that somatic cell nuclei may contain the major part of the cellular polymerase-α activity was shown when nuclei were isolated with nonaqueous or in calcium-containing media. Lynch et al.[23] demonstrated that most of the activity of hepatic polymerase-α was retained in nuclei of cells that were fractionated in calcium-containing isotonic media. Yet when Tris®-HCl buffer or inorganic salts were present in the fractionation medium, the activity of polymerase-α was reduced in isolated nucleus and increased in cytoplasm. A reasonable interpretation of these results is that polymerase-α readily leaks from the nucleus under most fractionation conditions, and this leaching is the basis of an artifactual cytoplasmic localization of that enzyme.[23]

This conclusion was challenged by Eichler et al.,[24] who found that calcium ions act as irreversible inactivators of cytoplasmic polymerase-α from KB cells. It was suggested that the low levels of polymerase-α in calcium-fractionated cytoplasm may be due to enzyme inactivation rather than to its absence from that subcellular compartment.[24] Matsukage et al.[25,26] countered that challenge by showing that calcium is indeed required for the retention of polymerase-α in nuclei, but its omission from the fractionation buffer changes the subcellular distribution of the polymerase without affecting total recovered activity. It was thus claimed that no significant fraction of the total cellular polymerase-α exists in the cytoplasm in vivo.[25,26] Two molecular forms of polymerase-α designated PI and PII, possessing high and lower affinity to DNA cellulose, respectively, were isolated from HeLa cells.[27] Upon fractionation in calcium-containing media, both forms were associated with the cell nucleus, but when fractionation was performed with aqueous buffers, form PII was found in the cytoplasm. It was implied, therefore, that form PII, which may be less tightly bound to DNA in chromatin, readily leaks from nuclei under aqueous fractionation conditions.[27] In a similar vein, correspondence was demonstrated between concentrations of salt, which are required to extract chromatin-bound polymerase-α subspecies, and the degree of their leakage into the cytoplasm.[26] That a major proportion of the total cellular activity of polymerase-α from various cell types is associated with chromatin was demonstrated by several laboratories.[17,28-32]

In addition to data obtained with polymerase-α, which was isolated from nuclei separated in calcium-containing media, other investigators observed that most of the cell polymerase-α activity was localized in nuclei that were obtained in nonaqueous media. Devising a nonaqueous scheme for the fractionation of animal cells,[33] Foster and Gurney[34] found that 83 to 89% of the total cellular activity of mammalian polymerase-α was confined to the cell nucleus.

In principle, the simplest procedure for designation of cellular localization of an enzyme would be to physically separate intact cellular components. However, with such procedures, discrepant results were obtained with different cell types and different experimental protocols. With sea urchin eggs, separation into nuclear and enucleate halves can be attained by centrifugation. In 9 of 11 experiments, DNA polymerase-α was detected in the nucleate fraction.[15] Martini et al.[35] isolated manually the germinal vesicles (nuclei) from oocytes of *Xenopus laevis* and found that virtually all the cellular activity of polymerase-α was restricted to the nuclei. Similarly, isolated germinal vesicles of fully developed *X. laevis* contained >80% of the activity of two separable molecular forms of polymerase-α.[36]

Other authors reported that 90% of the cellular activity of polymerase-α was detected in manually isolated nuclei of stages V and VI oocytes of *X. laevis*.[37] Nagano and associates[38] observed recently a similar association of polymerase-α with germinal vesicles of fully grown oocytes of the toad *Bufo bufo*. However, in enucleated mature oocytes, polymerase-α appeared in the cytoplasm, being associated with the rough endoplasmic reticulum therein. The most plausible explanation for this observation is that during oocyte maturation and breakdown of the nuclear membrane, polymerase-α is released into the cytoplasm.[38] Dissolution of nuclear membranes and leakage of polymerase-α into the cytoplasm may also be at the root of observations on cytoplasmic localization of polymerase-α in anucleate reticulocytes,[11] and in blood platelets.[12]

It could be argued again that the largely nuclear localization of polymerase-α in mechanically dissected sea urchin embryos and *X. laevis* eggs could be unique to embryogenesis, and that it does not necessarily reflect the situation in somatic cells. Experiments in which nuclei and cytoplasm of mammalian cells were separated by cytochalasin B-induced enucleation were conducted by several authors. Herrick et al.[39] dissociated nuclear karyoplastic fraction from cytoplasts of mouse L-929 cells that were treated with cytochalasin B. More than 85% of the cellular content of polymerase-α was associated with the karyoplasts, thus indicating nuclear localization of the enzyme. Reddy and Pardee[40] observed that a DNA polymerase was largely associated with cytochalasin B-generated karyoplasts from S-phase cells. However, in quiescent cells, that polymerase was located mainly within the cytoplasts (for further discussion of this work (see Section II.A.4 of this chapter). And last, Brown et al.[41] obtained crude karyoplasts and cytoplasts from cytochalasin B-treated bovine cells and further purified these subcellular fractions by Ficoll-Paque gradient centrifugation. DNA polymerase-α was detected in the separated nuclear and cytoplasmic fractions by immunofluorescent staining with polyclonal immunoglobulin raised against highly purified polymerase-α. The immunochemical detection assay indicated that all polymerase-α activity was discerned in either the cytoplasts or the perinuclear region of the crude nuclei.[41] It was estimated that nuclei contained <5% of the total cellular activity of polymerase-α.[41]

As gathered from the data, which was reviewed up to this point, the problem of the intracellular localization of polymerase-α was in a state of perplexing contradiction. Under some methods, notably aqueous fractionation of cells, polymerase-α appeared to be cytoplasmic, whereas use of other techniques for separation of nuclei from cytoplasm indicated that the enzyme resides in the cell nucleus. Moreover, use of the same fractionation procedure, i.e., cytochalasin B enucleation of cells, yielded opposing results in different laboratories.[39-41] Some of this discrepancy could reflect authentic different physiological processes in different animal cells and tissues. It is reasonable to assume that DNA polymerases are synthesized on cytoplasmic polysomes and are subsequently transported to the nucleus. One likely time for such relocalization is the period of dissolution of the nuclear membrane during the mitotic phase. Hence, in nondividing cells, polymerase may be retained in the cytoplasm for extended periods of time prior to DNA synthesis and inclusion of the polymerase within the nucleus. However, in rapidly dividing cells, one should expect the polymerase to be largely associated with the nucleus. The introduction of monoclonal antibodies to polymerase-α in the last few years provided independent evidence to show that polymerase-α is sequestered in the nucleus during the major part of the cell cycle.

Korn and colleagues[42] isolated a number of hybridoma cell lines that produced precipitating or neutralizing antibodies to KB cell polymerase-α. Four of these monospecific immunoglobulins were used to probe the intracellular localization of polymerase-α in WI38, BeWo, and KB cultured cells.[43] Fixed cells were labeled with the anti-polymerase-α monoclonal antibodies, using the immunoperoxidase staining procedure. About 95% of the immunolabeled polymerase-α was confined to the nuclei of all the examined cell types.[43] Nuclear compartmentalization of polymerase-α was observed throughout the cell cycle except for mitosis, during which the nuclear membrane dissolved and the polymerase was visualized in the cell cytoplasm. Examination of the finer substructure of the immunochemically labeled nuclei revealed that polymerase-α appeared in the midzonal nucleoplasmic region between unstained, centrally disposed nucleoli and peripherally located blocks of condensed chromatin.[43] Furthermore, immunodetection of polymerase-α in proliferating and quiescent cells indicated that higher levels of immunoreactive polymerase molecules were present in cells that actively replicated DNA (see Figure 1* and Section II.A.2.b below).

Matsukage et al.[44] raised monoclonal antibodies against chick cell polymerase-α. The intracellular compartmentalization of polymerase-α in cultured chick cells and in chick testis was probed by this fluorescently labeled antibody.[45] Polymerase-α was sequestered to the nucleus during the G_1, S, and G_2 phases of the cell cycle (see Figure 2*). Only during mitosis and dissolution of the nuclear membrane did polymerase-α appear in the cytoplasm.[45] Masaki et al.[46] raised monoclonal antibodies against the 10-S form of calf thymus DNA polymerase-α. That monoclonal immunoglobulin against calf thymus polymerase that cross reacted with human cell polymerase-α was used to study the intracellular localization of polymerase-α in lymphocytes and in cultured HeLa cells.[47] Immunofluorescence labeling of polymerase-α revealed that the enzyme was confined to the cell nucleus during both the S and G_1 phases of the cell cycle, but appeared in the cytoplasm during mitosis.[47] Immunolabeling of polymerase-α also indicated that

* Figures 1 and 2 appear after p. 110.

the amount of polymerase molecules in the cell was correlated to the proliferative activity of the cells (cf. Section II.A.2.a of this chapter).

The application of the highly specific monoclonal antibodies to the study of the intracellular localization of polymerase-α has provided an independent criterion for the designation of polymerase-α as a nuclear enzyme. In all the cells that were examined by several authors, and with the aid of independently isolated antibodies, polymerase-α was found to be associated with the cell nucleus throughout the cell cycle except for the short mitosis period during which the nuclear membrane disintegrates and the polymerase appears in the cytoplasm. This recurring result is in line with numerous previous investigations that applied various cell fractionation methods for the study of the intracellular distribution of polymerase-α (see above). It appears, however, that polymerase-α leaks into the cytoplasm under some conventional cell fractionation procedures.

It is not yet clear why bovine cell polymerase-α, which was probed with a polyclonal monospecific antibody, was found in the cytoplasm and the perinuclear region and not within the nucleus.[41] One unlikely possibility is that polymerase-α is sequestered to different subcellular compartments in different cell types. It may also be that the fixation procedures led to leaching of the bovine polymerase-α, or that some nuclear proteins masked a nuclear polymerase, which thus failed to be labeled with the antibody. Last, a more likely explanation is that the polyclonal antibody[41] was not monospecific to polymerase-α. Similar antibody was used to demonstrate cross antigenicity between polymerase-α and the cytoplasmic glucocorticoid hormone receptor.[48] However, upon reexamination of the reactivity of either hormone-charged or uncharged receptors with several different monoclonal antibodies against KB cell polymerase-α, no cross antigenicity between polymerase and receptor could be detected.[49] This lack of cross reactivity as evaluated with monoclonal antibody does not negate, however, other structural similarities between the glucocorticoid receptor and polymerase-α. Further, the possibility remains that polyclonal but not monoclonal antibody detects structural homology between the two proteins.

(2) DNA Polymerase-β

Under most conventional cell fractionation procedures, DNA polymerase-β from a wide variety of cells appears to be largely confined to and retained in the cell nucleus[4,23,34,50,51] and to be associated with chromatin therein.[32,52] Using highly specific polyclonal antibody against chick DNA polymerase-β, it was demonstrated by immunofluorescence labeling of chick cells that polymerase-β was mostly nuclear throughout the cell cycle (Figure 2).[45] Several studies suggested, however, that in selected cases polymerase-β may be localized in the cell cytoplasm. Isopycnic centrifugation separated unfertilized eggs of sea urchin into nucleate and anucleate halves, and most of the polymerase-β activity was detected in the cytoplasmic fraction.[21] Hobart and Infante[20] reported that cell fractionation in calcium-containing medium maintained mouse L cell polymerase-β in the nucleus. However, sea urchin polymerase-β appeared in the cytoplasm under identical fractionation conditions.[20] In manually enucleated oocytes of *X. laevis,* 90% of polymerase-α appeared in the nucleus, whereas only 20% of the cellular activity of polymerase-β was associated with it.[37] In both sea urchin eggs and in *X. laevis* oocytes, it is likely that polymerase-β is synthesized and stored in the cytoplasm prior to fertilization and before its activity in the nucleus is required.

(3) DNA Polymerase-γ

DNA polymerase-γ was initially identified as the sole mitochondrial DNA polymerase by Bolden et al.,[53] who demonstrated that mitochondria of mycoplasma-free HeLa cells contained a single DNA polymerase, which displayed properties of a γ-type enzyme. It was further calculated that mitochondria of HeLa and liver cells contained 29 and 16%, respectively, of the total cellular activity of polymerase-γ.[53] Shortly thereafter it was reported that DNA polymerase-γ was deposited within both mitochondria and nuclei of chick embryo cells, and the mitochondrial and nuclear enzymes were catalytically and physically indistinguishable.[54] In an elegant experiment, Hübscher and colleagues[55] separated brain cell nuclei and brain synaptosomes which were devoid of nuclei but contained mitochondria. Both nucleus and mitochondrion from brain tissue contained polymerase-γ activities that appeared to be structurally and enzymatically identical.[55] Association between mitochondria and polymerase-γ was indirectly suggested by demonstrations that the cellular activity of polymerase-γ peaked concomitantly with maximum synthesis of mitochondrial DNA in synchronized mouse cells.[56]

It was similarly demonstrated that the activities of mitochondrial cytochrome C oxidase and of DNA polymerase-γ were elevated with similar kinetics, following concanavalin A-induced stimulation of spleen lymphocyte proliferation.[57] That polymerase-γ is an authentic mitochondrial enzyme rather than an adventitiously bound extramitochondrial protein was shown by demonstrating that polymerase-γ was associated with the inner-membrane mitochondrial DNA replication complex (M-band).[58] Although in the majority of examined cell types polymerase-γ was found to be associated with both nucleus and mitochondrion, some investigators[37,39] reported that polymerase-γ was deposited in mitochondria only and was absent from nuclei of *X. laevis* oocytes and rat liver cells, respectively. Considering the size of the mitochondrial genome and proteins already known to be coded by it, it is unlikely that polymerase-γ is coded in mitochondrial DNA, and it is therefore nuclear DNA that is likely to contain the genetic information for that enzyme.

b. Association of Polymerase-α with the Nuclear Matrix

The nuclear matrix is a nucleoskeletal structure obtained by DNase digestion of the bulk of nuclear DNA, solubilization of the nuclear membrane with nonionic detergent, and removal of the large majority of nuclear proteins by extraction with 2 *M* salt.[60] The remaining nuclear residue, designated nuclear matrix, contains 1 to 2% of the total nuclear DNA and a limited set of several proteins.[60,61] Studies of the residual nuclear DNA indicated that it is associated with the matrix in the form of anchored supercoiled loops.[62-64] Furthermore, DNA that is closest to the matrix attachment sites was found to be enriched in newly replicated DNA.[65-67]

A topological model of replication has been offered wherein DNA loop domains move to be replicated by a matrix-bound replication complex.[66-68] Such a model requires that replicative enzymes are associated with the nuclear matrix. Smith and Berezney[69] showed that nuclear matrices from proliferating cells of regenerating rat liver contained measurable amounts of polymerase-α activity, but only trace amounts of polymerase-β. Activities of the two types of DNA polymerase were assessed in vitro by their characteristic sensitivities to the selective inhibitors aphidicolin, d_2TTP, and *N*-ethylmaleimide. Using the same assay system, it was subsequently shown that the proportion of matrix-bound nuclear polymerase-α in the regenerating liver correlated with DNA replication activity in the tissue.[70] Polymerase-α, which was associated with nuclear matrices, utilized endogenous DNA to synthesize fragment-sized products of 800 to 1200 polymerized nucleotides. In contrast to polymerase-α whose matrix-bound activity was elevated during active DNA replication, matrix-associated polymerase-β activity peaked prior to and immediately after maximum DNA replication occurred.[70] These data were interpreted as indicating that DNA polymerase-α is engaged in synthesis of DNA at the proposed site of active DNA replication, i.e., the nuclear matrix. In contrast, it was posited that polymerase-β is not involved in replication and that its function is supposedly to screen DNA for repair prior to and after replication.

A similar methodology for the determination of activities of DNA polymerases-α and -β has been applied for the examination of the intranuclear redistribution of DNA polymerase-α and -β prior to, during, and after DNA replication in the regenerating liver.[51] It was reported that a rise in matrix-bound polymerase-α was detectable at the prereplicative stage before the cellular activity of polymerase-α increased. Maximal proportion of cellular polymerase-α was bound to the matrix coincidentally with maximum in vivo rates of DNA replication. At the end of the replicative period, lower levels of the α-enzyme were detected in the high-salt-resistant nuclear matrix.[51] In contrast to the timed redistribution of polymerase-α within nuclei, activity of polymerase-β was not concentrated in any of the nuclear subfractions throughout the proliferative cycle.[51] Corroboration of these observations was provided recently by direct visualization of chick DNA polymerases-α and -β in nuclei and in isolated nuclear matrices with the aid of fluorescence-labeled monospecific antibodies against each of the two enzymes.[71] In intact nuclei of chick cells, anti-polymerase-α monoclonal antibodies stained intranuclear granules, whereas antibodies directed against polymerase-β labeled homogeneously the nuclei. Furthermore, after DNase and high-salt treatment of the nuclei, polymerase-α was detectable with the aid of monospecific antibodies in the residual nuclear matrices, whereas polymerase-β was not detectable in these structures.[71] It was suggested that the polymerase-α-containing granules in intact nuclei represent the nuclear matrix, and that a subfraction of polymerase-α molecules is tightly associated with these subnuclear structures. Analogous association of immunofluorescent-labeled polymerase-α with subnuclear granules of lymphocytes was also reported.[47] On the basis of previous works[25-27,72] and the above observations, it was proposed that several intranuclear forms of polymerase-α exist, which differ by their extractability from the nuclei with salt.[71] Supposedly, the form that is most avidly attached to DNA in the nucleus is that which was detected in the matrix.

The above described works[51,69,70,71] are particularly attractive in that they suggest dynamic replication-dependent intranuclear mobility of polymerase-α. Further, the association of polymerase-α with the nuclear matrix synchronously with DNA replication supports the proposal that the matrix is the site of active replication, and that polymerase-α is the replicative enzyme. It is not as yet fully clear, however, to what degree the association of polymerase-α with the matrix is a product of specific matrix preparative procedures. Thus for instance, while 10 to 50% of the activity of polymerase-α from mouse regenerating liver remains with a nonsoluble residue after a single extraction step with 0.4 *M* to 2.0 *M* NaCl, almost 100% of polymerase-α activity is solubilized after three consecutive extractions of nuclei with 0.4 *M* NaCl.[73] It also remains to be seen whether the matrix-bound polymerase-α represents a unique form of polymerase-α as suggested,[71] and whether indeed that form displays catalytic properties such as greater processivity and higher fidelity, and a requirement for ATP for activity as might be expected for a DNA replicating polymerase. Last, by operational definition the nuclear matrix is that residue that resists highly disruptive extraction procedure. Thus many controls are required to demonstrate that association of the polymerase with that residual structure is not an artifact of this disruptive isolation scheme.

2. Intracellular Levels of DNA Polymerase During Cell Cycling

One of the earliest arguments for the specific role of DNA polymerase-α in the replication of nuclear DNA is that the cellular activity of polymerase-α positively correlates with the DNA replication activity of the cell. Such correlation was demonstrated in various postmitotic cells, in quiescent cells that were stimulated to divide by a variety of means, and in synchronized cultured cells. Increased intracellular activity of polymerase-α has been consistently observed in all the proliferating cell systems that were examined. For a long time, only indirect studies with inhibitors of protein synthesis

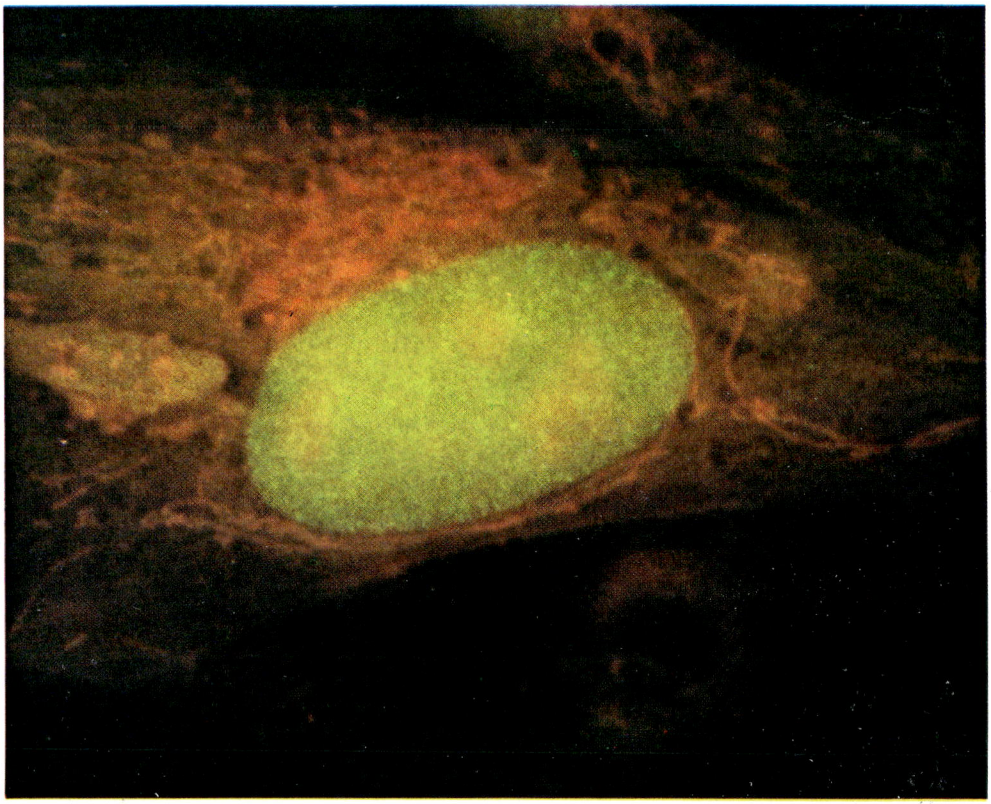

A.

FIGURE 1. Nuclear localization of DNA polymerases-α and -β and cell proliferation-dependent increase in immunochemically detected polymerase-α. Nuclear localization of chick cell DNA polymerases-α and -β. (A) *DNA polymerase-α: In situ* detection of polymerase-α was conducted with immunofluorescent monoclonal anti-polymerase-α immunoglobulin. (Magnification × 130.) (B) *DNA polymerase-β:* Polymerase-β was detected with immunofluorescent polyclonal monospecific antibodies. Detailed procedures for antibody preparation and *in situ* detection of polymerases in chick cells are described in Reference 45. (Figures A and B courtesy of Dr. Akio Matsukage, Aichi Cancer Research Institute, Nagoya, Japan.)

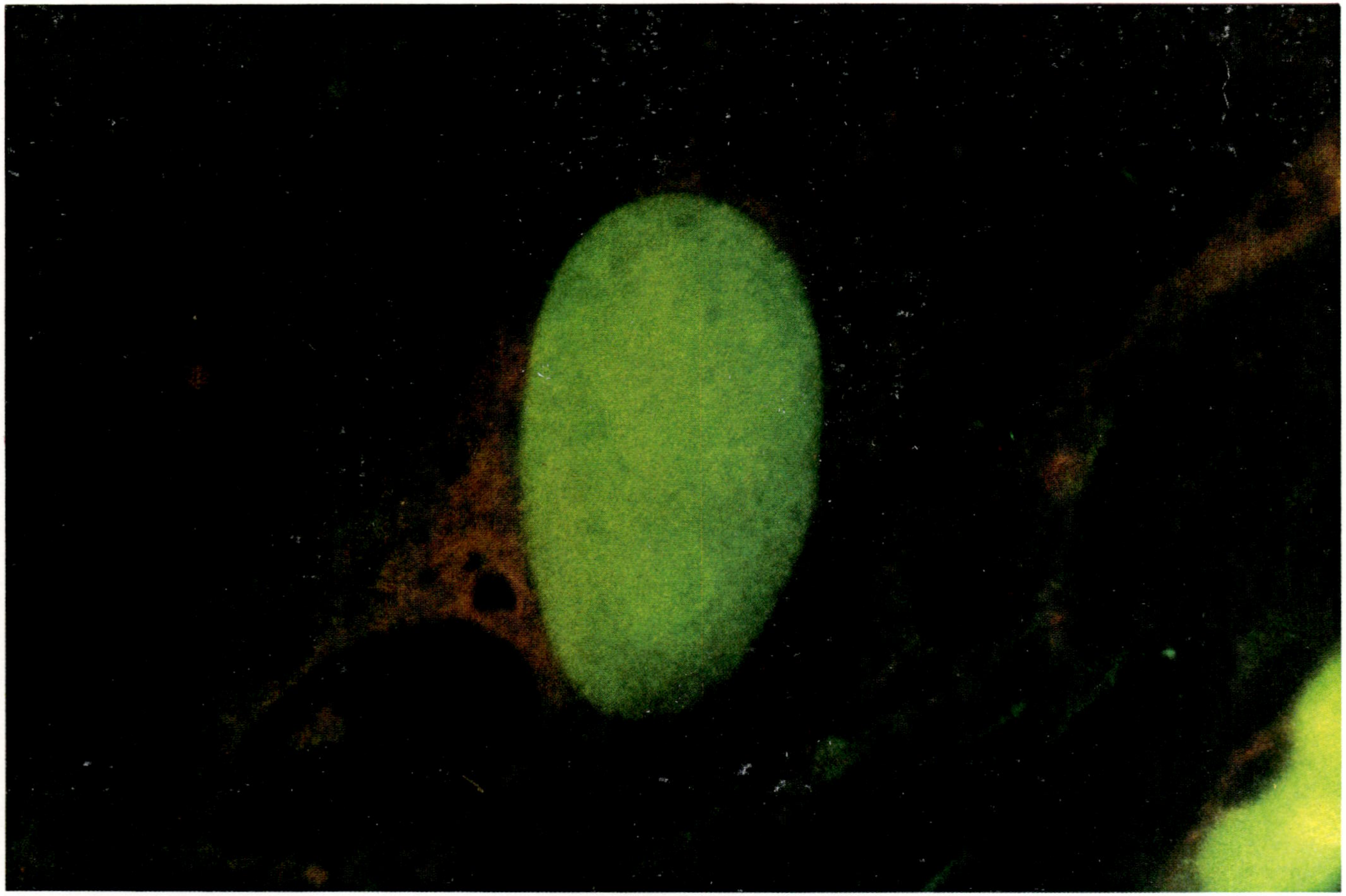

FIGURE 1B.

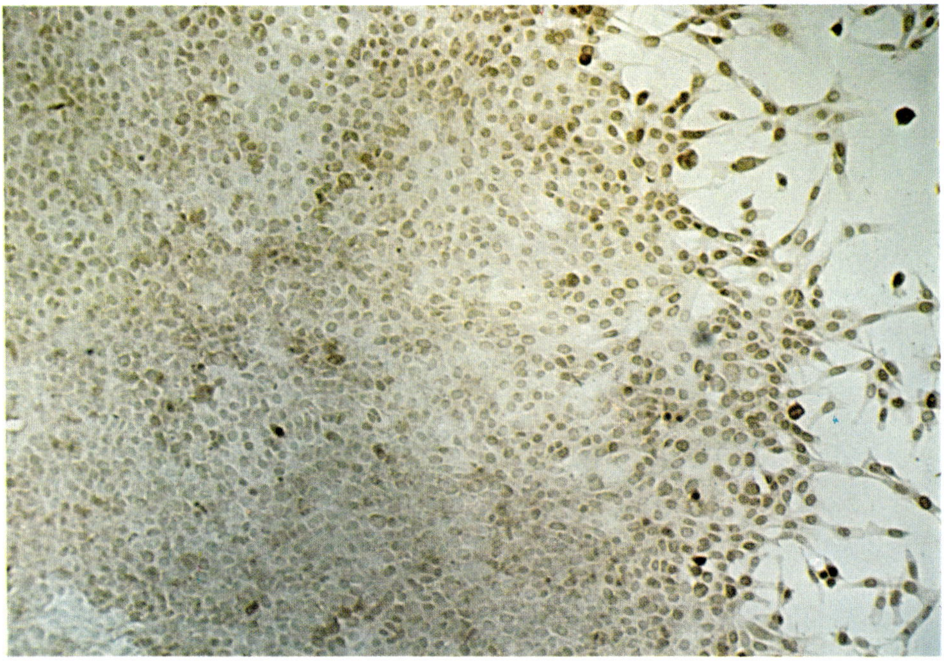

A.

FIGURE 2. Immunocytochemical demonstration of induction of DNA polymerase-α during "wound healing". 3T6 cells were seeded on a glass microscope slide chamber. Growth was in Eagle's minimum essential medium (MEM) for 3 days, during which they reached confluence, and a linear "wound" was then made with a scalpel blade. Cells attached to the slide were fixed in 4% paraformaldehyde solution. Slides were incubated with protein-A Sepharose purified monoclonal antihuman DNA polymerase-α antibody SJK 132-20 (30 $\mu g/m\ell$) for 30 min at 30° C. The slides were rinsed, incubated with rabbit antimouse IgG ("bridging antibody") followed by incubation with a soluble mouse peroxidase-antiperoxidase complex (PAP technique). The slides were then rinsed again, incubated for 8 min at 20°C in peroxidase substrate solution (pH 7.6) that contained diaminobenzidine, and rinsed again. Finally, the preparations were dehydrated in alcohol and xylene. (A) Sample prepared 20 hr after wounding. The wound at the right side of the photomicrograph is bounded by a collar of DNA polymerase-α antigen positive cells, some of which have already begun to divide and migrate into the wound. (B) Sample prepared 40 hr after wounding. The wounded area, at right, has been extensively filled in by cells that are DNA polymerase-α antigen-positive. (C) Sample prepared 60 hr after wounding. The wounded area, in the center of the photomicrograph, has been repaired and is filled with cells that are positive for DNA polymerase-α antigen. The zone of positivity merges on both sides with confluent cells that are negative for polymerase-α antigen. The magnification of all photomicrographs is × 130. (Figure 2 courtesy of Drs. David Korn and Teresa S. Wang, Stanford University, California.)

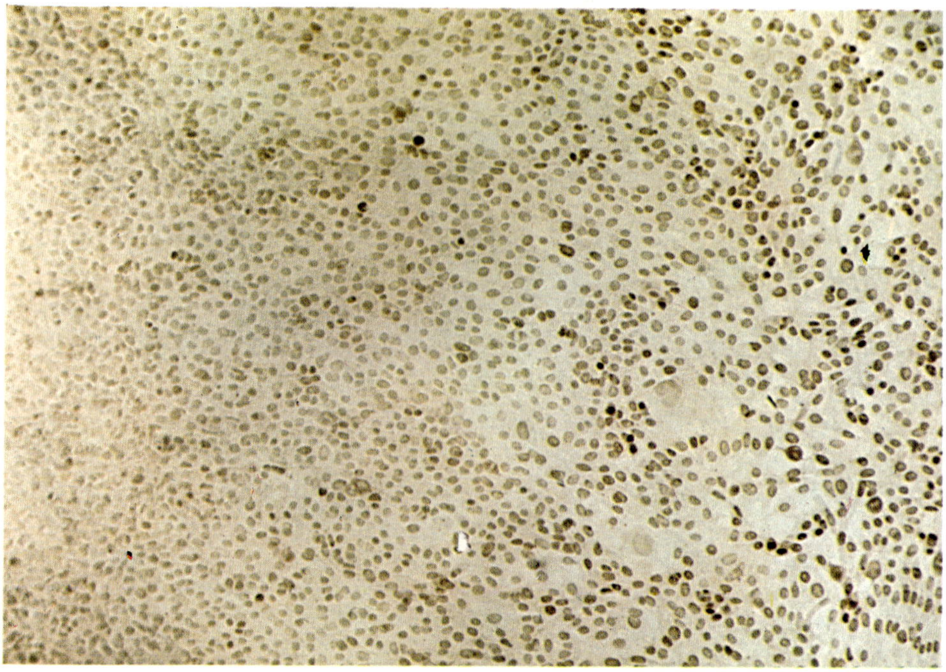

FIGURE 2B.

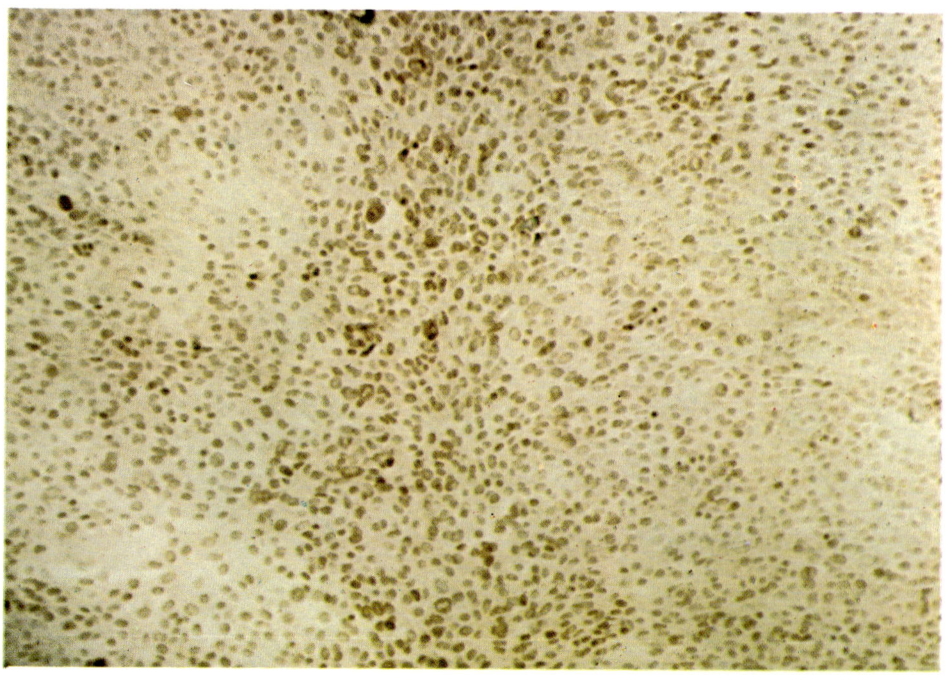

FIGURE 2C.

suggested that increased activity was due to *de novo* synthesis of polymerase-α instead of to postsynthetic activation of preexisting enzyme molecules. By use of *in situ* immunochemical labeling of polymerase-α with monospecific antibodies, evidence was presented that indeed, the number of polymerase-α molecules increases in dividing cells. Evidence for the specific increase in polymerase-α activity during cell proliferation is reviewed in this section.

a. *Evidence with Postmitotic Cells*

An obverse argument in favor of a central role for DNA polymerase-α in DNA replication is provided by the consistent observation that cells, which are maintained at an extended G_1 or G_0 phase, exhibit low levels of polymerase-α or are devoid of detectable activity of that polymerase. Mature liver that contains a low proportion of dividing cells does not contain detectable levels of DNA polymerase-α, whereas polymerase-β activity levels are similar in mature liver and in proliferating hepatocytes from a regenerating organ.[23,32,73,74] Postmitotic neurons were found to contain activities of β-type and γ-type DNA polymerases, but were devoid of DNA polymerase-α. Presumably, the presence of both DNA polymerase-β and -γ in these cells suggests that nuclear DNA repair synthesis and replication of mitochondrial DNA are conducted in postmitotic neuronal tissue.[75] Similarly, postmitotic sperm cells displayed polymerase-β as the only predominant DNA polymerase.[76-79] It was noted, however, that DNA polymerase-α activity persisted in nondividing differentiated cells at the terminal stages of spermatogenesis.[80-83] It was suggested, therefore, that polymerase-α may fill a role in processes additional to DNA replication (cf. Chapter 9, Section II). Matsukage et al.[45] used double immunofluorescence to stain *in situ* both α- and β-polymerases in chick testis with monospecific antibodies. Whereas polymerase-β was present in nuclei of all testicular cells, polymerase-α was detected only in nuclei of spermatogonia and spermatocytes, which are not terminally differentiated. In a recent survey of polymerase-α and -β activities in various organs of mice, considerable activity of polymerase-β, but very low levels of polymerase-α, were found in postmitotic tissues such as cardiac and skeletal muscle, kidney, and brain.[84]

The correlation between absence of DNA replication and low activity of polymerase-α is further strengthened by consistent observations on a progressive and parallel decline in both proliferative potential and polymerase-α activity as various cell types differentiate. For a discussion of differentiation and DNA polymerases in animal tissues, see Chapter 9, Section II.

b. *Studies on Quiescent Cells Stimulated to Divide*

When resting cells in tissues or quiescent cultured cells are stimulated to divide by a variety of means, the intracellular activity of DNA polymerase-α increases dramatically, whereas polymerase-β activity is usually minimally affected by the proliferative response.

In early studies, Chang and Bollum[85] demonstrated that liver regeneration after partial hepatectomy in the rat resulted in an increase in the activity of the high-molecular-weight polymerase-α, whereas the level of the smaller-sized polymerase-β remained unaltered. These results are illustrated in Figure 3. Very similar observations on regenerating liver were reported by Baril et al.[86,87] who also found that the proliferation-related increase in hepatic polymerase-α activity was abolished by administration of cycloheximide or actinomycin D. These studies were in accord with a number of earlier reports, which demonstrated abolishment of the increase in total DNA polymerase activity by actinomycin D and cycloheximide when quiescent animal cells are stimulated to divide.[92] It was implied, therefore, that *de novo* synthesis of DNA polymerase-α occurs as cells are induced to proliferate.

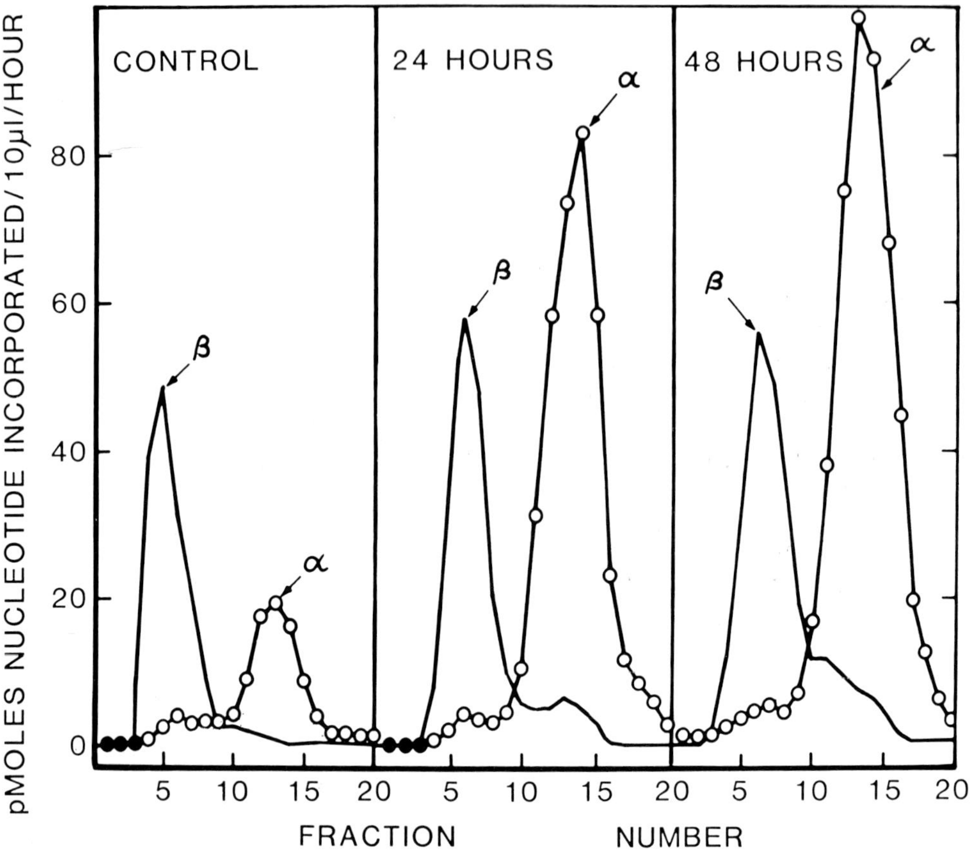

FIGURE 3. Increased activity of DNA polymerase-α and unaltered level of polymerase-β during hepatocyte proliferation in the regenerating rat liver. Liver homogenates (cytoplasm) obtained at the indicated times after partial hepatectomy of the rat were centrifuged in sucrose gradients. Activities of the low molecular weight polymerase-β and high molecular weight polymerase-α were assayed in fractions of the gradient. Note the increased activity of polymerase-α at 24 and 48 hr after partial hepatectomy, and the roughly constant activity of polymerase-β. (Figure adapted from Chang, L. M. S. and Bollum, F. J., *J. Biol. Chem.*, 247, 7948, 1972. With permission.)

Lynch et al.[23,74,88] reported that DNA polymerase-β was the only detectable polymerase in the mature rat liver, but partial hepatectomy and ensuing cell division were accompanied by appearance of polymerase-α in the cell nucleus. Smith and Berezney[70] followed the kinetics of DNA replication and changing α-and β-polymerase activities in regenerating livers of rats. The activity of rat nuclear matrix-bound polymerase-α increased concomitantly with DNA replication and declined as DNA replication ceased. Polymerase-β activity in the matrix did not parallel replicative activity, although it peaked prior to and after DNA synthesis occurred.[70] Fry et al.[32,73] studied the kinetics of cell division and changing polymerase activities during regeneration of liver in the mouse. Cytological analysis of the replicative response in the regenerating mouse liver revealed asynchrony of parenchymal and nonparenchymal cell division.[32,73] Activity of chromatin-associated DNA polymerase-β did not change during liver regeneration, whereas polymerase-α activity rose and declined with kinetics that were similar to but definitely not identical with the kinetics of DNA replication.[73] The nonidentity between the kinetics of DNA synthesis and changes in the level of chromatin-bound polymerase-α may reflect inadequacy of in vitro assessment of polymerase-α activity, or it may represent a true lack of full correlation between intracellular levels of polymerase-α and replicative activity.[73]

Several authors reported on specific increase in the activity of polymerase-α in blood cells that were stimulated to divide. Bollum and associates[89] noted that the activity of α-type polymerase was higher in immature, rapidly dividing leukemic human hematopoietic cells than in mature nondividing cells, whereas polymerase-β activity did not correlate with proliferative activity. Small quiescent leukemic cells of mice were separated from large proliferating cells by sedimentation at unit gravity.[90] The dividing cells contained, respectively, 12-fold and 4-fold higher activities of α- and β-polymerases than the resting cells.[90] Similar observations were also reported for erythropoietic cells obtained from mouse spleen.[91] Several authors reported on a specific increase in the activity of polymerase-α following stimulation of resting lymphoyctes by the mitogens concanavalin A or phytohemagglutinin.[30,92-95] Spadari et al.[96] monitored the activities of DNA polymerases-α, -β, and -γ in rabbit spleen lymphocytes that were stimulated by concanavalin A. It was found that polymerase-α activity increased in cells exposed to this mitogen, implying that the rise in polymerase-α is a prerequisite for DNA replication. In contrast, levels of polymerase-β did not change during cell division, and the activity of polymerase-γ declined.[96] Exposure of spleen cells to erythropoietin induced a fourfold increase in the activity of polymerase-α.[97] That increase was completely abolished by introduction of colcemid for 12 to 24 hr, suggesting that polymerase-α is rapidly turning over in vivo, and that its synthesis is dependent on continuous cell cycling.

Several authors reported on a severalfold specific increase in the activity of polymerase-α as cells in culture are stimulated to divide. When quiescent BHK-21/C13 cells are stimulated to proliferate by exposure to serum, their level of polymerase-α increases.[98,99] However, temporal discrepancy was noted between the rise and fall of polymerase-α activity and DNA synthesis (see below).[99] Similarly, during the accumulation of mouse FM3A cells at S phase after synchronization with 5-fluoro-2′-deoxyuridine, levels of nuclear polymerase-α increased threefold.[100]

The above-described observations on the specific increase in polymerase-α activity in both dividing cultured cells and proliferating tissues could not definitively distinguish *de novo* synthesis of polymerase-α following cell stimulation from postsynthetic activation of the enzyme. Abolishment of the increased activity of polymerase-α by inhibitors of protein synthesis suggested, but did not prove, that enhanced synthesis of enzymic protein occurred in dividing cells.[87,97] Use of the recently developed monoclonal antibodies against polymerase-α has provided an independent criterion to demonstrate that *de novo* synthesis of polymerase-α accompanies the induction of cell proliferation. Monoclonal antibodies directed against human cell polymerase-α stained 95% of the nuclei of highly proliferative transformed KB cells.[43] Cultured mammalian cells were labeled with anti-polymerase-α immunoglobulin in actively proliferating cell cultures, but in cultures containing contact-inhibited quiescent cells there was no apparent immunostaining, implying absence of polymerase-α protein (Figure 1).

Similarly, the level of immunofluorescence in nuclei of chick cells exposed to FITC-labeled monospecific antibodies against polymerase-α was greater in dividing cells than in contact-inhibited quiescent cells.[45] No difference in levels of immunolabeled polymerase-β was observed, however, in the two cell populations.[45] Immunofluorescence in lymphocytes exposed to fluorescent monoclonal antibody against polymerase-α stained <0.5 and 86.8% of the cell population at 19 and 74 hr, respectively, after exposure to the mitogen phytohemagglutinin.[47] The fluorescence-positive cells corresponded to those cells that were labeled with [^{3}H]-thymidine after exposure to the mitogen. However, arrest of cells at G_1 by hydroxyurea or aphidicolin, which inhibited DNA synthesis, did not lead to a decrease in immunofluorescence.[47] It seems, therefore, that high levels of polymerase-α persist at G_1, and thus a relatively long half-life of the enzyme is implied.

Previous study of the half lives of α-, β-, and γ-polymerases in HeLa cells that were exposed to inhibitors of protein synthesis indeed suggested an extended lifetime for polymerase-α.[101] In HeLa cell cultures, which were exposed to cycloheximide for 24 hr, the activity of polymerase-α decreased by 50%, and the residual enzyme behaved as a modified polymerase-α. In view of the plethora of results on partial proteolysis of polymerase-α (cf. Chapter 3) and loss of fidelity of the polymerase in contact-inhibited cells (Chapter 8), and for partially degraded enzyme (Chapter 9), further characterization of such modified subspecies of the polymerase may yield a clue to changes in the enzyme during cell cycling. Polymerase-β and -γ were found to be more labile, declining in activity by 30 to 40% after a 7-hr period of exposure to the protein synthesis inhibitor.[101]

c. Studies with Synchronized Cultured Cells

A detailed analysis of cell cycle-dependent modulations of DNA polymerase activities may be obtained by the study of tightly synchronized cultured cells. In an early report, Chang and Bollum[102] noted that the activity of the 3.4-S β-type polymerase remained unaltered during the different stages of the growth of cultured mouse L cells. In contrast, the level of the 6- to 8-S α-class polymerase varied by 5- to 12-fold in parallel with changes in DNA synthesis. Spadari and Weissbach[103] reported that the activity of synchronized HeLa cell polymerase-α increased continuously during S phase, attaining a twofold maximum increase at the end of the S period. Polymerase-β activity did not change during cell cycling, whereas the activity of polymerase-γ became elevated just prior to commencement of S phase and increased 2.5-fold during the first quarter of that stage.[103] Somewhat different results were obtained by Chiu and Baril[104] who reported that the activity of nuclear polymerase-α increased in synchronized HeLa cells between the G_1 phase and mid-S period, and declined 7- to 10-fold between late S and G_2. Again, the activity of polymerase-β remained unaltered throughout the cell cycle. Exposure of the synchronized HeLa cells to cycloheximide abolished the increase in polymerase-α activity but also reduced the levels of polymerase-β.[104] The interpretation of these studies is complicated by possible cell-cycle related changes in the avidity of polymerase-α to chromatin, which may affect its measured activity with exogenous DNA primer template. Another complication is the long half-life of polymerase-α, which may contribute to its extended activity after cessation of DNA replication.

3. DNA Replication in Cell Mutants with Modified DNA Polymerase-α

The availability of bacterial mutants defective at various loci, which control defined aspects of DNA replication, has greatly facilitated research on the molecular details of prokaryotic DNA replication.[1] The most definitive proof for the involvement of polymerase-α in replication of animal cell chromosomal DNA will be provided with the isolation of specific mutations at the polymerase locus which result in conditional changes in replication of DNA. Such cell mutants are now becoming available, and fragmentary information exists on aspects of DNA replication in these mutant strains. Tsai et al.[105] provided a first genetic clue for the role of DNA polymerase-α in DNA replication. These authors isolated a temperature-sensitive mutant (tsT244) of a mouse FM3A cell line that ceased to replicate its genome at a restrictive elevated temperature. The progressive decreasing rate of DNA synthesis at the nonpermissive temperature was concurrently accompanied by a decline in the intracellular activity of polymerase-α. In contrast, the activities of thymidine kinase and thymidylate kinase decreased at a rate lower than DNA replication. Although levels of β- and γ-polymerases were not monitored in the mutant, it was implied that heat lability of polymerase-α was responsible for the temperature sensitivity of the DNA replication process.[105] To our knowledge, the correlation between polymerase-α activity and DNA replication was not further presented for the tsT244 mutant FM3A cell. Moreover, temperature sensitivity for purified polymerase-α from this mutant has not been demonstrated, and the possibility exists that the heat lability lesion lies with another protein.

Liu and associates[106,107] isolated a series of mutants of Chinese hamster cultured V79 cell line that exhibited resistance to the polymerase-α inhibitor aphidicolin. DNA polymerase-α, purified from one such mutant designated Aph^r-4-2, was found to be resistant to aphidicolin. The mutation reduced tenfold the Km of the mutated polymerase-α for dCTP, thus increasing the K_i for aphidicolin.[107] The augmented affinity of the enzyme to dCTP provides a mechanistic explanation for its increased sensitivity to aphidicolin. A correlation between the genetic change in polymerase-α and DNA replication in vivo was possibly manifested in the extended division time of the mutant strain, which was not exhibited by wild type or revertant cells.[106] Other phenotypic aspects of the aphidicolin-resistant mutant cell strongly indicate that polymerase-α takes part in DNA repair synthesis (cf. Chapter 7, Section III.B).

It is clear from the above results that genetic evidence in support of the role of polymerase-α in DNA replication is still largely missing. Isolation of additional animal cell mutations in different polymerase genes will be important to unambiguously substantiate the role of each polymerase in different aspects of DNA metabolism.

4. Is DNA Polymerase-α a Component of a Multienzyme Replicative Complex?

Enzymes and proteins that act in concert in the replication of bacterial DNA are complexed together in a multiprotein complex that has been termed a replisome.[1] In phage T4-infected *Escherichia coli,* a series of enzymes involved in nucleotide processing and DNA synthesis are also associated in a high-molecular-weight aggregate.[108-110] Furthermore, dispersed reports suggested that multienzyme complexes for synthesis of DNA may be present in a variety of animal cells,[111-115] in cells infected with different types of viruses,[113,115-117] as well as in yeast cells.[118-120] The various observations on aggregates that direct DNA synthesis in eukaryotes have been sustained experimentally and put forth in a comprehensive model by Reddy and Pardee.[40,121-124] It has been proposed that DNA polymerase-α is complexed within the replicatively active cell nucleus with other enzymes that convert early precursors to immediate substrates for DNA synthesis, and that the polymerase acts in coordination with these other enzymes to replicate DNA during the S phase. In this section, results that support the concept of a multienzyme complex in animal cells are evaluated, and possible inadequacies of the experimental evidence for the complex are discussed.

Reddy and Pardee[40] presented results that suggest the existence of a multienzyme replicative complex in animal cells and coined the term *replitase* to denote that complex. Three elements were included in the initial observations made by these authors:[40] (1) cytochalasin B enucleation of CHEF/18 Chinese hamster cultured cells reveals that several enzymes involved in the metabolism of nucleotides and DNA such as thymidine kinase, dCMP kinase, ribonucleoside diphosphate reductase, dihydrofolate reductase, and DNA polymerase were confined mainly to the cytoplasm during the G_1 phase but relocated to nuclei toward the S period; (2) during the G_2 phase, this set of enzymes behaved as dissociated proteins as assessed by velocity sedimentation, whereas during the S period, they were isolated from karyoplasts as coaggregatants, and (3) activation of ribonucleoside diphosphate reductase by dithiothreitol led to preferential incorporation of labeled ribonucleoside diphosphates over deoxyribonucleoside triphosphates into alkali-resistant DNA. A model was codified to unify the above set of observations. It was hypothesized that enzymes, which are associated with synthesis of substrates for

DNA replication, as well as DNA polymerase, are sequestered to the cytoplasm during the G_1 phase but relocate to the nucleus toward the commencement of the S period and form together therein a functional multienzyme nuclear complex, the replitase. Futhermore, on the basis of observation (3), it was suggested that the replitase preferentially channels ribonucleosides into DNA by virtue of their efficient reduction to deoxyribonucleosides through the action of ribonucleoside diphosphate reductase.[40]

The initial suggestion of the replitase model was tested in subsequent experiments. It was reported that thymidylate synthetase activity increased in karyoplasts of hamster cells as they progressed from G_0/G_1 phase into the S period. Thus the confinement of that enzyme to the nucleus was due to its association with replitase.[121] In another work it was shown that lysolecithin-permeabilized Chinese hamster cells rapidly reduced [^{14}C]-labeled rCDP into dCDP, and the labeled precursor was incorporated into DNA.[122] The label that was incorporated into DNA was not diluted by excess unlabeled dCDP or dCTP, indicating preferential channeling of the ribonucleoside diphosphate precursor into DNA. Further, inhibitors of ribonucleoside reductase, dATP, or dTTP also inhibited appearance of the label from rCDP to DNA. It was suggested, therefore, that through the action of the replitase-associated reductase, kinase, and DNA polymerase, rCDP is channeled and efficiently converted into dCDP, dCTP, and DNA, respectively, without being admixed with the general cellular pool of dNDPs/dNTPs.[122]

In an ensuing work,[123] a multienzyme aggregate from sonicated nuclei of S phase cells was isolated by velocity sedimentation. The aggregate, which was considered to represent replitase, contained activities of DNA polymerase, DNA methylase, rNDP reductase, dihydrofolate reductase, and thymidylate synthetase, but was shown to be devoid of serendipitously associated mitochondrial enzymes or cytoplasmic proteins. Labeled rCDP was incorporated into DNA by concatenated steps in the isolated aggregate, and a large excess of unlabeled dCDP did not affect the rate of incorporation of the label in rCDP into DNA. Thus preferential incorporation of ribonucleoside precursors into DNA occurred as a result of reduction and subsequent polymerization by reductase and polymerase, respectively, acting in concert within the replitase.[123] Further, no exogenous DNA template was required for replitase-directed DNA synthesis, and it was, therefore, presumed that replitase is bound to fragments of nuclear DNA. Electron microscopy of the isolated aggregates revealed particles of 24 to 30 nm with an estimated molecular size of $\sim 5 \times 10^3$ kdaltons, a value which was also corroborated by acrylamide-agarose electrophoresis analysis of the replitase particles.[123]

Indirect evidence for the concerted interaction between enzymes which comprise the replitase particle was provided by demonstrating that specific inhibitors of ribonucleoside reductase, topoisomerase, and polymerase-α all inhibited thymidylate synthetase activity in nuclei of S-phase cells.[124] Hence hydroxyurea, novobiocin, and aphidicolin all suppressed the activity of thymidylate synthetase in intact nuclei of S-phase cells, but did not affect it in soluble cell extracts. It was suggested, therefore, that the physical association between ribonucleoside reductase, topoisomerase, polymerase-α, and thymidylate synthetase within the replitase results in allosteric interaction between the enzymic proteins, so that inhibition of one will suppress the activity of another.[124]

Independent evidence for the existence of a multienzyme precursor-processing and polymerizing replicative complex was provided in studies on human lymphoblastoid cells by Wickremasinghe et al.[125,126] It was found that gently lysed human HPB-ALL cells catalyzed incorporation into DNA of labeled distal precursors thymidine, dTMP, and UMP.[125] Kinetics of incorporation suggested that the precursors do not diffuse freely in the cell, but rather are channeled directly into DNA after being properly reduced and phosphorylated. Furthermore, presumed inhibition of polymerase-α by ara CTP abolished incorporation of labeled dTMP into DNA, thus suggesting that the flow of precursor into DNA is regulated by the activity of DNA polymerase. Gel filtration of cell extracts on Sepharose 6B[125] or Sephacryl® S200 yielded an aggregate of 230 to 440 kdaltons, which retained the capability to incorporate distal precursors into DNA.[126] Levels of [^{3}H]dTTP derived from [^{3}H]dTMP or [^{3}H]-thymidine exogenously added to the high-molecular-weight aggregate were low, implying rapid direct flow of precursors into DNA.[126]

The replitase model for the coordinated and controlled replication of DNA is highly attractive in that it provides a theoretical and empirical framework for cell cycle-regulated and concerted activity of precursor-processing and DNA polymerizing enzymes of animal cells. However, some aspects of the supporting experimental evidence for this model have been challenged. Using either Reddy and Pardee's original permeabilized CHEF/18 cells or 3T6 cells, Spyrou and Reichard[127] disputed the claim that rCDP is preferentially incorporated into DNA by showing that under the original experimental conditions, the precursor was incorporated into RNA that was mistakenly identified as DNA. Hence it was contended that inadequate alkali treatment of nucleic acids led to persistance of RNA, whereas full hydrolysis of RNA revealed that labeled rCDP was not found in residual DNA.[127] It was suggested that no single multienzyme complex in mammalian cells catalyzes both the synthesis of deoxynucleoside triphosphates and their polymerization into DNA.

Another aspect of the evidence for cell cycle-dependent formation of replitase is the coordinate intracellular relocalization of DNA polymerase and other components of the multienzyme complex toward the initiation of S phase.[40] However, evidence obtained recently by immunochemical methods indicates that DNA polymerase-α from various cell types persists in nuclei throughout the G_1, S, and G_2 periods, and only for the short period of nuclear membrane dissolution during mitosis is it found in the cytoplasm.[43,45,47] These results contradict the observed confinement of a major proportion of the activity of DNA polymerase to the cytoplasm of G_1-phase hamster cells.[40] These discrepant results suggest that further experimental work is required to establish the existence of replitase in animal cells and to elucidate the molecular details of the possible association of polymerase-α with such replicative complex.

Possible existence of (bacterial) replisome-like complexes in animal cells was recently suggested by Ottinger and Hübscher.[128] Using fresh calf thymus tissue and taking precautions to minimize proteolysis, these authors fractionated polymerase-α by phosphocellulose chromatography of extracts into four disparate peaks of activity. All four peaks behaved as high-molecular-weight aggregates displaying sedimentation coefficients >11.3 S, and all were capable of replicating in vitro long stretches of primed single-strand viral DNA. In addition to polymerase-α activity, the first peak included activities of double-strand DNA-dependent ATPase, topoisomerase type II, RNase H, and $3' \rightarrow 5'$ exonuclease. The other three peaks contained, in addition to polymerase-α and at various proportions, DNA primase, RNase H, topoisomerase type II, DNA methylase, and $3' \rightarrow 5'$ exonuclease. In light of the types of enzymes that comigrated with the first peak of polymerase-α, it was suggested that it represented a replisome-like complex, which was responsible for leading DNA strand synthesis. The other three peaks were suggested as being involved in lagging-strand replication.[128] These preliminary observations require careful analysis of each peak to investigate the nature of association between polymerase-α and the other enzyme activities, and to exclude the possibility of fortuitous comigration of unrelated activities. Most importantly the presence of each of the different components in stoichiometric amounts has to be established.

B. Participation of Polymerase-α in Genomic DNA Replication — Studies with Selective Inhibitors

1. Studies with Intact Cells

The selective inhibitor most advantageously used in the study of the role of DNA polymerase in the replication of nuclear DNA is aphidicolin. This antibiotic specifically inhibits the activity of polymerase-α, but does not affect polymerase-β and -γ. In animal cells that are permeable to it, aphidicolin also selectively and reversibly arrests replication of genomic DNA.[129-134] (Cf. also Chapter 3, Section VI.B for a discussion of the response of cells and isolated polymerase-α to aphidicolin). The link between the inhibition of polymerase-α activity and arrest of DNA replication in vivo was demonstrated by Pedrali-Noy et al.[131] It was shown that exposure of HeLa cells to aphidicolin blocked the cells on the brink of the S period, and subsequent removal of the drug allowed synchronous entry of the cell population into the replicative phase. In an autoradiographic study of DNA synthesis in concanavalin A-stimulated lymphocytes, it was further demonstrated that exposure of cells to aphidicolin led to inhibition of nuclear but not mitochondrial DNA synthesis.[134] Thus DNA polymerase-α was implicated in the replication of genomic DNA but not of mitochondrial DNA. This conclusion depends on the supposition that mitochondria are as permeable to aphidicolin as intact cells. Exposure of oocytes from the frog *X. laevis* to aphidicolin led to inhibition of amplification of the nucleolar extrachromasomal genes for ribosomal RNA.[135] It was thus concluded that polymerase-α of *X. laevis* participates in the amplification of ribosomal DNA. Exposure of cultured rat fibroblasts to aphidicolin at a concentration (1.4 μg/mℓ), which inhibited 90 to 95% of the incorporation of labeled thymidine into DNA, led to accumulation of DNA intermediate of nearly replicon size (2 to 5 × 10^4 kdaltons), whereas synthesis of Okazaki fragments continued.[136] It has been demonstrated that aphidicolin did not cause degradation of DNA, and that it inhibited synthesis of DNA in vivo by arresting replicon-size chain elongation.[136] Hence, polymerase-α was implicated in that reaction.

Butylanilinouracil[137] and butylphenylguanine[138] were shown to inhibit, at comparable concentrations, both purified HeLa cell DNA polymerase-α and in vivo replication of DNA. However, a study of the effects of these two analogs (cf. Chapter 3, Section VI.C) on growth of several animal cell types revealed that the cells were not uniformly sensitive to the drugs.[139] The variance in the potency of the drugs in inhibiting growth of different cells could be due to differences in cell permeability to the analogs, to different rates of in vivo inactivation of the drugs, or to variation in the in vivo sensitivity of polymerase-α to the inhibitors.[139]

2. Studies with Isolated Replicative Systems

To circumvent the problem of impermeability of intact animal cells to some selective inhibitors of DNA polymerases, several authors used subcellular systems that continue DNA polymerization and that are permeable both to labeled substrates for DNA synthesis and to inhibitory drugs.

In lysolecithin-permeabilized cultured hamster kidney cells, replicative synthesis of DNA is blocked, as is isolated polymerase-α, by *N*-ethylmaleimide, ara CTP, or a high concentration of salt.[140] Replication of DNA in lysates of HeLa cells and the activity of isolated polymerase-α are similarly resistant to d_2TTP at concentrations that were strongly inhibitory to polymerase-β or -γ.[141]

Isolated nuclei or nuclear extracts have been used as in vitro replicative systems to categorize replicative DNA polymerase. Several studies indicated close correlation between the sensitivity of purified polymerase-α to aphidicolin and the potency of that inhibitor in suppressing DNA synthesis in isolated nuclei.[142-144] DNA synthesis in nuclear extracts obtained from monkey CV-1 cells is completely inhibited by the polymerase-α inhibitors aphidicolin and ara CTP, and is as resistant to d_2TTP as is DNA polymerase-α. It was similarly found that ara CTP reduces the rate of elongation of primary DNA fragments in nuclei isolated from S-phase HeLa cells. [146] Subsequently, it was reported that DNA synthesis in isolated S-phase HeLa cell nuclei was as sensitive to ara CTP and as resistant to d_2TTP as DNA polymerase-α.[147] Burk et al.[148,149] studied effects of several inhibitory agents on synthesis of DNA in isolated nuclei, which were derived from cultured hamster kidney cells. It has been found that treatment of the nuclei with the detergent Brij® 58 inhibited by about 90% both nuclear DNA replication and activity of polymerase-α, whereas purified polymerase-β was slightly stimulated by the same concentrations of the detergent.[148]

Furthermore nuclear DNA synthesis and isolated polymerase-α were similarly inhibited by *N*-ethylmaleimide, pHMB, novobiocin, heparin, and phosphonoacetic acid. Partially purified polymerase-β was more resistant to each of these agents than either isolated polymerase-α or DNA synthesis in nuclei.[149] Although all the above studies indicate a strong correlation between the sensitivities of polymerase-α and DNA synthesis in isolated nuclei to a variety of inhibitors, caution is required in interpreting these results. First, it is doubtful whether DNA synthesis in isolated nuclei fully reflects in vivo replication of DNA. Isolated nuclei and cell lysates appear not to initiate new chains of DNA, but rather conduct limited elongation in vitro of strand whose synthesis was previously initiated in vivo.[148] In fact, it has been shown that DNA synthesis in nuclei isolated from proliferating cells of regenerating rat liver is not higher than synthesis in nuclei derived from mature undividing liver cells.[150] Furthermore, most of the synthetic activity in the isolated liver nuclei reflected mainly activity of chromatin bound polymerase-β. It has also been suggested that d_2TTP and *N*-ethylmaleimide may be inhibitory to DNA polymerase-β while complexed with chromatin in the nuclei.[150] Previous works[151,152] indeed suggested that the major polymerase activity, which was responsible for DNA synthesis in nuclei isolated from S-phase L cells, was DNA polymerase-β. Yet, extraction of isolated nuclei of HeLa cells with 0.2 *M* KCl indicated that residual, tightly bound polymerase-α may still be responsible for part of the synthetic activity of the isolated nuclei.[153] Hence it seems that DNA synthesis in isolated nuclei is a complex process whose relationship to DNA replication in vivo is not clear, and this complicates interpretation of results of inhibitor effects on nuclear systems.

An in vitro system, which has proven to be useful in demonstrating the role of DNA polymerase-α in replicative synthesis of DNA, is that of isolated SV40 or polyoma virus chromatin. A correlation between the replication of these viral genomes and DNA polymerase-α activity was first suggested by the observation that the onset of viral DNA synthesis in parvovirus-infected cells was accompanied by concomitant elevation in the activity of polymerase-α[154-156] Furthermore, inhibitor sensitivities of viral genome-associated polymerases indicated that DNA polymerase-α and -γ, but not -β, copurified with replicating SV40 minimicrosomes, and replicative activity was related to the level of chromatin-bound polymerase-α.[157-161] The association of DNA polymerase-γ, in addition to polymerase-α with replicating SV40 chromatin, led to a careful analysis of the polymerase that is responsible for replicative synthesis of viral DNA. By using *N*-ethylmaleimide, aphidicolin and d_2TTP, it was demonstrated that viral chromosome replication is inhibited similar to that of isolated polymerase-α and clearly different from either -β or -γ polymerases.[162] Furthermore, gradual salt extraction of polymerase-α from the viral chromatin indicated that the residual synthetic activity was related to the level of remaining chromatin-bound polymerase.[162] Other investigators indicated that inhibition of viral DNA synthesis by chloride ions resembled inhibiton of polymerase-α, whereas both viral chromatin replication and isolated polymerase-α were stimulated by acetate ions.[161] Interestingly, it has been shown recently that aphidicolin, which inhibits polymerase-α reversibly, arrests irreversibly the replication of

SV40 DNA in infected cells.[163] It appeared that extended exposure of SV40 chromosomes to aphidicolin both inhibited polymerase-α and selectively and irreparably impaired the replicating SV40 chromosome. Thus, although polymerase-α is the primary target of aphidicolin in inhibiting SV40 chromatin replication, that drug may also irreversibly prevent initiation of new rounds of viral DNA replication, perhaps by inducing irreversible changes in the viral chromatin.[163]

3. Probing DNA Replication with Monospecific Antibodies Against Polymerase-α

The successes in raising monospecific monoclonal antibodies to polymerase-α now provide a potentially powerful tool to delineate the role of polymerase-α in cellular metabolism of DNA. To our knowledge, at the time of this writing, in only one work was the functional role of polymerase-α in DNA replication demonstrated with the aid of monoclonal antibodies. Miller et al.[164] applied monoclonal antibodies against human KB cell polymerase-α to lysolecithin-permeabilized human HF cells, which conducted replicative DNA synthesis. DNA replication was inhibited specifically and in a dose responsive manner by the monospecific immunoglobulins. The antibody was shown by electron microscopy to be localized in the cell nucleus, and whereas a maximum inhibition of 80% of DNA synthesis was uniformly attained in all the cells, RNA synthesis was not affected by the antibody.[164] This immunochemical approach to the demonstration of the role of DNA polymerase-α in DNA replication is likely to be applied in the near future to the study of the participation of polymerase-α in repair synthesis of DNA, and for the investigation of roles of β- and γ- polymcrases in different aspects of DNA metabolism.

C. Correlating Enzymic Properties of Polymerase-α, -β, and -γ to Properties Required of a Replicative DNA Polymerase

In this section we review some of the properties of α-, β-, and γ- polymerases and speculate on their relevance to requirements for a replicative DNA polymerase. Enzymatic properties of DNA polymerase-α, -β, and -γ were reviewed in detail in Chapters 3, 4, and 5, respectively, and reference to pertinent sections in those chapters is made throughout this segment.

1. Relationships Between Unique Catalytic Properties of Polymerases -α, -β, and -γ and Properties Required of a Replicative DNA Polymerase

a. Processivity

A high degree of processivity is to be logically expected from a replicative DNA polymerase that catalyzes the polymerization of at least several hundred nucleotide-long stretches of DNA (Okazaki fragments). An even greater extent of processivity might be required for synthesis of the leading strand. Among the three major classes of animal cell DNA polymerases, it is DNA polymerase-γ that displays high processivity in vitro (cf. Chapter 5, Section I.B.2). DNA polymerase-β is distributive or very mildly processive (Chapter 4, Section IV.E), whereas polymerase-α is quasiprocessive, being able to polymerize 5 to 20 nucleotides per template binding event (cf. Chapter 3, Section V.F.1).

The high processivity of DNA polymerase-γ is requisite for its proposed role in the replication of mitochondrial DNA, but it could also reflect an unknown function of that enzyme in the replication of nuclear DNA (see below). It is the relatively low processivity of polymerase-α that is incompatible with its proposed central role in replicative synthesis of chromosomal DNA. Several authors observed that the inherent processivity of polymerase-α can be modified through its interaction with other proteins and cofactors. Thus it was reported that form A1 of calf thymus polymerase-α, which is composed of both catalytic and accessory subunits, was more processive than form C, which is devoid of the auxiliary subunit.[165] It was also demonstrated that the relatively low processivity of *Drosophila melanogaster* polymerase-α on primed phage M13 single strand DNA could be increased 20-fold in the presence of *E. coli* single-strand binding protein.[166] Hence it may be speculated that the inherently modest processiveness of polymerase-α is increased in vivo through its interaction with additional subunits/auxiliary proteins to satisfy requirements for processive replicative DNA synthesis. Furthermore, several authors reported that small molecules such as ATP, spermidine, and pyrophosphate affect the degree of processiveness of polymerase-α in vitro (cf. Chapter 3, Section V.F.1). It is clearly a future task of research on polymerase-α and DNA replication to identify and isolate proteins and cofactors that are able to modify the processiveness of replicative DNA synthesis. It should theoretically be possible to design an assay system for monitoring of processivity of polymerase-α during purification and in the presence of added cell proteins. Such an assay could in turn be used to identify and purify putative cellular cofactors that enhance processiveness.

b. Interaction with Template DNA

There is not much information on the interaction of polymerase-γ with template DNA, but existing experimental data suggest clear distinction between template preferences of DNA polymerases of the α and β classes. Steady-state kinetic analysis indicated that initial binding of polymerase-α to template DNA occurs at mandatory single strand regions, and that the enzyme shows preference for (dT)-enriched single-stranded sites (cf. Chapter 3, Section V.A). In contrast, DNA polymerase-β has no affinity for single-stranded DNA structure, but it strongly binds to nicked duplex DNA, which does not serve as template for polymerase-α (cf. Chapter 4, Section IV.A). One is tempted to speculate that following melting of DNA in vivo at dA-dT-rich regions, polymerase-α binds to the single-strand DNA to initiate synthesis from an RNA primer and to elongate it by polymerizing deoxynucleoside triphosphates. The clear preference of polymerase-α for (dT)-rich single-strand DNA is in line with such a model. The propensity of polymerase-β to nicked DNA duplex is more in line with repair function (see below and Chapter 7, Section V).

c. Interaction with Primer

Among the three classes of animal cell DNA polymerases, the α-type enzymes are the only ones that are able to extend natural RNA primers at a significant rate (cf. Chapter 3, Section V.B). The capability of polymerase-α to utilize RNA primer is supportive of its role in the in vivo replicative extension of primase-synthesized RNA. The tight and specific association of intact α-polymerases from various sources with DNA primase also sustains the contention that it is involved in replicative elongation of RNA primer in vivo (see below).

d. Gap Filling and Long-Stretch DNA Synthesis

Study of the reactivity of DNA polymerases-α and -β with DNA that contained gaps of defined sizes reveal differences in the preference of these two enzymes for gap length and in their ability to fill gaps to completion. Whereas DNA polymerase-β maximally utilizes gaps that are one to ten nucleotides and is able to fill them completely (Chapter 4, Section IV.C). DNA polymerase-α is usually unreactive with small gaps, utilizes gaps of 30 or less nucleotides, but fails to fill them to completion (Chapter 3, Section V.G). These in vitro properties of DNA polymerases-α and -β may be germane to their ability to participate in short- and long-patch repair synthesis of DNA (Chapter 7, Section V.A). However the failure of polymerase-α to fill gaps in DNA to completion is intriguing. It may be that under certain in vivo conditions, repair and/or replicative synthesis of DNA requires the concerted activity of polymerase-α and -β or γ-polymerases, which complete DNA synthesis started by the former. Hurwitz and associates[167] showed that α-polymerases extended RNA primer on ϕX174 DNA by 230 to 290 nucleotides only, but polymerase-β could further elongate those products to full-length RF structure of 4500 nucleotides. Similarly, Matsukage and associates[168] demonstrated that mouse cell polymerase-β was able to extend DNA chains initiated by polymerase-α, thus achieving higher product length than that obtained by polymerase-α alone. It should be noted, however, that several authors reported recently on successful replication of several thousand nucleotides long viral DNA by the action of unassisted polymerase-α (reviewed in Chapter 3, Section V.G). It is thus not clear at present whether concerted action of polymerase-α and other types of DNA polymerase are required during some stages of the in vivo replication of DNA.

e. Summary

In undertaking to determine which of the several classes of animal cell DNA polymerase take part in different aspects of DNA replication, the catalytic properties of these enzymes may be compared in an attempt to resolve which set of characteristics is consonant with those of a replicative enzyme. Yet such an approach may be premature in some respects. First catalytic parameters were by necessity determined for purified DNA polymerases under in vitro assay conditions whose resemblance to in vivo DNA replication conditions is unknown. Second, polymerase-α in particular, and possibly polymerase-γ as well, are multimeric proteins comprised of several subunits. Some of the studies conducted on these enzymes were performed with polymerase preparations that might have included only a partial complement of subunits. Moreover, polymerase-α is extremely sensitive to partial proteolytic digestion, and clearly numerous determinations of the catalytic properties of that enzyme were performed with partially cleaved enzyme preparations. Last, polymerases-α and -β and possibly polymerase-γ, are associated in vivo with auxiliary proteins and cofactors whose total number and functions are not known at present. Many of the properties that were described for these enzymes were ascribed to polymerases that were devoid of part or all ancilliary proteins. It is clear, therefore, that our present conception of the catalytic behaviour of α-, β-, and γ-polymerases is at best only a rough surmise of their enzymatic behaviour in the cell. Model systems that mimic partial replicative reactions using animal cell components have not yet been assembled. Catalytic properties of polymerases that comprise part of such multienzyme systems should be more closely related to in vivo catalysis of DNA polymerization.

2. Association of Polymerase-α with Auxiliary Protein

The central role of DNA polymerase-α in the replicative synthesis of nuclear DNA is strongly corroborated by the specific association of α-type polymerases with replicative proteins and cofactors. That replication of DNA in vivo is initiated in animal cells by short RNA primer segments is amply documented (for review see Chapter 3, Section IV.B). In recent years, it has been repeatedly shown that primer RNA is synthesized in animal cells by a unique RNA polymerase, designated DNA primase, which is distinguishable from the conventional RNA polymerases I, II, and III of eukaryotes. Furthermore, an intimate association between DNA primase and polymerase-α has been demonstrated in a large variety of cell types. As described in Chapter 3, Section IV.B and summarized in Table 2 of that chapter, in most of the cases examined, DNA primase could not be physically dissociated from DNA polymerase-α. However, whereas Hübscher[169] found that primase activity possibly resided within the catalytic core of calf thymus polymerase-α, Lehman and associates[170] showed that primase constituted a subunit of *D. melanogaster* polymerase-α, which was distinguishable from the catalytic polypeptide. Other investigators[171-175] also demonstrated in various cell types a tight association between polymerase-α and DNA primase. Only in the case of mouse hybridoma cells is it possible to completely dissociate DNA primase independantly from polymerase-α.[176] The intimate and specific association between polymerase-α and DNA primase strongly argues for the involvement of the former in the extension of RNA primers during DNA replication in the cell nucleus.

A correlation between intracellular levels of diadenosine-5′,5‴,-P^1,P^4-tetraphosphate (Ap_4A) and DNA replication in vivo has been established in a variety of eukaryotes. Hence, levels of Ap_4A in S phase cells are higher by two to three orders of magnitude than those in G_1/G_0 cells. Furthermore, exposure of permeabilized quiescent cells to Ap_4A-induced replicative-type DNA synthesis in the G_1 cells (for experimental evidence see Chapter 3, Section IV.D).[177-183] It has been suggested, therefore, that Ap_4A may act as a positive pleiotypic affector of DNA replication in animal cells.[183,184] Some experimental evidence links Ap_4A directly to DNA polymerase-α. First, highly purified preparations of polymerase-α exhibit Ap_4A binding activity, which appears to be mediated by a polymerase subunit or ancilliary protein.[177-181] Second, Ap_4A was reported to act in some cases as a primer for in vitro DNA synthesis catalyzed by polymerase-α[181,182] Although, as pointed out in Chapter 3, the nature of interaction between polymerase-α and Ap_4A and the mechanism of Ap_4A mediated stimulation of replication are not yet clear, an intriguing possible regulatory role for Ap_4A in activating polymerase-α in DNA replication is suggested.

An intimate association between polymerase-α and primer-recognition proteins C1C2 has been demonstrated recently.[185-187] These factors increase the in vitro affinity of polymerase-α to the primer stems at low primer to template ratio.[185] It is tempting to speculate that after polymerase-α becomes initially associated with single-stranded DNA at the opened replication fork, primer recognition cofactors enable the enzyme to rapidly and efficiently bind to relatively sparse RNA primers.

III. REPLICATION OF MITOCHONDRIAL DNA BY DNA POLYMERASE-γ

A. Compartmentalization of Polymerase-γ in Mitochondria

The initial conception that mitochondrial DNA polymerase constitutes a fourth separate class of polymerase in animal cells[188] was revised when Bolden et al.[53] found that DNA polymerase-γ was the only detectable polymerase in mitochondria that were purified from mycoplasma-free HeLa cells or from hepatocytes. Subsequently, it has been shown[54] that polymerase-γ was localized in both mitochondria and nuclei of chick cells. Hübscher et al.[55] showed that isolated brain cell synaptosomes, which did not contain nuclei but had mitochondria, displayed activity of polymerase-γ that was indistinguishable from the nuclear γ enzyme. Further, the mitochondrial authenticity of polymerase-γ was indicated by its association with the inner-membrane mitochondrial replication complex (M-band).[58] Indirect evidence was also provided to show correlation between mitochrondrial proliferation and mitochondrial enzyme activity and levels of polymerase-γ.[56,57] Fox et al.[35] and Tanaka and Koike[59] failed to detect polymerase-γ in nuclei, and found all its activity in mitochondria of *X. laevis* oocytes and rat hepatocyte, respectively (see also Section II.A.1 of this chapter). It is clear, therefore, that a considerable portion of the cellular activity of DNA polymerase-γ is deposited in the mitochondrion.

B. Participation of Polymerase-γ in Synthesis of Mitochondrial DNA Studies on the Effect of Selective Inhibitors and on Catalytic Properties

In a preliminary work it was reported that aphidicolin, as well as d_2TTP, inhibited the in vivo replication of mitochondrial DNA.[189] That observation suggested that in addition to the d_2TTP-sensitive polymerase-γ, aphidicolin-susceptible polymerase-α may also take part in the replication of mitochondrial DNA. That observation was, however, disputed by Zimmerman et al.[190] who showed that aphidicolin had no effect on the replication of DNA in purified mitochondria. Similarly, Genskens et al.[134] used autoradiography of intact cells to show that incorporation of labeled thymidine into mitochondrial DNA was not inhibited by aphidicolin, although nuclear DNA synthesis was suppressed by that drug. These reports argue against a role for DNA polymerase-α in the replication of mitochondrial DNA. The high processivity of DNA polymerase-γ and its possible involvement in strand displacement mode of parvovirus DNA synthesis were documented (see Chapter 5, Section I.B.2). It is reasonable to assume that polymerase-γ alone is able to extend a primer on one template strand of the mitochondrial genome with concurrent displacement of the other strand, which eventually forms a single-stranded D loop. After about two thirds of the first strand is copied, the primed displaced strand is copied as well. The recent report that bovine polymerase-γ is able to fully replicate singly primed parvovirus DNA[191] supports the notion that polymerase-γ is capable of replicating the mitochondrial genome. Although it has been assumed that mitochondria fail to repair damage to their DNA, there is adequate experimental evidence to support only lack of UV damage repair in mitochondria.[192] Hence, a role for polymerase-γ in other modes of repair synthesis cannot be ruled out.

IV. CONCLUDING REMARKS

None of the various methods that were employed to elicit roles of the different classes of animal cell DNA polymerases in the in vivo replication of cellular DNA could, by itself, point unambiguously to a specific polymerase as the nuclear or mitochondrial replicative enzyme. Yet the information accumulated by different approaches to the problem compellingly indicate that DNA polymerase-α plays a central role in the replication of nuclear DNA, while DNA polymerase-γ appears to be the only polymerase required for the replication of mitochondrial DNA.

Figure 4 represents an attempt to summarize present-day knowledge and conceptions of the place of DNA polymerase-α in the replicative machinery of animal cells. By necessity, the presented scheme is an oversimplification of complex, and often contradictory, data obtained from different cell types by different methods. In the center of the scheme, the catalytic core of polymerase-α is depicted in close association with primase, which may or may not constitute an integral part of the catalytic polypeptide (Section II.C.2). Other possible subunits whose precise number and function are obscure (Chapter 3, Section II) are not shown. The polymerase is associated with ancillary proteins, notably Ap_4A binding protein and primer-recognition cofactors (Section II.C.2). We speculate that polymerase-α may also be combined with additional auxiliary cofactors such as processivity-enhancing protein(s), template-sequence recognition factor(s), and fidelity-modifying protein(s). There is yet very little information to support the proposition that such cofactors do exist, but theoretical considerations, detailed at various parts of this book, merit a search for such ancillary factors.

Although compelling data indicate that DNA polymerase-α is the *major* replicative enzyme in eukaryotes, it is not clear what role, if any, is played by the nuclear polymerases-β and -γ in the replication of chromosomal DNA. It should be noted that in the special case of endoduplication of DNA in rat trophoblasts there is evidence that polymerase-β may serve as the major DNA duplicating enzyme (cf. Chapter 9, Section II for a full discussion).[193] We have, therefore, depicted these two polymerases as possible participants in the replicative complex. On the right side of Figure 3, some activities that may form a replisome-like particle with polymerase-α are listed. There is yet only preliminary evidence on the possible association between some of these activities

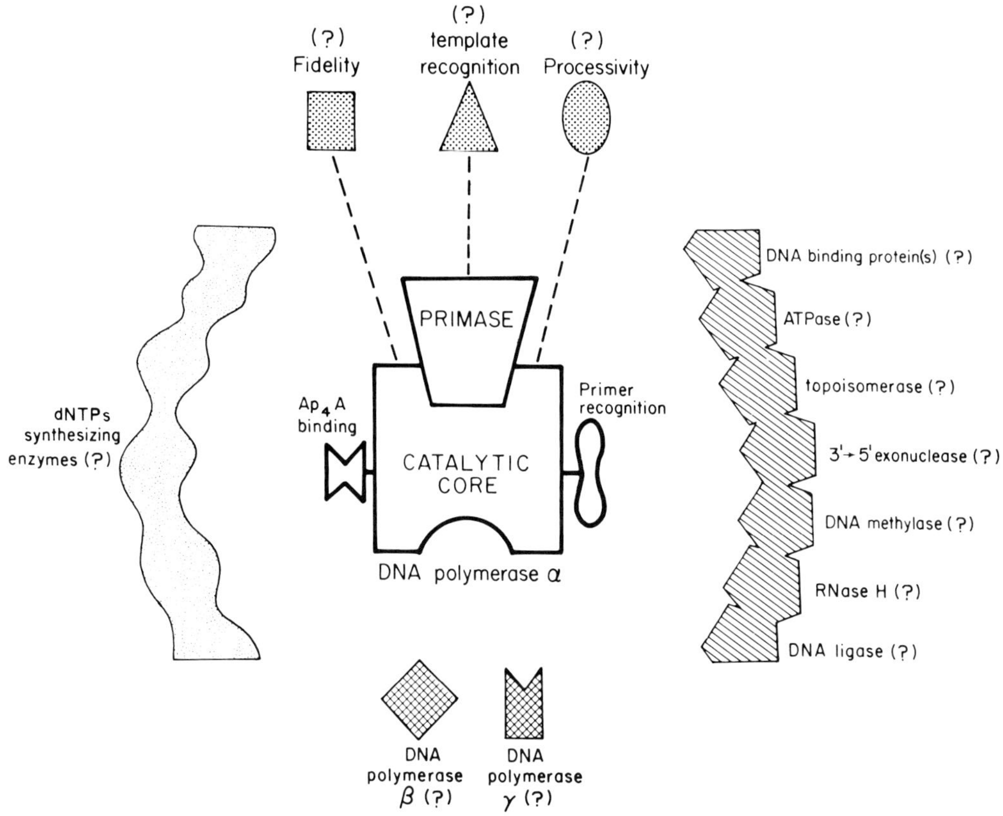

FIGURE 4. Hypothetical structure of animal cell DNA replication complex. A detailed discussion of the structure and possible components of the purported multienzyme complex is presented in the text.

and polymerase-*α*.[128] Also, there may be different replisome types that contain different sets of enzymes to conduct leading- and lagging-strand synthesis.[128] The specific association of polymerase-*α* with part or all of the listed eukaryotic replisome enzymes is only speculative at present. On the left side of Figure 3, we depict the possible association of polymerase-*α* with dNTP synthesizing enzymes to form a "replitase" complex. Data in support of the replitase model and contradictory to it were reviewed in Section II.A.4.

V. UPDATE

A. Association of DNA Polymerase-*α* with the Nuclear Matrix

Evidence was presented in support of the specific association of polymerase-*α* with the nuclear matrix of HeLa cells and the correlation between rates of DNA synthesis in vivo and activity of matrix-bound *α* -polymerase.[194] As reported in the past,[195,196] DNA replication during the S phase proceeds with biphasic kinetics that exhibit a distinct pause between two periods of synthesis. In concordance with these two periods of active replication, Foster and Collins[194] reported that activity of polymerase-*α* in the nuclear matrix also shows two similar peaks. Although attempts to immunologically detect changes in the level of matrix-associated polymerase protein have not succeeded, it has been suggested that differential binding of polymerase to the matrix during S phase is determining the rate of DNA synthesis.[194] Results obtained with nuclear matrices from Ehrlich ascites cells[197] are at variance with the reported association with nuclear matrix of DNA polymerase-α from rat hepatocytes[69,70] and Hela cell.[194] Nuclear matrix isolated from Ehrlich ascites cells contains DNA polymerase activity with unusually high heat stability conferred by its association with DNA and relative sensitivity to d_2TTP and resistance to MalNet that identify it as being mostly a β-class enzyme.[197] Moreover, analysis of DNA products indicates that the in vitro matrix-directed synthesis of DNA reflects repair rather than replicative synthesis.[197]

B. Intracellular Levels of DNA Polymerases During Cell Cycling

The change in the activities of DNA polymerases-α, -β, and -γ, during HeLa cell cycle have been reinvestigated in synchronized cultures by assaying DNA polymerization activities in cell extract under defined conditions that are presumably specific for each type of enzyme.[198] It has been reported that the activity of polymerase-α increases during S phase and declines as the cells enter G_2. The activity of polymerase-γ increases between the S and G_2 phases — a time period during which replication of mitochondrial DNA takes place. Surprisingly, however, the activity of polymerase-β was also found to increase during the early stages of the S phase.[198] The molecular events that are at the basis of this change in polymerase-β activity are a matter for speculation.

C. DNA Replication in Cell Mutants with Modified DNA Polymerase-α

A substantiation of the central role of polymerase-α in the replication of genomic DNA has been obtained with the isolation and characterization of a mouse FM3A cell mutant in which both DNA replication in vivo and polymerase-α became similarly temperature sensitive.[199] This mutant cell strain, designated tsFT20, was isolated from FM3A mammary carcinoma cultured cells that were mutagenized with *N*-methyl-*N'*-nitro-*N* nitrosoguanine. At 30°, growth of tsFT20 mutant cells was arrested at the G_1/S boundary and in S phase in close resemblance to wild type cells that were exposed to aphidicolin. Most importantly, polymerase-α from the mutant cell exhibited augmented sensitivity to exposure to 39°. That temperature sensitivity of DNA replication in tsFT20 was due to heat lability of α-polymerase was further demonstrated with the finding that in revertant cells of the tsFT20 strain, both DNA replication and DNA polymerase-α regained heat stability.[199] In addition to providing supporting evidence for the cardinal role of polymerase-α in DNA replication, temperature sensitive mutations in this enzyme provide a powerful tool for additional lines of investigations. Thus, for instance, tsFT20 strain provided important additional information for the identification of human chromosome X as the for polymerase-α gene (cf. Chapter 3, Section VII).

D. Is DNA Polymerase-α a Component of a Multienzyme Replication Complex?

A fundamental argument in favor of the "replitase" multienzyme complex is evidence presented by Reddy and Pardee[40,122] on the preferential channeling of ribonucleoside triphosphates into DNA synthesized by permeabilized cells. These reports were challenged by Spyrou and Reichard[127] who found that most of the radiolabeled rCDP that was supplied to permeabilized cells was incorporated into alkali labile nucleic acid that was identified as RNA rather that DNA. In a careful reassessment of this controversy, labeled rCDP has been shown to be incorporated in permeabilized hamster cells into low molecular weight double-stranded nucleic acid in which 60% of the precursor existed as 5'-dCMP and 40% as 5'-rCMP.[200] Specific inhibitors of ribonucleoside reductase (hydroxyurea) and polymerase-α (BuPGdR-TP) blocked the incorporation of rCDP into this hybrid product.[200] The molecular nature of the RNA-DNA hybrid is not clear, but it has been proposed that the RNA portion of the molecule may represent primer RNA chains.[200]

REFERENCES

1. Kornberg, A., *DNA Replication,* W. H. Freeman, San Francisco, 1980 (supplement 1982).
2. Lloyd, D., Poole, R. K., and Edwards, S. W., *The Cell Division Cycle,* Academic Press, New York, 1982.
3. Lewin, B., *Gene Expression,* Vol. 2, 2nd ed., Wiley & Interscience, New York, 1980.
4. Weissbach, A., Schlabach, A., Fridlender, B., and Bolden, A., DNA polymerases from human cell, *Nature (London) New Biol.,* 231, 167, 1971.
5. Chang, L. M. S., and Bollum, F. J., Low molecular weight deoxyribonucleic acid polymerase from rabbit bone marrow, *Biochemistry,* 11, 1264, 1972.
6. Garrett, R. J. B. and Bollum, F. J., DNA polymerase activity in agarose column fractions of rat liver cytoplasm, *Biochem. Biophys. Acta,* 312, 45, 1973.
7. Hecht, N. B., Interconvertability of mouse DNA polymerase activities derived from the nucleus and cytoplasm, *Biochem. Biophys. Acta,* 312, 471, 1973.
8. Craig, R. K. and Keir, H. M., Deoxyribonucleic acid polymerases of BHK-21/C-13 cells: heterogeneity, molecular asymmetry, and subcellular distribution of the enzymes, *Biochem. J.,* 145, 225, 1975.
9. Sedwick, W. D., Shu-Fong, T., and Korn, D., "Cytoplasmic" deoxyribonucleic acid polymerase. Structure and properties of the highly purified enzyme from human KB cells, *J. Biol. Chem.,* 250, 7045, 1975.
10. Roufa, D. J., Moses, R. E., and Reed, S. J., The DNA polymerases of Chinese hamster cells. Subcellular distribution and properties of two DNA polymerase, *Arch. Biochem. Biophys.,* 167, 547, 1975.
11. Byrnes, J. J., Downey, K. M., Jurmark, B. S., and So, A. G., Reticulocyte DNA polymerase, *Nature (London),* 248, 687, 1974.
12. Tarrago-Litvak, L., Desgranges, C., Araya, A., and Litvak, S., DNA polymerase of anucleated cells. Isolation and characterization of DNA polymerases from human platelets, *Eur. J. Biochem.,* 93, 271, 1977.
13. Mazia, D. and Hinegardner, R. T., Enzymes of DNA synthesis in nuclei of sea urchin embryos, *Proc. Natl. Acad. Sci. U.S.A.,* 50, 148, 1963.
14. Loeb, L. A., Fansler, B., Williams, R., and Mazia, D., Sea urchin DNA polymerase. I. Localization in nuclei during rapid DNA synthesis, *Exp. Cell Res.,* 57, 298, 1969.
15. Fansler, B. and Loeb, L. A., Sea urchin nuclear DNA polymerase. II. Changing localization during early development, *Exp. Cell Res.,* 57, 305, 1969.
16. Loeb, L. A. and Fansler, B., Intracellular migrationof DNA polymerase in early developing sea urchin embryos, *Biochim. Biophys. Acta,* 217, 50, 1970.
17. Loeb, L. A., Molecular association of DNA polymerase with chromatin in sea urchin embryos, *Nature (London),* 226, 448, 1970.
18. Fansler, B. and Loeb, L. A., Sea urchin nuclear DNA polymerase. IV. Reversible association of DNA polymerase with nuclei during the cell cycle, *Exp. Cell Res.,* 75, 433, 1972.
19. Loeb, L. A., Purification and properties of deoxyribonucleic acid polymerase from nuclei of sea urchin embryos, *J. Biol. Chem.,* 244, 1672, 1969.
20. Hobart, P. M. and Infante, A. A., Persistent cytoplasmic location of DNA polymerase-β in sea urchins during development, *Biochem. Biophys. Acta,* 607, 256, 1980.
21. Shioda, M., Nagano, H., and Mano, Y., Cytoplasmic location of DNA polymerase-α and -β of sea urchin eggs, *Biochem. Biophys. Res. Commun.,* 78, 1362, 1977.
22. Shioda, M., Nagano, H., and Mano, Y., Transition of DNA polymerase-α and endoplasmic reticulum during gastrulation of the sea urchin, *Dev. Biol.,* 91, 111, 1982.
23. Lynch, W. E., Surrey, S., and Lieberman, I., Nuclear deoxyribonucleic acid polymerases of liver, *J. Biol. Chem.,* 250, 8179, 1975.
24. Eichler, D. C., Fisher, P. A., and Korn, D., Effect of calcium on the recovery and distribution of DNA polymerase-α from cultured human cells, *J. Biol. Chem.,* 252, 4011, 1977.
25. Nishioka, N., Matsukage, A., and Takahashi, T., On the binding of DNA polymerase-α to nuclear structure in mouse myeloma MOPC 104E, *Cell Struct. Funct.,* 2, 61, 1977.
26. Matsukage, A., Nishioka, N., Nishizawa, M., and Takahashi, T., Differential extraction of DNA polymerase from the nucleus of the mouse myeloma cell, *Cell Struct. Funct.,* 4, 295, 1979.
27. Ono, Y., Enomoto, T., Hanaoka, F., and Yamada, M., Separation of two forms of HeLa DNA polymerase α with different binding affinity to DNA, *Gann,* 69, 207, 1978.
28. Knopf, K. W. and Weissbach, A., A study of DNA synthesis in chromatin isolated from HeLa cells, *Biochemistry,* 16, 3190, 1977.
29. Kaftory, A. and Fry, M., Highly efficient copying of single stranded DNA by eukaryotic cell chromatin, *Nucl. Acids Res.,* 5, 2679, 1978.
30. Schlaeger, E. J., van Telgen, H.-J., Klempnauer, K.-H., and Knippers, R., Association of DNA polymerase with nucleosomes from mammalian cell chromatin, *Eur. J. Biochem.,* 84, 95, 1978.

31. Yagura, T. and Seno, T., DNA synthesis in isolated chromatin. Nature of activities and relationship to kinetics of DNA polymerase release from chromatin DNA, *Biochem. Biophys. Acta*, 608, 277, 1980.
32. Fry, M., Silber, J., Martin, G. M., and Loeb, L. A., Delayed and reduced cell replication and diminishing levels of DNA polymerase α in regenerating liver of aging mice, *J. Cell Physiol.*, 118, 225, 1984.
33. Gurney, T., Jr. and Foster, D. N., Nonaqueous isolation of nuclei from cultured cells, *Methods Cell Biol.*, 16, 45, 1977.
34. Foster, D. N. and Gurney, T., Jr., Nuclear location of mammalian DNA polymerase activities, *J. Biol. Chem.*, 251, 7893, 1976.
35. Martini, G., Tato, B., Attardi-Gandini, D., and Tocchini-Valenti, G. P., Nuclear localization of DNA polymerase α in *Xenopus laevis* oocytes, *Biochem. Biophys. Res. Commun.*, 72, 875, 1976.
36. Grippo, P., Taddei, C., Locorontondo, G., and Gambino-Guiffrida, A., Cellular localization of DNA polymerase activities in full grown oocytes and embryos of *Xenopus laevis*, *Exp. Cell Res.*, 109, 247, 1977.
37. Fox, A. M., Breaux, C. B., and Benbow, R. M., Intracellular localization of DNA polymerase activities within lage oocytes of the frog *Xenopus laevis*, *Dev. Biol.*, 80, 79, 1980.
38. Nagano, H., Okamo, K., Ikegami, S., and Katagiri, C., Changes in the intracellular location of DNA polymerase-α during oocyte maturation of the toad *Bufo bufo japonicus*, *Biochem. Biophys. Res. Commun.*, 106, 683, 1982.
39. Herrick, G., Spear, B. B., and Veomett, G., Intracellular localization of mouse polymerase-α, *Proc. Natl. Acad. Sci. U.S.A.*, 73, 1136, 1976.
40. Reddy, G. P. V. and Pardee, A. B., Multienzyme complex for metabolic channeling in mammalian DNA replication, *Proc. Natl. Acad. Sci. U.S.A.*, 77, 3312, 1980.
41. Brown, M., Bollum, F. J., and Chang, L. M. S., Intracellular localization of DNA polymerase-α, *Proc. Natl. Acad. Sci. U.S.A.*, 78, 3049, 1981.
42. Tanaka, S., Hu, S. Z., Wang, T. S. F., and Korn, D., Preparation and preliminary characterizattion of monoclonal antibodies against human DNA polymerase-α, *J. Biol. Chem.*, 257, 8386, 1982.
43. Bench, K. G., Tanaka, S., Hu, S. Z., Wang, T. S. F., and Korn, D., Intracellular localization of human DNA polymerase α monoclonal antibodies, *J. Biol. Chem.*, 257, 8391, 1982.
44. Matsukage, A., Yamaguchi, M., Tanabe, K., Nishizawa, M., Takahashi, T., Seto, M., and Takahashi, T., Establishment of hybridoma clones which produce antichick embryo DNA polymerase-α monoclonal antibodies, *GANN*, 73, 850, 1982.
45. Matsukage, A., Yamamoto, S., Yamaguchi, M., Kuskabe, M., and Takahashi, T., Immunochemical localization of chick DNA polymerase α and β+, *J. Cell. Physiol.*, 117, 266, 1983.
46. Masaki, S., Shiku, H., Kaneda, T., Koiwai, O., and Yoshida, S., Production and characterization of monoclonal antibody against 10S DNA polymerase α from calf thymus, *Nucl. Acids Res.*, 10, 4703, 1982.
47. Nakamura, H., Morita, T., Masaki, S., and Yoshida, S., Intracellular localization and metabolism of DNA polymerase-α in human cells visualized with monoclonal antibodies, *Exp. Cell Res.*, 151, 123, 1984.
48. Schmidt, T. J., Bollum, F. J., and Litwack, G., Correlations between the activities of DNA polymerase α and the glucocorticoid receptor, *Proc. Natl. Acad. Sci. U.S.A.*, 79, 4555, 1982.
49. Lan, N. C., Wang, T. S.-F., Johnson, L. K., and Korn , D., Monoclonal antibodies against human DNA polymerase α do not cross-react with glucocorticoid receptors, *J. Steroid Biochem.*, 20, 251, 1984.
50. Chang, L. M. S., Low molecular weight deoxyribonucleic acid polymerase from calf thymus chromatin, *J. Biol. Chem.*, 248, 251, 1984.
51. Smith, H. C. and Bereznay, R., Dynamic domains of DNA polymerase α in regenerating rat liver, *Biochemistry*, 22, 3042, 1983.
52. Fry, M., Shearman, C. W., Martin, G. M., and Loeb, L. A., On the accuracy of DNA synthesis directed by chromatin associated with DNA polymerase β, *Biochemistry*, 19, 5937, 1980.
53. Bolden, A., Pedrali-Noy, G., and Weissbach, A., DNA polymerase of mitochondria is γ-polymerase, *J. Biol. Chem.*, 252, 3351, 1977.
54. Bertazzoni, U., Scovassi, A. I., and Brun, G. M., Chick embryo DNA polymerase γ. Identity of γ-polymerases purified from nuclei and mitochondria, *Eur. J. Biochem.*, 81, 237, 1977.
55. Hübscher, U., Kuenzle, C. C., and Spadari, S., Identity of DNA polymerase γ from synaptosomal mitochondrial and rat brain nuclei, *Eur. J. Biochem.*, 81, 249, 1977.
56. Radsak, K. and Schütz, E., Changes of mitochondrial DNA polymerase-γ activity in synchronized mouse cell cultures, *Eur. J. Biochem.*, 89, 3, 1978.
57. Hardt, N., De Kegal, D., Vanheule, L., Villani, G., and Spadari, S., DNA polymerase γ, cytochrome C oxidase and mitochondrial integrity in rabbit spleen lymphocytes stimulated with concanvalin A, *Exp. Cell Res.*, 127, 269, 1980.

58. Adams, W. J. and Kalf, G. F., DNA polymerase isolated from the mitochondrial chromosome appears to be identical to DNA polymerase-γ, *Biochem. Biophys. Res. Commun.*, 95, 1875, 1980.
59. Tanaka, S. and Koike, K., DNA polymerase-γ is localized in mitochondria, *Biochem. Biophys. Res. Commun.*, 81, 791, 1978.
60. Berezney, R., Effect of protease inhibitors on matrix proteins and the association of replicating DNA, *Exp. Cell Res.*, 123, 411, 1979.
61. Berezney, R. and Coffey, D. S., Nuclear matrix: isolation and characterization of a framework structure from rat liver nuclei, *J. Cell. Biol.*, 73, 616, 1977.
62. Wanka, F., Mullenders, L. H. F., Bekers, A. G. M., Pennings, L. J., Aelen, J. M. A., and Eygensteyn, J., Association of nuclear DNA with a rapidly sedimenting structure, *Biochem. Biophys. Res. Commun.*, 74, 739, 1977.
63. Vogelstein, B., Pardoll, D. M., and Coffey, D. S., Supercoiled loops and eukaryotic DNA replication, *Cell*, 22, 79, 1980.
64. Berezney, R. and Buchholtz, L. A., Isolation and characterization of rat liver nuclear matrices containing high molecular weight deoxyribonucleic acid, *Biochemistry*, 20, 4995, 1981.
65. Berezney, R. and Coffey, D. S., Nuclear protein matrix: association with newly synthesized DNA, *Science*, 189, 291, 1975.
66. Dijkwel, P. A., Mullenkers, L. H. F., and Wanka, F., Analysis of the attachment of replicating DNA to a nuclear matrix in mammalian interphase nuclei, *Nucl. Acids. Res.*, 6, 219, 1979.
67. Padroll, D. M., Vogelstein, B., and Coffey, D. S., A fixed site for DNA replication in eukaryotic cells, *Cell*, 19, 527, 1980.
68. Berezney, R. and Buchholtz, L. A., Dynamic association of replicating DNA fragments with the nuclear matrix of regenerating liver, *Exp. Cell Res.*, 132, 1, 1981.
69. Smith, H. C., and Berezney, R., DNA polymerase α is tightly bound to the nuclear matrix of actively replicating liver, *Biochem. Biophys. Res. Commun.*, 97, 1541, 1980.
70. Smith, H. C. and Berezney, R., Nuclear matrix-bound deoxyribonucleic acid synthesis: an *in vitro* system, *Biochemistry*, 21, 6751, 1982.
71. Yamamoto, S., Takahashi, T., and Matsukage, A., Tight association of DNA polymerase α with granular structures in the nuclear matrix of chick embryo cell: immunochemical detection with monoclonal antibody against DNA polymerase α, *Cell Struct. Funct.*, 9, 83, 1984.
72. Enomoto, T., Tanuma, S.-I., and Yamada, M.-A., Purification and characterization of two forms of polymerase α from HeLa cell nuclei, *Biochemistry*, 22, 1134, 1983.
73. Fry, M., Renart, Z., Weisman-Shomer, P., Lapidot, J., and Sharf, R., Factor D — a template selective stimulatory protein of DNA polymerases: unchanged level in chromatin during liver regeneration, submitted, 1985.
74. Lynch, W. E., Short, J., and Lieberman, I., The 7.1S nuclear DNA polymerase and DNA replication in intact liver,
75. Hübscher, U., Kuenzle, C. C., Limacher, W., Shcerrer, P., and Spadari, S., Functions of DNA polymerases α, β, and γ in neurons during development, *Cold Spring Harbor Symp. Quant. Biol.*, 43, 625, 1978.
76. Chevaillier, Ph. and Phillippe, M., Activité DNA polymérase nucléaire dans le spermatozoides de souris et évolution de cette activité an cours de la spermatogenése, *Exp. Cell Res.*, 99, 237, 1976.
77. Phillipe, M. and Chevaillier, Ph., Further characterization of DNA polymerase activity in mouse sperm nuclei, *Biochem. Biophys. Acta*, 447, 188, 1976.
78. Grippo, P., Geremia, R., Locorotondo, G., and Monesi, V., DNA dependant DNA polymerase species in male germ cells of the mouse, *Cell Diff.*, 7, 237, 1978.
79. Chavaillier, Ph. and Philippe, M., *In situ* detection of a DNA polymerase activity in the nuclei of mouse spermatazoa, *Chromosoma*, 54, 33, 1979.
80. Hecht, N. B., Farrel, D., and Davidson, D., Changing DNA polymerase activities during the development of the testis in the mouse, *Dev. Biol.*, 48, 56, 1976.
81. Hecht, N. B., Farrel, D., and Williams, J. L., DNA polymerase in mouse spermatogenic cells separated by sedimentation velocty, *Biochem. Biophys. Acta*, 561, 358, 1979.
82. Phillippe, M. and Chevaillier, Ph., Study of dogfish *(Scylidrinus caniculus)* deoxyribonucleic acid polymerases α and β. Extraction, characterization and changes during spermatogenesis, *Biochem. J.*, 189, 635, 1980.
83. Hecht, N. B. and Parvinien, M., DNA synthesis catalyzed by endogenous templates and DNA-dependent DNA polymerases in spermatogenic cells from rat, *Exp. Cell Res.*, 135, 103, 1981.
84. Ono, K., Barré-Sinoussi, F., Rey, F., Jasmin, C., and Chermann, J. C., Organ distribution of reverse transcriptase and cellular DNA polymerases in Friend virus-infected mice: analysis by phosphocellulose column chromatography, *Cell. Mol. Biol.*, 30, 303, 1984.
85. Chang, L. M. S. and Bollum, F. J., Variation of deoxyribonucleic acid polymerases activities during rat liver regeneration, *J. Biol. Chem.*, 247, 7948, 1972.

86. Baril, E. F., Jenkins, M. D., Brown, O. E., Laszlo, J., and Morris, H. P., DNA polymerase I and II in regenerating rat liver and Morris hepatoma, *Cancer Res.*, 33, 1187, 1973.
87. Davis, P. B., Laszlo, J., and Baril, E. F., Induction of DNA polymerase-α during liver regeneration in rats on controlled feeding schedules, *Cancer Res.*, 36, 432, 1976.
88. Lynch, W. E. and Lieberman, I., A DNA polymerase in liver nuclei whose activity rises with DNA synthesis after partial hepatectomy, *Biochem. Biophys. Res. Commun.*, 52, 843, 1973.
89. Coleman, M. S., Hutton, J. J., and Bollum, F. J., DNA polymerases in normal and leukemic human hematopoietic cells, *Blood*, 44, 19, 1974.
90. Barr, R. D., Sarin, P., Sarna, G., and Perry, S., The relationship of DNA polymerase activity to cell cycle stage, *Eur. J. Cancer*, 12, 705, 1976.
91. Roodman, G. D., Hutton, J. J., and Bollum, F. J., DNA polymerase, thymidine kinase and DNA synthesis in erythropoietic mouse spleen cells separated on bovine serum albumin gradients, *Biochem. Biophys. Acta*, 425, 478, 1976.
92. Agarwal, S. S. and Loeb, L. A., Studies on the induction of DNA polymerase during transformation of human lymphocytes, *Cancer Res.*, 32, 107, 1972.
93. Coleman, M. S., Hutton, I. J., and Bollum, F. J., Terminal riboadenylate transferase in human lymphocytes, *Nature (London)*, 248, 407, 1974.
94. Mayer, R. J., Smith, R. G., and Gallo, R. C., DNA metabolizing enzymes in normal human lymphoid cells. VI. Induction of DNA polymerase α, β, and γ following stimulation with phythohemaglutinin, *Blood*, 46, 509, 1975.
95. Bertazzoni, U., Stefanini, M., Noy, G. P., Giulotto, E., Nuzzo, F., Falaschi, A., and Spadari, S., Variations of DNA polymerase-α and -β during prolonged stimulation of human lymphocytes, *Proc. Natl. Acad. Sci. U.S.A.*, 73, 785, 1976.
96. Spadari, S., Villani, G., and Hardt, N., DNA polymerase α, β, and γ activities in rabbit spleen cell populations stimulated by various doses of concanavalin A, *Exp. Cell Res.*, 113, 57, 1978.
97. Roodman, G. D., Hutton, J. J., and Bollum, F. J., DNA polymerase activities during erythropoiesis. Effects of erythropoietin, vinblastin, colcemid, and daunomycin, *Exp. Cell Res.*, 91, 269, 1975.
98. Craig, R. K., Costello, P. A., and Keir, H. M., Deoxyribonucleic acid polymerase of BHK-21/C13 cells. Relationship to the physiological state of the cells to synchronous induction of synthesis of deoxyribonucleic acid, *Biochem. J.*, 145, 233, 1975.
99. Keir, H. M., Craig, R. K., and McLennon, A. G., Variation of deoxyribonucleic acid polymerases in the cell cycle, *Biochem. Soc. Symp.*, 42, 37, 1977.
100. Hanaoka, F., Nagata, K., Watanbe, Y., Enomoto, T., and Yamada, M.-A., Accumulation of S phase populations of FM3A cells growing in the mouse due to administration of 5-flouro-2-deoxyuridine, *Cell Struct. Funct.*, 6, 357, 1981.
101. Pedrali-Noy, G. and Weissbach, A., HeLa cells DNA polymerase: the effect of cycloheximide *in vivo* and detection of a new form of polymerase α, *Biochem. Biophys. Acta*, 477, 70, 1977.
102. Chang, L. M. S. and Bollum, F. J., Induction of DNA polymerase in mouse L cells, *J. Mol. Biol.*, 74, 1, 1977.
103. Spadari, S. and Weissbach, A., The interaction between DNA synthesis and various DNA polymerase activities in synchronized HeLa cells, *J. Mol. Biol.*, 86, 11, 1974.
104. Chiu, R. W. and Baril, E. F., Nuclear DNA polymerases and the HeLa cell cycle, *J. Biol. Chem.*, 250, 7951, 1975.
105. Tsai, Y. J., Hanaoka, F., Nakano, M. M., and Yamada, M. A., A mammalian DNA mutant decreasing nuclear DNA polymerase α activity at nonpermissive temperature, *Biochem. Biophys. Res. Commun.*, 91, 1190, 1979.
106. Chang, C.-C., Boezi, J. A., Warren, S. T., Sabourin, C. L. K., Liu, P. K., Glatzer, L., and Trosko, J. E., Isolation and characterization of a UV-sensitive hypermutable aphidicolin-resistant Chinese hamster cell line, *Somat. Cell Genet.*, 7, 235, 1981.
107. Liu, P. K., Chang, C.-C., Trosko, J. E., Dube, D. K., Martin, G. M., and Loeb, L. A., Mammalian mutator mutant with an aphidicolon-resistant DNA polymerase α, *Proc. Natl. Acad. Sci. U.S.A.*, 80, 797, 1983.
108. Tomich, P. K., Chiu, C. S., Worcha, M. G., and Greenberg, G. R., Evidence for a complex regulating the *in vivo* activities of early enzymes induced by bacteriophage T4, *J. Biol. Chem.*, 249, 7613, 1974.
109. Reddy, G. P. V., Singh, A., Stafford, M. E., and Mathews, C. K., Enzyme associations in T4 phage DNA precursor synthesis, *Proc. Natl. Acad. Sci. U.S.A.*, 74, 3152, 1977.
110. Allen, J. R., Reddy, G. P. V., Lasser, G. W., and Mathews, C. K., T4 ribonucleotide reductase. Physical and kinetic linkage to other enzymes of deoxyribonucleotide biosynthesis, *J. Biol. Chem.*, 255, 7583, 1980.
111. Baril, E., Baril, B., Elford, H., and Luflig, R. G., Aggregated mammalian enzymes in deoxyribonucleotide and DNA replication, in *Mechanisms and Regulation of DNA Replication*, Kolber, A. R. and Kohiyama, M., Eds., Plenum Press, New York, 1973, 275.

112. Genta, V. W., Kaufman, D. G., Kaufman, W. K., and Gerwin, B. I., Eukaryotic replication complex, *Nature (London),* 259, 502, 1976.
113. Yamashita, T., Arens, M., and Green, M., Adenovirus deoxyribonucleic acid replication. Isolation of a soluble replication system and analysis of the *in vitro* DNA product, *J. Biol. Chem.,* 252, 7940, 1977.
114. Crabb, W. D., Firshein, W., and Infante, A. A., A potential DNA replication complex isolated from sea urchin embryos by renografin gradient centrifugation, *Biochem. Biophys. Acta,* 609, 456, 1980.
115. Müller, W. E. G., Zahn, R. K., Arendes, J., Schroder, H. C., and Falke, D., DNA replication complex from cells infected with herpes virus, *J. Biochem.,* 115, 159, 1981.
116. Su, R. T. and DePamphilis, M. L., *In vitro* replication of simian virus 40 DNA in a nucleoprotein complex, *Proc. Natl. Acad. Sci. U.S.A.,* 73, 3466, 1976.
117. Gourlie, B. B., Krauss, M. R., Buckler-White, A. J., Benbow, R. M., and Pigiet, V., Polyoma virus minichromosomes: a soluble *in vitro* replication system, *J. Virol.,* 38, 805, 1981.
118. Jazwinski, S. M. and Edelman, G. M., Protein complexes from active replicative fractions associated *in vitro* with the replication origins of yeast 2 μm DNA plasmid, *Proc. Natl. Acad. Sci. U.S.A.,* 79, 3428, 1982.
119. Jazwinski, S. M., and Edelman, G. M., *In vitro* association of a replication complex with a yeast chromosomal replicator, *J. Biol. Chem.,* 258, 2754, 1983.
120. Jazwinski, S. M., and Edelman, G. M., Evidence for participation of a multiprotein complex in yeast DNA replication *in vitro, J. Biol. Chem.,* 259, 6852, 1984.
121. Reddy, G. P. V., Catalytic functions of thymidilate sythetase is confined to S phase due to its association with replitase, *Biochem. Biophys. Res. Commun.,* 109, 908, 1982.
122. Reddy, G. P. V., and Pardee, A. B., Coupled ribonucleoside diphosphate reduction, channeling and incorporation into DNA of mammalian cells, *J. Biol. Chem.,* 257, 12526, 1982.
123. Noguchi, H., Reddy, G. P. V., and Pardee, A. B., Rapid incorporation of label from ribonucleoside diphosphates into DNA by cell-free high molecular weight fraction from animal cell nuclei, *Cell,* 32, 443, 1983.
124. Reddy, G. P. V. and Pardee, A. B., Inhibitor evidence for allosteric interaction in the replitase multenzyme complex, *Nature (London),* 303, 86, 1983.
125. Wickremasinghe, R. G., Yaxley, J. C., and Hoffbrand, A. V., Solubilization and partial characterization of a multienzyme complex of DNA synthesis from human lyphoblastoid cells, *Eur. J. Biochem.,* 126, 589, 1982.
126. Wickremasinghe, R. G., Yaxley, J. C., and Hoffbrand, A. V., Gel filtration of a complex of DNA polymerase and DNA precursor synthesizing enzymes from human lymphoblastoid cell line, *Biochem. Biophys. Acta,* 740, 243, 1983.
127. Spyrou, G. and Reichard, P., Ribonucleotides are not channeled into DNA in permeabilized mammalian cells, *Biochem. Biophys. Res. Commun.,* 115, 1022, 1983.
128. Ottiger, H.-P., and Hübscher, U., Mammalian DNA polymerase α holoenzyme with pssible functions at the leading and lagging strand of the replicative fork, *Proc. Natl. Acad. Sci. U.S.A.,* in press, 1984.
129. Ikegami, S., Taguchi, T., Ohashi, M., Oguro, M., Nagano, H., and Mano, Y., Aphidicolon prevents mitotic cell division by interfering with the activity of DNA polymerase α, *Nature (London),* 275, 458, 1978.
130. Pedrali-Noy, G. and Spadari, S., Effect of aphidicolin on viral and human DNA polymerases, *Biochem. Biophys. Res. Commun.* 88, 1194, 1979.
131. Pedrali-Noy, G., Spadari, S., Miller-Faures, A., Miller, A. O. A., Kruppa, J., and Koch, G., Synchronization of HeLA cultures by inhibition of DNA polymerase α with aphidicolon, *Nucl. Acids Res.,* 8, 377, 1980.
132. Ichikawa, A., Negishi, M., Tomita, K., and Ikegami, S., Aphidicolin: a specific inhibitor of DNA synthesis in synchronous mastocytoma P-815 cells, *Jpn. J. Pharmcol.,* 30, 301, 1980.
133. Pedrali-Noy, G., Belvedere, M., Crepaldi, T., Focher, F., and Spadari, S., Inhibition of DNA replication and growth of several human and murine neoplastic cells by aphidicolin without detectable effect upon synthesis of immunoglobulins and HeLa antigens, *Cancer Res.,* 42, 3810, 1982.
134. Genskens, M., Hardt, N., Pedrali-Noy, G., and Spadari, S.,An autoradiographic demonstration of nuclear DNA replication by DNA polymerase α and of mitochondrial DNA synthesis by DNA polymerase α, *Nucl. Acids Res.,* 9, 1599, 1981.
135. Zimmerman, W. and Weissbach, A., Participation of deoxyribonucleic acid polymerase α in amplification of ribosomal deoxyribonucleic acid in *Xenopus laevis, Mol. Cell Biol.,* 1, 680, 1981.
136. Yagura, T., Kozu, T., and Seno, T., Arrest of chain growth of replication-sized intermediates by aphidicolin during rat fibroblast cell chromosome replication, *Eur. J. Biochem.,* 123, 15, 1982.
137. Wright, G. E., Baril, E. F., and Brown, N. C., Butylanilinouracil: a selective inhibitor of HeLa cell DNA synthesis and HeLa cell DNA polymerase α, *Nucl. Acids Res.,* 8, 99, 1980.

138. Wright, G. E., Baril, E. F., Brown, V. M., and Brown, N. C., Design and characterization of N^2-arylaminopurines which selectively inhibit replicative DNA synthesis and replication-specific DNA polymerases: guanine derivatives active on mammalian DNA polymerase alpha and bacterial DNA polymerase III, *Nucl. Acids Res.*, 10, 4431, 1982.
139. Rochowska, M., Siedlecki, J., Shurzak, H., Wright, G., and Zmudzka, B., Inhibition of calf thymus DNA polymerase α and of normal and cancer cell growth by butylanilinouracil and butylphenylguanine, *Biochem. Biophys. Acta*, 699, 67, 1982.
140. Castellot, J. J., Jr., Miller, M. R., Lehtomaki, D. M., and Pardee, A. B., Comparison of DNA replication and repair enzymology using permeabilized baby hamster kidney cells, *J. Biol. Chem.*, 254, 6904, 1979.
141. Waqar, M. A., Evans, M. J., and Huberman, J. A., Effect of 2′-3′-dideoxythymidine-5′-triphosphate on HeLa cell *in vitro* DNA synthesis: evidence that DNA polymerase α is the only polymerase required for DNA replication, *Nucl. Acids Res.*, 5, 1933, 1978.
142. Wist, E. and Prydz, H., The effect of aphidicolin on DNA synthesis in isolated HeLa nuclei, *Nucl. Acids Res.*, 6, 1583, 1979.
143. Oguro, M., Shioda, M., Nagano, H., Mano, Y., Hamoka, F., and Yamada, M., The mode of action of aphidicolin on DNA synthesis in isolated nuclei, *Biochem. Biophys. Res. Commun.*, 92, 12, 1980.
144. Krokan, H., Wist, E., and Krokan, K. H., Aphidicolin inhibits DNA synthesis DNA polymerase α and isolated nuclei by a similar mechanism, *Nucl. Acids Res.*, 9, 4709, 1981.
145. Krokan, H., Schaffer, P., and DePamphilis, M. L., Involvement of eukaryotic deoxyribonucleic acid polymerase α and γ in the replication of cellular and viral deoxyribonucleic acid, *Biochemistry*, 18, 4431, 1979.
146. Wist, E., Krokan, H., and Prydz, H., Effect of 1-β-D-arabinofuranosylcytosine triphosphate on DNA synthesis in isolated HeLa nuclei, *Biochemistry*, 15, 3647, 1976.
147. Wist, E., The role of DNA polymerase α, β and γ in nuclear DNA synthesis, *Biochem. Biophys. Acta*, 562, 62, 1979.
148. Burk, J. F. and Pearson, C. K., Deoxyribonucleic acid synthesis in isolated nuclei from baby hamster kidney cells (BHK-21/C13), *Biochem. J.*, 178, 613, 1979.
149. Burk, J. F., Duff, P. M., and Pearson, C. K., Effect of drugs on deoxyribonucleic acid synthesis in isolated mammalian nuclei. Comparison with purified deoxyribonucleic acid polymerases, *Biochem. J.*, 178, 621, 1979.
150. Zoncheddu, A., Accomando, R., and Badaracco, G., Comparison of the effects of differential inhibitors of eukaryotic DNA polymerases on DNA synthesis in isolated rat liver nuclei, *Int. J. Biochem.*, 15, 337, 1983.
151. Butt, R. T., Wood, W. M., and Adams, R. L. P., Role of deoxyribonucleic acid polymerase β in nuclear deoxyribonucleic acid synthesis, *Biochem. Soc. Trans.*, 4, 807, 1976.
152. Butt, R. T., Wood, W. M., McKay, E. L., and Adams, R. L. P., Involvement of deoxyribonucleic acid polymerase β in nuclear deoxyribonucleic acid synthesis, *Biochem. J.*, 173, 309, 1978.
153. Tanuma, S.-I., Enomoto, T., and Yamada, M.-A., Characterization of DNA synthesis in salt-treated HeLa cell nuclei, *Cell Struct. Funct.*, 5, 27, 1980.
154. Wintersberger, U. and Wintersberger, E., DNA polymerases in virus-infected mouse kidney cells, *J. Virol.*, 16, 1095, 1975.
155. Mêchali, M., Girard, M., and de Recondo, A.-M., DNA polymerase activities in growing cells infected with simian virus 40, *J. Virol.*, 23, 117, 1977.
156. Wintersberger, U. and Wintersberger, E., Nuclear localization of DNA polymerase α and DNA synthesis in polyoma virus infected mouse cells, *Oncology*, 34, 190, 1977.
157. Edenberg, H. J., Anderson, S., and DePamphilis, M. L., Involvement of DNA polymerase α in simian virus 40 DNA replication, *J. Biol. Chem.*, 253, 3273, 1978.
158. Otto, B. and Fanning, E., DNA polymerase α is associated with replicating SV40 nucleoprotein complex, *Nucl. Acids Res.*, 5, 1715, 1978.
159. Tsubota, Y., Waqar, M. A., Burke, J. F., Milavitz, B. I., Evans, M. J., Kowalski, D., and Huberman, J. A., Association of enzymes with replicating and nonreplicating simian virus 40 chromosomes, *Cold Spring Harbor Symp. Quant. Biol.*, 43, 693, 1978.
160. Richter, A., Otto, B., and Knippers, R., Replication of SV40 chromatin in extracts from eggs of *Xenopus laevis*, *Nucl. Acids Res.*, 9, 3793, 1981.
161. Richter, A., Schen, R., and Otto, B., Replication of simian virus 40 chromatin *in vitro* depends on the amount of DNA polymerase α associated with replicating chromatin, *Eur. J. Biochem.*, 109, 67, 1980.
162. Waqar, M. A., Evans, M. J., Burke, J. F., Plummer, J. M., and Huberman, J. A., *In vitro* synthesis of an α-like DNA polymerase bound to replicating simian virus 40 chromosomes, *J. Virol.*, 48, 304, 1983.
163. Dinter-Gottlieb and Kaufmann, G., Aphidicolin arrest irreversibly impairs replicating simian virus 40 chromosomes, *J. Biol. Chem.*, 258, 3809, 1983.

164. Miller, M. R., Ulrich, R. G., Wang, T. S.-F., and Korn, D., Monoclonal antibodies against human DNA polymerase-α inhibit DNA replication in permeabilized human cells, *J. Biol. Chem.*, 260, 134, 1985.
165. Hockensmith, J. W. and Bambara, R. A., Kinetic characteristics which distinguish two forms of calf thymus DNA polymerase α, *Biochemistry*, 29, 227, 1981.
166. Villani, G., Fay, P. J., Bambara, R. A., and Lehman, I. R., Elongation of RNA-primed DNA templates by DNA polymerase α from *Drosophila melanogaster, J. Biol. Chem.*, 256, 8202, 1981.
167. Ikeda, J.-E., Longiaru, M., Horwitz, M. S., and Hurwitz, J., Elongation of primed DNA templates by eukaryotic DNA polymerase, *Proc. Natl. Acad. Sci. U.S.A.*, 77, 5827, 1980.
168. Matsukage, A., Nishizawa, M., Takahashi, T., and Hozumi, T., *In vitro* synthesis of short DNA pieces by DNA polymerase α from mouse myeloma, *J. Biochem.*, 88, 1869, 1980.
169. Hübscher, U., The mammalian primase is part of a high molecular weight DNA polymerase α polypeptide, *EMBO J.* 2, 133, 1983.
170. Kaguni, L. S., Rossignol, J.-M., Conaway, R. C., Banks, G. R., and Lehman, I. R., Association of DNA primase with the β/γ subunits of DNA polymerase α from *Drosophila melanogaster* embryos, *J. Biol. Chem.*, 288, 9037, 1983.
171. Shioda, M., Nelson, E. M., Bayne, M. L., and Benbow, R. M., DNA primase activity associated with DNA polymerase α from *Xenopus laevis* ovaries, *Proc. Natl. Acad. Sci. U.S.A.*, 79, 7209, 1982.
172. Yagura, T., Kozu, T., Seno, T., Saneyoshi, M., Hiraga, S., and Nagano, H., Novel form of DNA polymerase α associated with DNA primase activity of vertebrates, *J. Biol. Chem.*, 258, 13070, 1983.
173. Yagura, T., Tanaka, S., Kozu, T., Seno, T., and Korn, D., Tight association of DNA primase with a subspecies of mouse DNA polymerase α, *J. Biol. Chem.*, 258, 6698, 1983.
174. Wang, T. S.-F., Hu, S.-Z., and Korn, D., DNA primase from KB cells. Characterization of a primase tightly associated with immunoaffinity-purified DNA polymerase α HeLa cells, *J. Biol. Chem.*, 259, 1854, 1984.
175. Gronostajski, R. M., Field, J., and Hurwitz, J., Purification of a primase activity associated with DNA polymerase α from HeLa cells, *J. Biol. Chem.*, 259, 9479, 1984.
176. Tseng, B. Y. and Ahlem, C. N., A DNA primase from mouse cells. Purification and partial characterization, *J. Biol. Chem.*, 258, 9845, 1983.
177. Grummt, F., Waltl, G., Jantzen, H.-M., Hamprecht, K., Hübscher, U., and Kuenzle, C. C., Diadenosine 5′,5‴-P^1,P^4-tetraphosphate, a ligand of the 57 kilodalton subunit of DNA polymerase α, *Proc. Natl. Acad. Sci. U.S.A.*, 76, 6081, 1979.
178. Rapaport, E., Zamecnik, P. C., and Baril, E. F., HeLa cell DNA polymerase α is tightly associated with tryptophanyl t-RNA synthetase and diadenosine 5′,5‴-P^1, P^4-tetraphosphate binding activities, *Proc. Natl. Acad. Sci. U.S.A.*, 78, 838, 1981.
179. Baril, E. F., Bonin, P., Burstein, D., Mara, K., and Zamecnik, P., Resolution of the diadenosine 5′,5‴-P^1, P^4-tetraphosphate binding subunit from a multiprotein form of HeLa cell DNA polymerase α, *Proc. Natl. Acad. Sci. U.S.A.*, 80, 4931, 1983.
180. Rapaport, E. and Feldman, L., Adenosine (5′) tetraphosphate (5′) adenosine-binding protein of calf thymus, *Eur. J. Biochem.*, 138, 111, 1984.
181. Rapaport, E., Zamecnik, P. C., and Baril, E. F., Association of diadenosine 5′,5‴-P^1, P^4-tetraphosphate binding protein with HeLa cell DNA polymerase α, *J. Biol. Chem.*, 256, 12148, 1981.
182. Zamecnik, P. C., Rapaport, E., and Baril, E. F., Priming of DNA synthesis by diadenosine 5′,5‴-P^1, P^4-tetraphosphate with a double stranded octamer as template and DNA polymerase α, *Proc. Natl. Acad. Sci. U.S.A.*, 79, 1791, 1982.
183. Zamecnik, P., Diadenosine 5′,5‴-P^1, P^4-tetraphosphate (Ap_4A): its role in cellular metabolism, *Anal. Biochem.*, 134, 1, 1983.
184. Varshavsky, A., Diadenosine 5′,5‴-P^1, P^4-tetraphosphate: a pleiotropically acting alarmone?, *Cell*, 34, 711, 1983.
185. Lamothe, P., Baril, B., Chi, A., Lee, L., and Baril, E. F., Accessory proteins for DNA polymerase α activity with single strand DNA templates, *Proc. Natl. Acad. Sci. U.S.A.*, 78, 4723, 1981.
186. Pritchard, C. G. and DePamphilis, M. L., Preparation of DNA polymerase α by reconstituting DNA polymerase α with its specific stimulatory cofactors C_1C_2, *J. Biol. Chem.*, 258, 9801, 1983.
187. Pritchard, C. G., Weaver, D. T., Baril, E. F., and DePamphilis, M. L., DNA polymerase α cofactors C_1C_2 function as primer recognition proteins, *J. Biol. Chem.*, 258, 9810, 1983.
188. Weissbach, A., Baltimore, D., Bollum, F. J., Gallo, R. C., and Korn, D., Nomenclature of eukaryotic DNA polymerases, *Science*, 190, 401, 1975.
189. McLennan, A. G., Effect of aphidicolin and 2′-3′-dideoxythymidine on mitochondrial DNA replication, *Biochem. Biophys. Res. Commun.*, 94, 116, 1980.
190. Zimmerman, W., Chen, S. M., Bolden, A., and Weissbach, A., Mitochondrial DNA replication does not involve DNA polymerase α, *J. Biol. Chem.*, 255, 11847, 1980.
191. Robertson, A. T., Bates, R. C., and Stout, E. R., Purified DNA polymerase α replicates bovine parvovirus DNA to a unit-length product, *Biochem. Biophys. Res. Commun.*, 117, 580, 1983.

192. Clayton, D. A., Doda, J. N., and Friedberg, E. C., The absence of pyrimidine dimer repair mechanism in mammalian mitochondria, *Proc. Natl. Acad. Sci. U.S.A.*, 71, 2777, 1974.
193. Siegel, R. L. and Kalf, G. F., DNA polymerase β involvement in DNA endoreduplication in rat giant trophoblast cells, *J. Biol. Chem.*, 257, 1786, 1982.
194. Foster, K. A. and Collins, J. M., The interrelation between DNA synthesis rates and DNA polymerases bound to the nuclear matrix in synchronized HeLa cells, *J. Biol. Chem.*, 260, 4229, 1985.
195. Holmquist, G. P., Gray, M., Porter, T., and Jordan, J., Replication characterization of Giemsa dark- and light-band DNA, *Cell*, 31, 121, 1982.
196. Goldman, M. A., Holmquist, G. P., Gray, M. C., Caston, L. A., and Nag, A., Replication timing of genes and middle repetitive sequences, *Science*, 224, 686, 1984.
197. Chanpu, S., Muller, M., and Werner, D., DNA polymerases in isolated "nuclear matrix" of Ehrlich ascites cells, *Biochem. Biophys. Res. Commun.*, 118, 609, 1984.
198. Delfini, C., Alfani, E., DeVenezia, V., Oberholtzer, G., Tomasello, C., Eremenko, T., and Volpe, P., Cell-cycle dependence and properties of the HeLa cell DNA polymerase system, *Proc. Natl. Acad. Sci, U.S.A.*, 82, 2220, 1985.
199. Murakami, Y., Yasuda, H., Miyazawa, H., Hanaoka, F., and Yamada, M.-A., Characterization of a temperature-sensitive mutant of mouse FM3A cell defective in DNA replication, *Proc. Natl. Acad. Sci. U.S.A.*, 82, 1761, 1985.
200. Reddy, G. P. V., Klinge, E. M., and Pardee, A. B., A mixed DNA-RNA polymer is synthesized from ribonucleotides by permeabilized S phase Chinese hamster cells, *J. Biol. Chem.*, submitted 1985.

Chapter 7

ROLES OF ANIMAL CELL DNA POLYMERASES IN DNA REPAIR SYNTHESIS

Every cause produces more than one effect.

Herbert Spencer

I. INTRODUCTORY REMARKS

It has become increasingly apparent in recent years that DNA, like all other cellular components, undergoes biologic turnover. This turnover results from resynthesis of DNA after removal of altered pieces of DNA following damage by both endogenous and exogenous agents. Efficient and faithful mechanisms for the repair synthesis of DNA are required for genetic continuity of species, as well as for stringent control of genetic diversity. The availability of mutant cells that are defective in different aspects of DNA repair in bacteria has permitted a detailed genetic analysis of DNA repair pathways. In eukaryotes, however, modes of DNA repair are less well understood.

In all organisms, extensive resynthesis of DNA is an obligatory step in repair of DNA, with the exception of repair by adaption, photoreactivation, and possibly insertion. With eukaryotic cells, only adaption — the removal of small alkylated adducts from the O^6 position of purine bases, is a significant pathway of repair of damaged DNA without extensive resynthesis. Thus essentially all damage to DNA in animal cells requires considerable resynthesis of DNA by DNA polymerase. For each damaged nucleotide removed, the patch size of resynthesized DNA can range between a few and several hundred nucleotides.

Pathways of DNA repair that require resynthesis can be classified into excision repair, postreplication, and induced error-prone repair.[1-6] By analogy with repair modes in bacterial cells, excision repair involves the cleavage of DNA at damaged sites, exonuclease-catalyzed hydrolysis of a segment of DNA, and resynthesis by a DNA polymerase followed by ligation of the patched segment. Recently, this scheme has been revised to include endonucleolytic removal of an oligonucleotide containing the damaged adduct.[7] In contrast, postreplication repair results from blockage of replication by unrepaired bulky regions and subsequent resynthesis of the gap by undefined mechanisms that involve strand exchange or *de novo* initiation of synthesis. It is not clear whether or not a mechanism for bypass of bulky adducts occurs in animal cells that is analogous to induced error prone repair in bacteria. Excellent reviews have summarized the experiments on DNA repair in both bacteria and eukaryotic cells.[1-6] The types of chemical lesions in DNA and their removal is the subject of a recent comprehensive treatise by Singer and Grunberger.[8]

Discussion in this chapter is limited by necessity to the roles of DNA polymerases-α, -β, and -γ in DNA repair and thus only to the stage of polymerase-catalyzed DNA synthesis at the DNA lesion. Several experimental approaches were employed in an effort to elucidate specific roles for the different types of DNA polymerases in DNA repair synthesis. Early attempts to correlate intracellular levels of DNA polymerases with repair activity met with only limited success because of both the variety and complexity of repair processes in animal cells, and chiefly because unlike replication, the induction of DNA repair does not necessarily involve significant elevation in DNA polymerase activity. So far, attempts to correlate hereditary deficiencies in repair with altered levels of DNA polymerase(s) have not yielded clues to the roles of polymerases in repair. Three other experimental approaches were more fruitful and led to the present conception that both polymerase-α and polymerase-β participate in the diverse modes of DNA repair. First, use of selective inhibitors of DNA polymerases in cell systems that perform repair of damaged DNA compellingly indicates that DNA polymerase-α takes part in some modes of DNA repair, and that polymerase-β participates in the same and in additional repair pathways. Second, a recently isolated hamster cell mutant strain with a mutated polymerase-α exhibits a distinct hypermutable phenotype. The finding that a genetic defect in polymerase-α leads to spontaneous and induced hypermutability strongly indicates that the α enzyme participates in DNA repair. Last, correlating the singular enzymic properties of DNA polymeraase-α and -β with their possible role in repair allowed the construction of in vitro systems in which polymerase-α and -β and other repair enzymes together perform repair of defined damage to DNA. In this chapter, the evidence for the participation of both α- and β-polymerases in repair synthesis of DNA is presented.

II. PHYSIOLOGY OF DNA POLYMERASES-α AND -β IN CELLS CONTAINING DAMAGED DNA

Monitoring intracellular levels of DNA polymerase-α, -β and -γ in postmitotic and cycling cells provided initial clues for the central role of DNA polymerase-α in DNA replication (cf. Chapter 6, Section II). In an analogous attempt to determine which classes of polymerase take part in repair synthesis of DNA, levels of polymerase-α, -β, and -γ were monitored in cells after their exposure to DNA-damaging agents. Two categories of cells were employed in these studies:

- Postmitotic cells: Levels of DNA polymerases were measured in G_0-arrested cells after DNA damage or in terminally differentiated tissues that do not carry out DNA replication and by definition perform predominantly excision repair of DNA;[5] analysis of repair polymerases in nondividing cells is further simplified in that in some noncycling cells α-polymerase activity is undetectable and thus polymerase-β and -γ are the remaining obvious candidates for repair synthesis.
- Dividing cells: Such cells exhibit a more complex repair response, including postreplication repair,[5] and they usually contain all three major types of DNA polymerases; hence, in proliferating cells it is difficult to correlate any particular mode of repair with activity of a defined class of polymerase.

A. Postmitotic Cells

Neurons in adult animal brain are believed to never divide and thus can be considered the paradigm for a postmitotic cell. DNA polymerase-β is the predominant, if not the only, DNA polymerase activity detected in these cells.[9,10] Exposure of isolated rat brain neurons to UV irradiation results in seven- to tenfold increase in the rate of thymidine incorporation into DNA.[9] Since DNA polymerase-β is virtually the only DNA polymerase present in these neurons, Hubscher et al.[9] proposed that this polymerase was responsible for the stimulation of DNA synthesis that presumably resulted from repair of UV damage. It was found, however, that *N*-methyl-*N*-nitrosourea did not induce a similar DNA repair synthesis.[9] This lack of repair synthesis was interpreted as being due to the inability of neurons to excise alkylated bases from DNA as demonstrated in earlier work.[11,12] It had been subsequently shown that β- and γ-polymerases comprised 99.2 and 0.8%, respectively, of the total nuclear polymerase activity of rat cortex neurons.[13] Although DNA polymerase-β was thus the overwhelmingly major nuclear activity, and although UV repair synthesis was suppressed by d_2TTP, suggesting that the β enzyme was responsible for repair synthesis, a role of DNA polymerase-γ in repair synthesis could not be excluded.

An alternative approach to the preparation of quiescent cell populations is to deprive proliferating cells of essential growth stimulants and thus render them postmitotic. This approach assumes that nutrient deprivation does not alter the cellular pathways for DNA repair. Rossignol et al.[14] deprived confluent cultured Syrian hamster cells of arginine, thus completely suppressing replicative DNA synthesis. The G_0-arrested cells were exposed to *N*-ethyl-*N*-methyl-*N*′-nitro-*N*-nitrosoguanidine (MNNG), to induce unscheduled DNA synthesis. Sucrose gradient analysis revealed DNA polymerase-β as the only detectable polymerase in nuclei and cytosol of the quiescent cells, both before and after exposure to MNNG.[14] It was concluded, therefore, that polymerase-β was responsible for repair synthesis of MNNG-induced DNA damage. A similar conclusion was drawn by the finding that nuclei, which were isolated from confluent cultured mouse L cells, contained only polymerase-β and exhibited stimulated repair synthesis after being exposed to DNase I.[15]

The results of the experiments with postmitotic cells strongly suggest that polymerase-β acts as the only synthetic repair polymerase under the tested conditions. However, the generality of such a conclusion should be qualified. First, since very low levels of DNA polymerase are required for repair synthesis,[5] it could be argued that DNA polymerase-γ or trace levels of polymerase-α, which could not be detected by the analytic methods which were employed, do also take part in DNA repair synthesis. If in a typical experiment, damage to DNA is assumed to provide 1000 sites per genome for repair and each site involves the synthesis of 100 nucleotides, then the total number of nucleotides polymerized during repair synthesis is 10^5 per cell. This corresponds to 0.003% of the human genome. If activity of polymerase-α in the cell is proportional to the stretch of DNA synthesized, then only 0.003% of the level of polymerase-α that is required for genome replication will suffice for repair. Such low levels of activity are well below detectability, and thus absence of detectable activity of polymerase-α does not provide strong argument against its participation in repair. Second, only a limited variety of DNA-damaging agents was used in these studies, and it is not clear whether DNA polymerase-β alone would have been able to repair other untested types of damage to DNA.

B. Cycling Cells

In cycling eukaryotic cells, it may be that both excision repair and postreplication repair occur concomitantly. Several investigators undertook to measure intracellular levels of DNA polymerases-α, -β, and -γ after exposure of dividing animal cells to DNA-damaging agents, with the expectation of establishing a correlation between DNA repair activity and a specific increase in the activity of a particular type of polymerase. These experiments have not been definitively instructive in assigning functions for the different DNA polymerases in repair synthesis. Bertazzoni et al.[16] induced DNA replication in populations of human peripheral lymphocytes by stimulation with phytohemagglutinin. This resulted in an increase in polymerase-α activity, which paralleled the rise in thymidine incorporation into DNA (cf. Figure 2, Chapter 2). Following UV irradiation of the mitogen-stimulated lymphocytes, two maxima of DNA synthesis were observed, one corresponding to a replicative stage with an increase in polymerase-α activity and the other correlated with a peak of increased DNA polymerase-β activity. In contrast, Wicker et al.[17] found that repair of DNA and recovery of biological function after UV irradiation of CV-1 cultured monkey cells did not involve a measurable variation in the intracellular levels of polymerases-α, -β, or -γ. In dividing HeLa cells exposed to MNNG, the cellular activities of all three classes of DNA polymerase increased with discernable kinetics after treatment with that carcinogen.[18] In contrast, activity of polymerase-α decreased, and levels of polymerase-β and polymerase-γ remained unaltered following high doses of UV irradiation of human lymphocytes and heteroploid cultured cells.[19] In vivo studies on levels of polymerase-β and -γ in hepatocytes of rats, which were fed with the DNA damaging carcinogen diethylnitrosamine, were carried out by Craddock and Ansley.[20,21] It was reported that treatment with the carcinogen led to a regenerative response, which was accompanied by an increase in levels of hepatic polymerase-α. However, chronic exposure to the DNA damaging agent, which supposedly leads to continuous repair activity, was marked by elevated activity of polymerase-β.[20] Correlation between DNA replication and repair activities and respective increases in levels of hepatic polymerase-α and polymerase-β was also observed after exposure of the rats to *N*-acetylaminofluorene.[21]

By using selective inhibitors activities of DNA polymerases-α and -β were determined in nuclear matrices isolated from hepatocytes in the course of liver regeneration.[22] Whereas levels of nuclear matrix-bound polymerase-α rose during DNA replication, activity of polymerase-β in nuclei and matrices reached maxima prior to replication and at late and postreplicative stages. It has been proposed[22] that the pre- and postreplicative elevation in levels of matrix-associated polymerase-β correspond to stages of genome screening by that polymerase, as originally hypothesized by Gupta and Sirover.[23] The interesting concept included in this hypothesis is that unrepaired DNA damage in animal cells must be excised prior to DNA replication to avoid errors at the time of genome duplication. To achieve that, cells conduct a prereplicative burst of DNA repair processes immediately prior to the onset of S phase.

The mixture of results obtained in the above-described studies do not permit the drawing of any firm conclusion on the role of particular DNA polymerases in repair synthesis. First, it may be naive to expect that repair synthesis necessarily entails an increase in intracellular levels of the repairing polymerase. As discussed in Chapter 6, replication of the complete genome is usually accompanied by a two- to fivefold increase in the activity of the replicative polymerase-α (cf. Chapter 6, Section II.A.2). It may thus be that the much more limited repair synthesis of DNA involves a very modest and possibly undetectable rise in the intracellular level of repair polymerase(s), or it may not require any change in the ambient level of the polymerase(s). Second, the complexity of DNA synthesis processes in cycling cells, which include both replication and repair, renders experimental results very hard to interpret. Last, dividing cells may employ several pathways of repair and DNA synthesis at the same time,[4,5] a possibility that may also add to the complexity of the interpretation of results obtained with such cells.

III. STUDIES ON MUTANT CELLS WITH DEFICIENT REPAIR OR MODIFIED DNA POLYMERASE-α

Well-characterized animal-cell mutants with deficient repair function(s) or with defined deficits in a particular DNA polymerase, should provide a powerful tool for the determination of roles of DNA polymerases in repair synthesis. Genetic analysis of DNA repair and DNA polymerases in animal cells is still at a primordial phase. In this section we survey recent attempts to use mutant cell strains with defective repair functions or with modified polymerase-α to elicit understanding of the roles of DNA polymerases in DNA repair synthesis.

A. Hereditary Diseases of Defective DNA Repair

Certain hereditary human diseases exhibit deficits in DNA repair, and cells derived from afflicted individuals maintain the defective phenotype in culture. Cells isolated from patients with xeroderma pigmentosum, the most extensively studied of these diseases, can be classified into seven or eight complementation groups. Presumably each of these groups involves a mutation of a different gene. Considering the limited number of proteins required for repair of UV damage in prokaryotes, it would seem likely that one of these complementation groups may involve a DNA polymerase that is required for resynthesis of DNA following damage by UV irradiation. At least two research groups compared levels of DNA polymerase-α, -β, -γ in fibroblasts from normal subjects to enzyme levels in cells derived from human patients afflicted with assorted diseases, including xeroderma pigmentosum, which affect DNA repair. In both studies, no measurable alteration in the level of any of the polymerase types was discerned for the various genetic defects.[24,25] Clearly, either none of the examined diseases involve DNA polymerases, or subtle changes in the enzymic properties of the polymerase could occur that were undetected in the above studies. Since many diseases exhibiting deficits in DNA repair are associated with increased incidence of malignancy, it may be that searching for decreased fidelity of DNA synthesis in individuals afflicted with diseases of deficient DNA repair will prove a fruitful area of investigation.

B. Mutant Cell with a Modified DNA Polymerase-α

Chang et al.[26] isolated and characterized a mutant strain of Chinese hamster V79 cell that is slow growing, UV sensitive, and aphidicolin resistant.[26-29] The mutant cell, designated Aph^r-4-2, possesses a genetically modified DNA polymerase-α that is resistant to aphidicolin and exhibits increased affinity to dCTP.[30] Initial characterization of the mutant phenotype revealed that it was hypersensitive to UV irradiation, and that it exhibited ninefold increase in the frequency of chromatid gaps and breaks after exposure to bromodeoxyuridine.[26] Subsequent comparison of UV irradiation-induced mutability of various genes in wild type cells and the Aph^r-4-2 mutant revealed that the mutant strain was hypermutable at the loci for resistance to ouabain, diphtheria toxin, and 6-thioguanine.[27] Additional experiments indicated that the rate of spontaneous mutagenesis at the ouabain, 6-thioguanine, and diphtheria toxin resistance loci was elevated in the aphidicolin-resistant cell.[28,30]

Although the Aph^r-4-2 strain was hypermutable for UV irradiation and MNNG, it showed normal response to X-ray irradiation and dimethyl sulfate.[29] The established mutator phenotype of the Aph^r-4-2 cell strain, and the presumed initial single-site mutation at the polymerase-α locus, provide the strongest argument for a central role of polymerase-α in DNA repair, provided that indeed, only the polymerase-α gene was mutated in the Aph^r-4-2 strain. The procedure for obtaining this mutant involved exposure to UV irradiation, as well as to BudR and black light. It is not unreasonable, therefore, that mutations in other DNA repair genes in the same cell may also have been produced and selected for. As noted above, whereas hypermutability of Aph^r-4-2 was observed after UV irradiation which entails long-patch DNA repair, the mutant cell did not display increased sensitivity to X-ray irradiation which results in short-patch repair.[29] Two explanations are offered. DNA polymerase-α may be involved only in long-patch repair and does not participate in short-patch repair synthesis[29] (see below for further discussion). Alternatively, the mutation in the polymerase-α gene, which increases the enzyme's affinity to dCTP, renders it a mutator polymerase as a result of more frequent substitution of dCTP in place of any complementary nucleotide. Hence, extended synthesis by mutant polymerase-α during long-patch repair results in apparent hypermutability, whereas short-patch repair is too limited to bring about a measurably higher mutation rate.[29] The recent success in introduction of the gene for the mutated polymerase-α into recipient cells[31] opens the way for isolation and molecular cloning of that gene. The isolated gene may be used eventually for an unequivocal proof that introduction of a mutated polymerase-α gene into normal recipient cells leads to hypermutable repair synthesis of DNA.

IV. INVOLVEMENT OF POLYMERASES-α AND -β IN REPAIR: STUDIES WITH SELECTIVE INHIBITORS

One of the most actively employed approaches to the elucidation of roles of DNA polymerases in DNA repair has been the use of selective inhibitors of particular polymerase types to assess their relative contribution to repair synthesis. This approach is by necessity somewhat simplistic. First, inhibitors of DNA polymerases are never completely effective in arresting the activity of any polymerase. Therefore, lack of inhibitory effect on DNA repair could be due to residual activity of the suppressed enzyme, rather than to the polymerase not having a role in repair. It should also be noted that effectiveness and mode of inhibition by several agents such as aphidicolin, butylanilinouracil, and butylphenylguanine differ for purified polymerase-α and in intact cells or nuclei (cf. Chapter 3, Sections VI.B and C). Furthermore, different polymerase classes may operate in concert in some pathways of DNA repair and may even be interchangeable. Such complexity is difficult to unravel by use of selective inhibitors. Yet in recent years, studies on the effectiveness of selective inhibitors on the repair of DNA damage have provided evidence that both DNA polymerase-α and -β take part in repair synthesis. It is still debatable, however, at what specific pathways of repair each of the polymerases participates, to what extent they may be interchangeable, and whether they act in concert in some modes of DNA repair synthesis. Furthermore, a possible role for DNA polymerase-γ in repair was neither confirmed nor repudiated by inhibitor studies, and its place in DNA repair in either mitochondria or nuclei has not been delineated. Similarly, a role for polymerase-δ in repair synthesis has yet to be investigated.

A. Evidence for the Participation of Polymerase-α in Repair Synthesis

Aphidicolin, a highly specific inhibitor of polymerases of the α category, has been used to test whether polymerase-α takes part in repair synthesis. Early work suggested that only aphidicolin-resistant polymerase(s) participate in repair synthesis. However, later developments established that some modes of repair are suppressed by aphidicolin and other inhibitors of DNA polymerase-α.

1. Early Observations

Seki et al.[32] found that although aphidicolin effectively arrested DNA replication in permeabilized mouse cells, it had no effect on unscheduled DNA synthesis that occurred after exposure of the cells to the radiomimetic drug bleomycin. Furthermore, d_2TTP inhibited DNA replication in the permeabilized cell system with a K_i of 8.0 μM, whereas repair of bleomycin damage was inhibited at $K_i = 0.5$ μM.[33] It was thus concluded that an aphidicolin-resistant d_2TTP-sensitive, non-α-polymerase conducts repair synthesis of bleomycin-damaged DNA.[32] As learned later, the choice of bleomycin was unfortunate since, rather uniquely, polymerase-α appears not to take a major part in repair of damage by that agent (see below). Lack of participation of polymerase-α in repair was also implied by Pedrali-Noy and Spadari,[34] who found that aphidicolin inhibited replication of chromosomal DNA in cultured human cells, but did not affect the enhanced unscheduled synthesis of DNA elicited by UV irradiation. Subsequently, autoradiographic analysis was used to reaffirm and sustain the absence of effect of aphidicolin on UV-induced unscheduled DNA synthesis.[35,36]

In contrast to these initial results, may authors have subsequently demonstrated that repair of damage exerted by various agents on DNA of different cell types requires the participation of an aphidicolin-sensitive polymerase-α. Hanaoka et al.[37] minimized replicative DNA synthesis in HeLa cells by exposure of cultures to hydroxyurea and ara C and then irradiated the cells with UV to elicit repair response. Aphidicolin was as effective in inhibiting repair synthesis as it was in suppressing replication.[37] The conclusion that polymerase-α indeed participates in repair synthesis was supported by an independent demonstration that ara CTP, a selective inhibitor of polymerase-α, depresses repair synthesis in HeLa cell lysates.[38]

Similarly, Berger and associates[39] reported that aphidicolin, as well as phosphonacetic acid and *N*-ethylmaleimide, effectively blocked repair synthesis in permeable human lymphocytes that were exposed to either MNNG or to UV irradiation. Further, indications for the role of polymerase-α in DNA repair synthesis could be deduced in osmotically lysed human fibroblasts from the resistance of their repair synthesis to d_2TTP and their sensitivity to aphidicolin.[40] Snyder and Regan[41] quantitated repair activity in UV-irradiated human fibroblasts by direct velocity sedimentation analysis of damaged and repaired DNA, and by measurements of amounts of chromatographically separated pyrimidine dimers. Both criteria indicated that aphidicolin inhibited DNA repair synthesis. In a subsequent work, it was further demonstrated that although aphidicolin effectively blocked DNA repair in confluent cultures of UV irradiated human fibroblasts, it did not interfere with repair in rapidly cycling UV-damaged cells.[42] It was suggested that previous failures to observe suppression of repair synthesis by aphidicolin[33-36] resulted from use of dividing cells, which may have higher intracellular pools of dNTPs and are thus capable of overcoming aphidicolin inhibition. Subsequently, other authors also found that aphidicolin was less effective in inhibiting repair synthesis in proliferating cells (see below).

2. Cell Type Variance in Response of Repair to Aphidicolin

A possible source for variance in the response of DNA-repairing cells to inhibition of polymerase-α activity by aphidicolin is cell-type variability. This could result from cell-type differences in nuclear levels of aphidicolin, different intracellular levels of deoxypyrimidines, or differences in the relative contribution of polymerase-α to repair synthesis in different cells. Variations between cell types must be taken into account when assessing the effectiveness of aphidicolin in inhibiting repair synthesis. In a systematic study, Seki et al.[43] found that aphidicolin partially inhibited unscheduled DNA synthesis in a large variety of human cell types after their exposure to bleomycin. A combination of aphidicolin and d_2TTP was more effective than aphidicolin alone in inhibiting repair in permeabilized cells, suggesting that both polymerase-α and polymerase-β participated in repair synthesis. Use of various combinations of aphidicolin, d_2TTP, and *N*-ethylmaleimide indicated that polymerase-γ had no role in repair of bleomycin-induced damage to DNA. Interestingly, the relative inhibitory effectiveness of aphidicolin and d_2TTP varied in various human cell types, and in rodent cells aphidicolin did not suppress repair of bleomycin damage. It was proposed, therefore, that different cell types employ either polymerase-β alone or both polymerase-α and polymerase-β in variable proportions to perform repair synthesis.[43]

Collins et al.[44] established a system of quiescent human fibroblasts, which were maintained at confluence in stale (spent) medium, in which semiconservative DNA replication was not performed, but unscheduled repair synthesis that was conducted was inhibited by the polymerase-α inhibitors novobiocin and aphidicolin. Subsequently, previous reports[34,36] on the lack of effect of aphidicolin on repair synthesis in UV irradiated HeLa cells were reexamined in that cell system.[45] It was found that whereas 2 μg/mℓ aphidicolin sufficed to induce DNA break accumulation in normal human fibroblasts, 10 μg/mℓ were required to elicit a similar effect on repair in HeLa cells. It was argued that susceptibility of repair synthesis to aphidicolin may differ in different cell types, and furthermore, repair should be rigorously monitored by assessing breakage and sealing of DNA rather than by the more ambiguous measurement of unscheduled incorporation of labeled thymidine.[45] Van Zeeland and colleagues[46] confirmed previous contrasting observations on the inhibitory effect of aphidicolin on repair in confluent human fibroblasts[40] and the lack of effect of that inhibitor on UV repair in proliferating HeLa cells.[34] However, aphidicolin did inhibit repair in HeLa cells in the presence of hydroxyurea, suggesting that different dCTP pool sizes in the two cycling and quiescent cell types were responsible for the discrepant effect of aphidicolin on repair.[46] Clearly, the initial claim that aphidicolin has no effect on DNA repair was premature, and careful analysis indicates that under defined conditions the participation of both DNA polymerase-α and -β in repair synthesis can be discerned. Further evidence for the complex polymerase response to DNA damage is discussed below.

3. Further Evidence for the Function of Polymerase-α in Repair Synthesis

That DNA polymerase-α does take part in repair synthesis that is induced by a variety of agents has been repeatedly demonstrated by numerous authors. Waters[47] reported that single-strand DNA breaks are accumulated in UV-irradiated human skin fibroblasts in the presence of aphidicolin. Such an accumulation of single-strand breaks was not observed, however, in excision-deficient xeroderma pigmentosum fibroblasts. It was proposed, therefore, that polymerase-α participates in excision-repair synthesis of DNA.[47] Aphidicolin and ara C were also found to partially inhibit repair of DNA in X-ray irradiated normal and xeroderma pigmentosum human fibroblasts, suggesting that polymerase-α takes part in some of the repair synthesis of X-ray-induced damage.[48] Exposure of intact cultured rat hepatocytes to methylmethanosulfonate (MMS) induced DNA strand breakage and ensuing unscheduled DNA synthesis, which led to eventual sealing of the repaired DNA patch.[49] Aphidicolin blocked unscheduled DNA synthesis, and the fragmented DNA patch remained unrepaired in its presence.[49] It was reported that aphidicolin was also effective in inhibiting unscheduled DNA synthesis in nuclei isolated from hepatocytes of rats which were exposed to dimethylnitrosamine.[50]

Yamada et al.[51] studied the role of polymerase-α in repair of cellular DNA exposed to the alkylating agent MNNG. Lysates were prepared from HeLa cells that were treated with hydroxyurea and ara C to arrest replicative DNA synthesis. Exposure of the lysates to MNNG induced unscheduled DNA synthesis, which was in turn 90% inhibited by aphidicolin, but not by d_2TTP. It was suggested, therefore, that in this mode of base excision repair of alkylated DNA, polymerase-α is the major synthetic enzyme, whereas polymerase-β plays a minor role in that type of repair synthesis.[51]

In a subsequent work, Hanaoka et al.[52] studied the effect of polymerase-α inhibition on the long-patch repair mode of UV irradiation-induced damage to HeLa cell DNA. Lysates of G_1-arrested HeLa cells were UV-irradiated and exposed to aphidicolin alone or to combinations of aphidicolin and hydroxyurea or aphidicolin and d_2TTP. In the presence of aphidicolin, and in particular with aphidicolin and hydroxyurea which reduces the level of dCTP, there was an accumulation of single-strand DNA breaks. Removal of aphidicolin restored unscheduled DNA synthesis and joining of the DNA fragments. The presence of d_2TTP did not significantly inhibit repair synthesis, nor did it potentiate aphidicolin, and thus the role of DNA polymerase-β in the repair of UV damage was deemed minor under the described experimental conditions.[52]

4. Role of Polymerase-α in Repair of Damage Caused by Different Agents

Miller and associates[53-55] conducted a series of extensive studies on the effect of several inhibitors of DNA polymerase-α and -β on repair of damage to DNA elicited by different agents. Repair of CHO cell DNA, which was damaged by either MNNG or *N*-nitrosomethylurea (NMU), was prevented by aphidicolin or ara C, both effective inhibitors of polymerase-α.[53] In contrast, repair of DNA that was damaged by bleomycin or neocarzinostatin was not similarly affected by aphidicolin. Thus it appears that the nature of damage determines which polymerase class will be involved in the repair synthesis reaction.[53] In a subsequent work, the effect of d_2TTP, and inhibitor of β- and γ-polymerases, but not of polymerase-α, on repair of DNA was studied in lysolecithin-permeabilized human cells.[54] Neither long-patch repair of damage of UV irradiation or MNU, nor short-patch repair of γ irradiation-induced damage was affected by d_2TTP, and it was thus concluded that the uninhibited polymerase-α is the central repairing polymerase in these repair modes.

In contrast, repair of damage inflicted by bleomycin and neocarzinostatin was inhibited by d_2TTP, and thus polymerase-β was deemed to involved in repair synthesis following exposure to these two agents.[54] It was suggested that size of the repaired patch is not a determining factor in the choice of polymerase. Rather, it was proposed that the nature of damage to DNA is the decisive factor in the choice of a repairing polymerase. A hypothesis was advanced that γ or UV irradiation, as well as MNNG and MNU, create damaged sites in DNA that are recognized by incision nuclease(s) which generate usable 3′ and 5′ termini. In turn, these termini are recognized by repair nucleases, which form long or short gaps that are filled by polymerase-α. On the other hand, bleomycin and neocarzinostatin break DNA directly, thus creating sugar moieties with blocked 3′ and 5′ termini, respectively. Such ends at the breakage site may be recognizable only by a polymerase-β-associated nuclease (DNase V), and thus the β enzyme is responsible for repair of these damaged loci.[54]

In a subsequent work, the effects of d_2TTP, *N*-ethylmaleimide, and aphidicolin on repair synthesis were studied in permeabilized human fibroblasts and in rodent CHO cells.[55] Since repair of damage inflicted by various agents was not equally sensitive to both d_2TTP and *N*-ethylmaleimide, it was concluded that polymerase-γ does not play a major role in repair synthesis. In both cell types, DNA repair following UV irradiation and MNNG and MNU treatments was suppressed by aphidicolin or araCTP. Yet a discrepancy was noted between the concentrations of both drugs that were required for inhibition of purified polymerase-α and repair synthesis *in situ*.[55]

5. Effect of the Extent of Damage on Mobilization of Polymerases for Repair Synthesis

Dresler and Lieberman[56-59] conducted a series of studies on the effects of inhibitors of DNA polymerases on DNA synthesis in permeabilized cells. A protocol was developed for the conduct of in vitro repair synthesis, which resembles in vivo repair activity. Quiescent cultured cells were permeabilized by mechanical shear and exposed to DNA-damaging agents. Repair synthesis was conducted at low dNTP concentrations that selectively depressed replication of DNA without affecting repair, which in turn was monitored by both autoradiography and DNA density-shift analysis.[56,57] Under these assay conditions, repair of UV irradiation damage was 90% inhibited by aphidicolin, whereas high levels of d_2TTP that completely blocked purified DNA polymerase-β did not prevent repair synthesis. Furthermore, novobiocin, an inhibitor of both topoisomerase II and polymerase-α also partially arrested repair activity in the permeabilized cell system.[56] In a subsequent publication, it was reported that aphidicolin was also 60 to 70% inhibitory to repair of MNU and *N*-acetoxy-2-*O*-acetylaminofluorene (NAAAF) damage to DNA, whereas repair synthesis following exposure to bleomycin was only 20 to 25% inhibited by aphidicolin.[57] An additional element of complexity of the repair response was suggested when it was found that the inhibitory effect of aphidicolin on repair synthesis was dependent on the degree of damage inflicted on the DNA.

Thus, following X-ray irradiation or MNU treatment at low doses, repair was only 0 to 20% depressed by aphidicolin, but at high levels of damage, repair became 40 to

70% sensitive to aphidicolin. Although it is difficult to assess damage to DNA at low doses of UV irradiation and NAAAF, it was suggested that the degree of participation of polymerase-α in repair of damage inflicted by these two agents was also dose responsive. In contrast, even at high doses of bleomycin, only a maximum of 50% of repair synthesis was blocked by aphidicolin.[57] These observations on dose-dependence of the repairing polymerase-α were subsequently expanded in a study on repair synthesis following exposure to UV irradiation, NAAAF, MNU, and bleomycin.[58] The general conclusion is that at low levels of damage to DNA, polymerase-β is the major repairing polymerase, whereas at high levels of damage, polymerase-α is progressively mobilized for repair synthesis. These findings are further discussed in Section V.B of this chapter.

Recently, Dresler[59] performed a comparative study on the effect of aphidicolin on DNA replication and on repair of UV-induced damage to DNA in permeabilized dividing human fibroblasts. Both repair and replication of DNA were found to be inhibited by aphidicolin, yet a 20-fold higher concentration of the antibiotic was required to inhibit repair than to suppress replication under identical salt and dNTP substrate concentrations. Both repair and replication were inhibited by aphidicolin competitively with dCTP and with similar K_i values. However, K_m values for dCTP were 0.17 μM and 2.0 μM for repair and replication, respectively, and K_m values for all four dNTPs were 0.07 μM and 30.0 μM for repair and replication, respectively.[59] The lower K_i of aphidicolin for repair synthesis could reflect increase in the intracellular levels of dNTPs following damage to DNA (see Chapter 8). Yet, it may be that different subspecies of polymerase-α participate in the repair and replication processes. Alternatively, interaction of polymerase-α with components of the hypothetical replication complex and/or a putative repair complex might modify its affinity to dNTP substrates.

6. *Effect on Damaged DNA of the Arrest of Polymerase-α Activity*

Several authors addressed the question of molecular consequences of the arrest of DNA repair synthesis by aphidicolin. In an early work, Bender and Preston[60] studied the influence of aphidicolin on the frequencies of X-ray-induced chromatid and chromosomal aberrations in human cells irradiated at either the G_0 or G_2 phases of the cell cycle. Exposure of cells to aphidicolin during the G_2 period increased the frequencies of chromatid breaks and exchanges, whereas treatment of irradiated G_0 cells with aphidicolin enhanced both chromosomal and chromatid aberrations.[60]

These preliminary studies did not reveal, however, the molecular mechanism of induction of chromosomal and chromatid breaks by aphidicolin. Cleaver[61] showed recently that ara C, aphidicolin, and d_2TTP each partially repressed unscheduled DNA synthesis and enhanced accumulation of single-strand breaks in DNA of UV-irradiated human fibroblasts. Calculations were presented suggesting that shorter than normal repair patches were formed in the presence of ara C or aphidicolin (6 to 17 nucleotides), whereas without inhibitors repaired patches consisted of 25 to 75 nucleotides.[61] It was suggested that in the presence of inhibitors of polymerase-α, a single quasi-processive cycle of DNA polymerization is allowed, and then repair synthesis ceases. Removal of aphidicolin allowed completion of the patch until DNA ligase could seal it.[61] These results suggested that repair synthesis was only partially sensitive to inhibitors of polymerase-α, and in their presence shorter than normal repair patches were accumulated.

Discrepant results were reported recently by Smith and Okumoto,[62] who investigated the molecular consequences of exposure of UV-irradiated quiescent human cells to aphidicolin. Under their conditions, repair synthesis was inhibited 70 to 80% by either aphidicolin or ara C. Direct cesium chloride density gradient analysis of repair patches made in the presence of either drug revealed that they were not shorter than normal, but rather that their number was drastically reduced. Patches made in the presence of inhibitors of polymerase-α were found to be degradable by S1 nuclease, thus indicating that they were created through strand-displacement synthesis, and that they remained unligated for an extended period of time. Removal of aphidicolin led to enhanced completion of patch synthesis to full sealability by DNA ligase.[62] It was therefore suggested that in the presence of inhibitors of polymerase-α, strand-displacement synthesis by polymerase-β took place at a slow rate to fill gaps left uncompleted by the arrested polymerase-α.[62] The finding that fewer, rather than shorter, patches are made in the presence of aphidicolin contradicts the notion that the inhibitor merely slows the rate of DNA polymerization at incision sites near lesions. Rather it is suggested that blocking of the DNA polymerase-α-directed repair pathway leads to utilization of an alternate repair mode that most probably involves strand displacement synthesis by polymerase-β (see below).

With an aphidicolin-resistant mutant of *Herpes simplex* virus, experiments indicated that aphidicolin markedly restricted plaque formation by a UV-irradiated virus.[63] The drug had no effect, however, on the plating efficiency of nonirradiated virus. These results suggested that polymerase-α was involved in the repair of UV-irradiated viral DNA. This is somewhat in contrast to an earlier finding at the same laboratory[64] that repair synthesis of viral and cellular DNA at the late stages of infection with nonirradiated *Herpes simplex* virus was resistant to aphidicolin and to phosphonacetic acid, thus implicating polymerase-β rather than polymerase-α in repair of DNA in virus-infected cells. Hence it was suggested that polymerase-α and -β may be involved in the repair of UV-evoked and virus-induced damage to DNA, respectively.[64]

B. DNA Polymerase-β

DNA polymerase-β may be the major or only repair polymerase responsible for resynthesis of DNA after damage by certain agents, in postmitotic cells that are devoid of detectable polymerase-α (cf. Section II.A of this chapter). However, the question of which polymerases are mobilized to repair damaged DNA in cells that contain DNA polymerases of all three major classes remains still largely unanswered. In this section we attempt to categorize factors that may contribute to the final relative extent of DNA polymerase-β participation in repair of DNA.

First, the growth rate of cells may affect the mobilization of polymerase-β for repair synthesis. Several authors reported that inhibitors of polymerase-α such as aphidicolin and ara C were less effective in inhibiting repair synthesis in dividing cells than in their quiescent counterparts.[42,62,65] A plausible interpretation of this observation is that proliferating cells contain higher intracellular levels of dCTP, which competitively overcomes the effects of either aphidicolin or ara C.[42] However, an alternative explanation is that as cells proceed into S phase, repair functions are switched from polymerase-α to polymerase-β.[65] To our knowledge, neither of these possibilities has yet been decisively confirmed or refuted.

That polymerases-α and -β may operate in concert in DNA repair synthesis was suggested by results of numerous experiments that showed partial inhibition of repair by selective inhibition of either polymerase-α or polymerase-β, and a higher degree of arrest of repair by simultaneous inhibition of both polymerase types. For example, aphidicolin and d_2TTP were each partially inhibitory to repair synthesis in HeLa cells after UV irradiation.[66] Furthermore, d_2TTP microinjected into nuclei or cytoplasm of UV irradiated cells depressed DNA repair synthesis as did ara C. Hence polymerase-α and -β may conduct repair synthesis together, and may perhaps be interchangeable. A similar conclusion was reached in studies on the inhibitory effects of d_2TTP and ara C on repair following UV irradiation or MNNG damage to human carcinoma cell DNA.[67] Neither of the two inhibitors alone completely blocked repair, but their combination totally repressed repair as measured by sedimentation of restored supercoiled DNA nucleoids. Tyrrel[68] studied the effects of aphidicolin on UV-irradiated normal human fibroblasts and on xeroderma pigmentosum variant cells that were capable of normal excision repair. Aphidicolin was found to be cytotoxic to UV-irradiated fibroblasts. The difference in survival between UV-irradiated cells in the presence of aphidicolin and in its absence yielded an indirect estimate of the proportion of sites repaired by polymerase-α, which was 67%. The cytotoxic effect of aphidicolin suggests that at some damaged sites, polymerase-α is not replaceable by polymerase-β. However, other evidence indicates that alternate pathways for repair involving DNA polymerase-β do operate when polymerase-α is blocked.[62]

Arguments were presented to show that DNA polymerase-β may be a major repair enzyme in rodent cells. DNA polymerase-β from a variety of cell types has been shown to be more thermosensitive than polymerase-α.[69-70] Dewey and associates[71-73] established a good correlation between the extent of heat inactivation of polymerase-β and thermal killing of CHO cells after X-ray irradiation. The fit between cell survival and residual polymerase-α activity was less rigorous. It is difficult, however, to interpret these results in terms of the relative role of each polymerase type in repair, since threshold intracellular levels of α- and β-polymerases that are required for DNA repair are not known. Furthermore, heat inactivation of other repair enzymes and proteins may also contribute to the complexity of the cellular response to hyperthermia. Nevertheless, these observations could be particularly important because of the potentiating effects of hyperthermia on cell killing after X irradiation and the selective killing of certain tumor cells by hyperthermia.

Dresler and Lieberman[58] put forward the argument that the extent of damage to DNA determines to what extent polymerase-β is joined by polymerase-α in repairing DNA. Hence, at low doses of damage by UV irradiation, NAAAF and NMU, repair synthesis of DNA was mainly mediated by an aphidicolin- and *N*-ethylmaleimide-resistant and d_2TTP-sensitive activity, presumably polymerase-β. At higher doses of the damaging agent, the proportion of aphidicolin/*N*-ethylmaleimide-sensitive repair increased until a maximum 70 to 80% of the repair synthesis was catalyzed by polymerase-α.[58] Previous failures to detect dose responsiveness of polymerase-α mobilization for repair[55] were explained as resulting from use of saturating concentrations of DNA-damaging agents.[58]

As discussed at length in Section V.A, Miller and associates[53-55] provided evidence to show that a major factor in determining the relative contribution of polymerase-α polymerase-β to DNA repair polymerization is the nature of damaging agent employed. That this may indeed be the case was first demonstrated by Castellot et al.[74] who showed that ara CTP, *N*-ethylmaleimide, and high concentrations of KCl — all inhibitors of DNA replication and of polymerase-α — minimally affected repair synthesis after exposure of lysolecithin-permeabilized hamster cells to bleomycin. Several authors have subsequently demonstrated that repair of bleomycin-evoked damage was relatively resistant to inhibitors of polymerase-α activity.[32,43,53,57] Similarly, repair of damage elicited by neocarzinostatin appeared also to be conducted mainly by polymerase-β. In contrast, as discussed in Section V.A, repair of damage induced by UV and X irradiation and of damage elicited by NAAAF, MNNG, and NMU is blocked more efficiently by inhibitors of polymerase-α.[53-55,57] It was thus argued that the recognizability of the damaged site by specific repair nucleases, rather than the patch size, determines whether polymerase-β or polymerase-α will be the major repairing polymerase.[54] This implies, therefore, that repair involves the concerted action of nucleases and one of the DNA polymerases.

V. ENZYMATIC PROPERTIES OF DNA POLYMERASES AND THEIR CONCORDANCE WITH ROLES IN DNA REPAIR

A. Catalytic Properties of DNA Polymerase-α and -β

Although properties of isolated DNA polymerases may allude to their ability to repair certain lesions in DNA, it should be kept in mind that the repair process involves a full set of enzymes and proteins that are responsible for recognition of the lesion, removal of adducts, and hydrolysis of the DNA at the damaged site. Thus characteristic enzymic properties of high purified α- and β-polymerases suggest their possible role in DNA repair only in a tentative manner. Two catalytic parameters appear to be relevant to repair synthesis: gap-size preference of α- and β-polymerases, and the ability of polymerase-β to perform limited-strand-displacement synthesis of DNA. Also, the distributive nature of DNA polymerization by polymerase-β is often cited as an indication for its proposed participation in repair synthesis of relatively short stretches of DNA. Although lack of processiveness may not hinder repair synthesis of short patches, it is hard to see why it should be an advantage for a repairing polymerase.

The mechanism of creation of a long gap in animal cell DNA after excision of bulky adducts is not known. Based on recent studies in bacteria, it seems likely that the adduct is removed by simultaneous endonucleolytic breaks that are several nucleotides apart, and that encompass the site of damage.[7] Thereafter, elongation is mediated by exonucleolytic activity, the identity of which has not yet been established, although DNase V is an attractive candidate.

The order of the long-patch repair reaction is not known at present, and several possibilities can be envisioned. First, DNA synthesis could precede exonucleolytic digestion, thus advancing by a strand-displacement mechanism. In this case, the ability of polymerase-β to conduct strand-displacement synthesis[75-77] makes it a better-adapted enzyme for this mode of repair than polymerase-α, which appears not to perform strand-displacement DNA polymerization. Furthermore, whereas polymerase-α is not reactive with nicks and single-nucleotide gaps in DNA,[76,77] polymerase-β has high affinity to nicks and fills single-base gaps to completion.[77,78] An alternate route is for DNA synthesis and exonucleolysis to proceed coordinately. This last alternative is attractive, since it does not leave highly vulnerable single-stranded segments of DNA exposed for unpropitious hydrolysis. Again, because of the reactivity of polymerase-β with incised sites and its capacity to extend DNA strands by displacement, it is the best-fitted enzyme to perform repair synthesis by that mechanism. Third, it may be that excision is completed before synthesis starts. Here, a gap is created in DNA, and both polymerase-α and polymerase-β are able to participate in gap-filling. Yet, the two enzymes exhibit different preferences for gaps of different size. As discussed in Chapter 3, Section V.G, polymerase-α prefers gaps of 30 to 60 nucleotides, and it usually does not fill them to completion. In contrast, polymerase-β potentially utilizes shorter gaps and is able to fill them to completion.[76-79] These properties appear to indicate that both polymerase types may take part together in excision repair.

The ability of polymerase-β to utilize single-base gaps[76,77] contrasts with the failure of polymerase-α to interact in most examined cases with such nicks.[76,77] Hence polymerase-β appears to fit better as a repairing enzyme for small gaps. It should be stressed again, however, that properties of highly purified polymerases may not fully reflect their potential for repair synthesis in vivo. Interaction of polymerase(s) with other proteins in a putative repair complex might alter their catalytic properties and enable them to conduct repair synthesis that an isolated enzymes cannot perform.

B. In Vitro Utilization of Damaged DNA Templates by Polymerase-α and -β

To gain insight into modes of repair by animal cell polymerases, several authors used

DNA damaged by defined agents as templates for isolated polymerases. Siedlecki et al.[80] reported that repaired DNA derived from BudR-treated *Escherichia coli* was copied more efficiently by purified calf thymus polymerase-β than by polymerase-α. Subsequently, Nowak et al.[81] showed that calf thymus polymerase-β but not polymerase-α was able to fill gaps created by micrococcal correnuclease in UV-irradiated ColE1 DNA. In investigating the in vitro copying of duplex *E. coli* DNA damaged by thymine starvation or by MMS, it was found that the synthesis by polymerase-β exceeded the extent of gaps available in the template.[82] That result was interpreted as reflecting strand-displacement synthesis by polymerase-β.

Philippe et al.[83] compared the ability of highly purified human cell DNA polymerase-α and -β to copy UV-irradiated hook polymer: poly(dT) primed with $(dA)_{15}$ covalently attached to the 3′ end of the template. Surprisingly, in vitro copying of that template was enhanced when it was UV irradiated. In the presence of Mn^{2+}, polymerase-β produced full-length copies of either unirradiated or irradiated template. In marked contrast, polymerase-α copied only short segments of the unirradiated template, but produced full-length copies of poly(dT) irradiated at an appropriate UV dose.[83] The relevance of these observations to in vivo repair of UV damage is not clear. It is possible that poly(dT) exists in solution as polystranded structures that are only partially available for copying by the polymerases. The presence of thymidine dimers generated by UV irradiation may make the template available for polymerase-catalyzed replications through a primer slippage mechanism.[83] The ability of animal cell DNA polymerases to bypass in vitro apurinic/apyrimidinic lesions in DNA, and to copy DNA opposite adducts of NAAAF, were investigated by Loeb and colleagues and by Strauss and associates. The results of these experiments pertain directly to the mutagenic potential of DNA polymerases, and are discussed at length in Chapter 8.

C. Accessory Proteins for DNA Polymerases and the Hypothetical Multienzyme Repair Complex

In deliberating the adaptability of DNA polymerase-α and -β to repair synthesis of DNA, the specific association of these enzymes with auxiliary proteins should be considered. Neither highly purified DNA polymerase-α nor -β are associated with nucleases (cf. Chapter 3 and 4, respectively). However, crude polymerase-β from Novikoff hepatoma and from HeLa cells copurified with a specific bidirectional nuclease, DNase V, which appears to form a specific complex with the polymerase (cf. Chapter 4, Section III.A). It has been suggested [84,85] that polymerase-β and DNase V, which stimulates its activity, together form part of a repair complex that performs excision repair through the concerted action of both nuclease and polymerase. In vitro functional reconstruction of this putative repair complex was reported recently by Mosbaugh and Linn see Section VI of this chapter for discussion).[77] Thomas and Meyer[86] isolated a DNA-dependent ATPase that stimulates the activity of polymerase-β particularly ungapped duplex DNA. The properties of the enzyme, designated ATPase III, and its specific stimulatory effect on polymerase-β raised the possibility that it, too, is a component of a repair complex formed with polymerase-β (cf. Chapter 4, Section III.B).[86] A possible fourth component in the multiprotein repair complex is a 48-kdalton DNA-binding protein from Novikoff hepatoma cells that specifically stimulated homologous DNA polymerase-β.[86,101]

Data on an aphidicolin-resistant, hypermutable mutant rodent cell (cf. Section IV.A) clearly indicates that polymerase-α takes part in some modes of DNA repair. It is not known, however, whether other proteins interact in vivo with polymerase-α during the performance of DNA repair synthesis. Several proteins have been shown to serve as ancillary cofactors for polymerase-α (cf. Chapter 3, Section IV). Some of these accessory proteins appear to function together with polymerase-α in its role as a replicative enzyme. Yet none of these proteins exhibit properties that specifically mark it as a possible component in a repair complex. Bose et al.[87] and Mosbaugh and Linn[79] constructed multienzyme in vitro repair systems in which polymerase-*α* functioned in repair synthesis in vitro (see Section VI below). However, it is not known whether polymerase-*α*, acting as a repair enzyme, is part of a multiprotein repair complex in vivo. The tantalizing possiblity that two types of polymerase-*α*-containing multiprotein complexes exist in the cell, one responsible for DNA replication and the other for repair, was hinted at by the finding that some catalytic properties of polymerase-*α* differ when it functions in replication and in repair (cf. Section IV above).[59]

VI. RECONSTRUCTED IN VITRO MODEL SYSTEMS FOR REPAIR OF DAMAGED DNA

As noted above, recent new direction in the quest for elucidation of roles of DNA polymerases in DNA repair is the construction of multienzyme systems that conduct in vitro repair of defined lesions in DNA. Some of these systems contained polymerase-*α* as the synthetic enzyme, and others polymerase-*β*.

Bose et al.[87] constructed an in vitro system to mimic short-patch excision repair of alkali-labile apurinic/apyrimidinic lesions created in phage T7 DNA by methylmethane sulfonate. Nuclease preparations consisting of both apurinic nuclease and a $5' \rightarrow 3'$ exonuclease excised only one to four nucleotides per lesion. Human lymphoblastoid cell polymerase-*α* was able to fill the gaps thus created, and DNA ligase was able to subsequently seal the small repair patch.[87] Hence, under the defined in vitro conditions of this experiment, it was demonstrated that purified polymerase-*α* was able to function in small-patch repair of DNA. Mosbaugh and Linn[79] created gaps of defined mean length in partially depurinated phage PM2 DNA by a combination of HeLa cell apurinic/apyrimidinic endonuclease II and HeLa DNase V. It was found that HeLa polymerase-*α* utilized gaps of 20 to 63 nucleotides, but synthesis was terminated as gaps were reduced to ~15 bases. It was concluded, therefore, that in an in vitro system, HeLa polymerase-*α* was neither able to utilize small gaps nor could it fill larger gaps to completion. These properties contrast with those of polymerase-*β*, which interacted with large and small gaps and filled them to completion (see below).[77,79] It is not clear why lymphoblastoid cell[87] and HeLa cell[77,79] polymerases differ in their ability to perform short-patch repair. The possibility that polymerase-*α* is able to function in repair of patches a few nucleotides in length, and that it functions together with polymerase-*β* to resynthesize longer patches is suggested by the two described in vitro systems.

Mosbaugh and Linn[77] examined in detail the ability of DNA polymerase-*α* and -*β* to repair gaps of only a single nucleotide in DNA. HeLa cell apurinic/apyrimidinic endonuclease II created nicks on the 5′ side of an apurinic site in DNA, leaving a 3′ hydroxyl terminus, and polymerase-*β* was found to be able to carry out a limited strand-displacement synthesis from that nicked site. When *E. coli* endonuclease III incised at the 3′ side of the apurinic site, and apurinic/apyrimidinic endonuclease II removed the baseless ribose phosphate, a single nucleotide gap was created and polymerase-*β* was able to fill it. In contrast, polymerase-*α* was unable to react with either nicks or single-base gaps.[77] Concerted action of the above nucleases, together with DNase V and polymerase-*β*, led to nick translation from single incisions at apurinic/apyrimidinic sites. A system was constructed to repair in vitro UV-induced pyrimidine dimers. The components of that system are depicted in Figure 1. DNA polymerase-*β*, together with DNase V and DNA ligase, could repair lesions caused by UV irradiation after T4 UV endonuclease and HeLa cell apurinic endonuclease II incised the DNA at the thymidine-dimer site. DNase V removed the pyrimidine dimers, primarily as dinucleotides, and then promoted nick-translational DNA synthesis by polymerase-*β* that

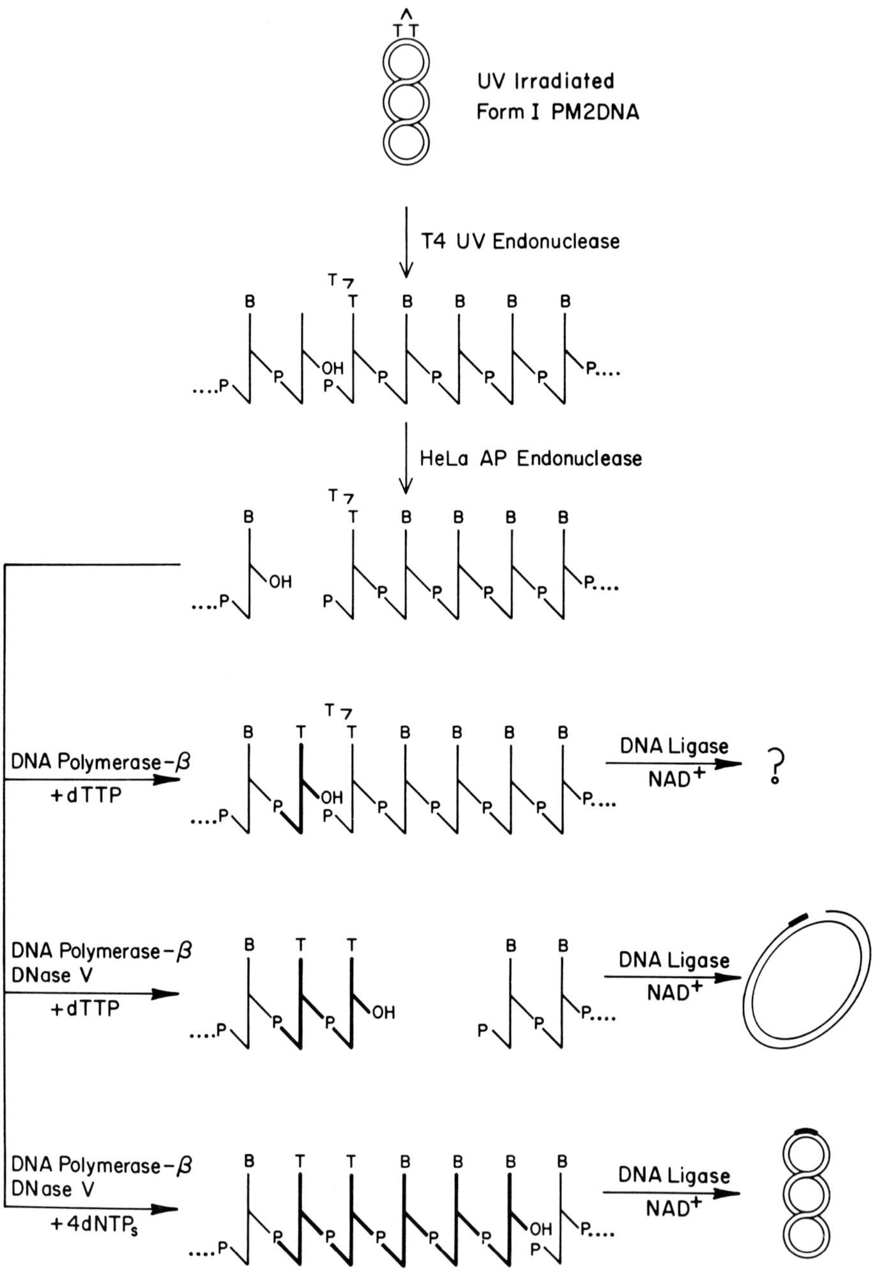

FIGURE 1. A multienzyme system for the in vitro excision repair of UV-induced pyrimidine dimers. Form I PM2 DNA labeled with [^{3}H]-thymidine is UV-irradiated to generate ∼1.5 pyrimidine dimers per duplex DNA circle. The UV-irradiated DNA is incised by T4 UV endonuclease, which produces nicks with 3′ deoxyribose termini and thymine-thymidylate cyclobutane dimer 5′ termini. The 3′ apyrimidinic site is in turn removed by HeLa cell apurinic/apyrimidinic endonuclease II, and a single-nucleotide gap is created that contains a 3′ hydroxyl end, which supports DNA synthesis. Enzymes at different combinations are added with [^{32}P]dTTP to follow repair synthesis. Ligation of the repair patch is monitored by restoration of closed circles, as assessed by alkaline sucrose gradient sedimentation of the repaired DNA. With DNA polymerase-β alone, <10% of the molecules are sealed by ligase. Similarly, with polymerase-β and DNase V present but with dTTP as the only substrate, minimal sealing is observed. However, when polymerase-β, DNase V, and all four dNTPs are added, excision repair takes place and DNA ligase seals the repair patch so that [^{32}P][^{3}H]-labeled covalently closed DNA is formed. The amount of [^{32}P]dTMP incorporated into DNA suggests that the nick-translated patch is roughly 33 nucleotides long. (Adapted from Mosbaugh, D. W. and Linn, S., *J. Biol. Chem.*, 258, 108, 1983. With permission.)

was completed when DNA ligase sealed the repair patch Figure 1.[77] It appears, therefore, that under these experimental conditions, DNA polymerase-β participates in both long- and short-patch repair, whereas polymerase-α was unable to repair single-nucleotide gaps. That polymerase-β can utilize both long and short gaps in DNA was also shown in a system described above,[79] in which polymerase-β was able to fill to completion both short and long gaps, although it was more reactive with the former.

A model system for repair synthesis in chromatin rather than naked DNA was constructed recently by Evans and Linn.[88] Minichromosomes of SV40 were UV-irradiated and treated with T4 UV endonuclease and HeLa cell apurinic endonuclease II to incise DNA at the dipyrimidinic lesion. Excision was subsequently conducted with DNase V, and synthesis was carried out by DNA polymerase-β that could fill the gap so the DNA ligase seals it.[88] In the chromatin system, DNA polymerase-β did not perform strand-displacement synthesis as in nicked naked DNA, but rather filled excised gaps to completion.

It is not clear, of course, whether the reconstructed repair systems reflect accurately the components and mechanisms of action of an in vivo repair complex, but they are certainly instructive as to the potential of polymerase-α and -β to function in defined modes of DNA repair.

VII. CONCLUDING REMARKS

The accumulated body of data on the ability of both DNA polymerase-α and polymerase-β to participate, separately or together, in different modes of repair clearly indicates that, as in bacteria, eukaryotes do not have a singular DNA repair polymerase. However, much work is required to reveal the specific role of each class of polymerase in any repair pathway, the extent of interchangeability between the polymerases, and their specific interaction with additional components of postulated repair complexes. Because of the diversity of chemical agents that damage DNA and the multiplicity of cellular pathways for DNA repair, our understanding of roles of DNA polymerases in repair synthesis in eukaryotes has lagged behind the elucidation of their part in replicative synthesis of DNA.

VIII. UPDATE

A. Involvement of DNA Polymerases-α and -β in Repair Synthesis: Studies with Selective Inhibitors

Aphidicolin and d_2TTP have been used to distinguish between repair synthesis which is catalyzed by polymerase-α and polymerase-β after exposure of HeLa cells to several different damaging agents.[89] Direct layering of the cells on alkali sucrose density gradients permitted lowering of the level of nonspecific shearing of the DNA and allowed detection of a single lesion per 10^6 kdaltons of DNA. It appeared that after UV irradiation only polymerase-α was involved in repair synthesis. By contrast, following damage by neocarzinostatin or bleomycin polymerases-α and -β operated on different lesions and after exposure to dimethyl sulfate both polymerases participated in repair of the same lesions.[89]

REFERENCES

1. Hanawalt, P. C., Cooper, A. K., Ganesan, A. K., and Smith, C. A., DNA repair in bacteria and mammalian cells, *Annu. Rev. Biochem.,* 48, 783, 1979.
2. Hall, J. D. and Mount, D. W., Mechanisms of DNA replication and mutagenesis in ultraviolet-irradiated bacteria and mammalian cells, *Prog. Nucl. Acid Res. Mol. Biol.,* 25, 53, 1981.
3. Grossman, L., Enzymes involved in the repair of damaged DNA, *Arch. Biochem. Biophys.,* 211, 511, 1981.
4. Lindahl, T., DNA repair enzymes, *Annu. Rev. Biochem.,* 51, 61, 1981.
5. Sarasin, A., Rossignol, J.-M., and Philippe, M., DNA polymerases and DNA repair in eucaryotic cells, in *New Approaches in Eukaryotic DNA Replication,* de Recondo, A. M., Ed., Plenum Press, New York, 1983, 313.
6. Walker, G. C., Mutagenesis and inducible responses to deoxyribonucleic acid damage in *Escherichia coli, Microbiol. Rev.,* 48, 60, 1984.
7. Sancar, A. and Rupp, W. D., A novel repair enzyme: UVRABC excision nuclease of *Escherichia coli* cuts a DNA strand on both sides of the damaged region, *Cell,* 33, 249, 1983.
8. Singer, B. and Grunberger, D., *Molecular Biology of Mutagens and Carcinogens,* Plenum Press, New York, 1983.
9. Hübscher, U., Kuenzle, C. C., and Spadari, S., Functional roles of DNA polymerases β and γ, *Proc. Natl. Acad. Sci. U.S.A.,* 76, 2316, 1979.
10. Kalluri, S. K., Martin, G. M., and Loeb, L. A., Fidelty of DNA polymerase-β in neurons from young and very aged mice, submitted.
11. Margison, G. P. and Kleihues, P., Chemical carcinogenesis in the nervous system. Differential accumulation of O^6-methylguanine in rat brain deoxyribonucleic acid during repetitive administration of *N*-methyl-*N*-nitrosourea, *Biochem. J.,* 148, 521, 1975.
12. Kleihues, P. and Buechler, J., Long term persistence of O^6-methylguanine in rat brain, *Nature (London),* 269, 625, 1977.
13. Waser, J., Hübscher, U., Kuenzle, C. C., and Spadari, S., DNA polymerase-β from brain neurons is a repair enzyme, *Eur. J. Biochem.,* 97, 361, 1979.
14. Rossignol, J.-M., Gentil, A., Lacharpagne, J., and de Recondo, A.-M., Involvement of DNA polymerase-β in a *N*-methyl-*N*-nitro-*N*-nitrosoguanidine (MNNG)-induced DNA repair process, *Biochem. Int.,* 1, 253, 1980.
15. Butt, R. T., Wood, W. M., McKay, E. L., and Adams, R. L. P., Involvement of deoxyribonucleic acid polymerase-β in nuclear deoxyribonucleic acid synthesis, *Biochem. J.,* 173, 309, 1978.
16. Bertazzoni, U., Stefanini, M., Noy, G. P., Giulotto, E., Nuzzo, F., Falaschi, A., and Spadari, S., Variations of DNA polymerase-α and -β during prolonged stimulation of human lymphocytes, *Proc. Natl. Acad. Sci. U.S.A.,* 73, 785, 1976.
17. Wicker, R., Scovassi, A. I., and Nocentini, S., DNA polymerase-α, β, and γ activities in ultraviolet irradiated CV-1 monkey cells, *Nucl. Acids Res.,* 6, 1591, 1979.
18. Suzuki, K., Miyaki, M., Akamatsu, N., and Ono, T., Increase of DNA polymerase activities in HeLa cells treated with *N*-methyl-*N*-nitro-*N*-nitrosoguanidine, *FEBS Lett.,* 119, 150, 1980.
19. Stefanini, M., Scovassi, A. I., and Bertazzoni, U., Functions of DNA polymerases α, β, and γ in DNA replication and repair, in *Lymphocyte Stimulation,* Castellani, A., Ed., Plenum Press, New York, 1980, 33.
20. Craddock, V. M. and Ansley, C. M., Sequential changes in DNA polymerases α and β during diethylnitrosamine-induced carcinogenesis, *Biochim. Biophys. Acta,* 564, 15, 1979.
21. Craddock, V. M., DNA polymerases in replication and repair of DNA during carcinogenesis induced by feeding *N*-acetylaminofluorene, *Carcinogenesis,* 2, 61, 1981.
22. Smith, H. C. and Berezney, R., Nuclear matrix deoxyribonucleic acid synthesis: an *in vitro* system, *Biochemistry,* 21, 6751, 1982.
23. Gupta, P. K. and Sirover, M. A., Sequential stimulation of DNA repair and DNA replication in normal human cells, *Mutat. Res.,* 72, 273, 1980.
24. Parker, V. P. and Lieberman, M. W., Levels of DNA polymerases α, β, and γ in control and repair of deficient human diploid fibroblasts, *Nucl. Acids Res.,* 4, 2029, 1977.
25. Bertazzoni, U., Scovassi, A. I., Stefanini, M., Giulotto, E., Spadari, S., and Pedrini, A., DNA polymerases α, β and γ in inherited diseases affecting DNA repair, *Nucl. Acids Res.,* 5, 2189, 1978.
26. Chang, C.-C., Boezi, J. A., Warren, S. T., Sabourin, C. L. K., Liu, P. K., Glatzer, L., and Trosko, J. E., Isolation and characterization of a UV-sensitive hypermutable aphidicolin-resistant Chinese hamster cell line, *Somat. Cell Genet.,* 7, 235, 1981.
27. Liu, P. K., Trosko, J. E., and Chang, C.-C., Hypermutability of a UV-sensitive aphidicolin-resistant mutant of Chinese hamster fibroblasts, *Mutat. Res.,* 106, 333, 1982.
28. Liu, P. K., Chang, C.-C., and Trosko, J. E., Association of mutator activity with UV sensitivity in an aphidicolin-resistant mutant of Chinese hamster V79 cells, *Mutat. Res.,* 106, 317, 1982.

29. Liu, P. K., Chang, C.-C., and Trosko, J. E., Evidence for mutagenic repair in V79 cell mutant with aphidicolin-resistant DNA polymerase-α, *Somat. Cell Mol. Genet.*, 10, 235, 1984.
30. Liu, P. K., Chang, C.-C., Trosko, J. E., Dube, D. K., Martin, G. M., and Loeb, L. A., Mammalian mutator mutant with an aphidicolin-resistant DNA polymerase-α, *Proc. Natl. Acad. Sci. U.S.A.*, 80, 797, 1983.
31. Liu, P. K. and Loeb, L. A., Transfection of DNA polymerase-α gene, *Science*, 226, 833, 1984.
32. Seki, S., Oda, T., and Ohashi, M., Differential effects of aphidicolin on replicative DNA synthesis and unscheduled DNA synthesis in permeable mouse sarcoma cells, *Biochim. Biophys. Acta*, 610, 413, 1980.
33. Seki, S., and Oda, T., Effects of 2′,3′-dideoxythymidine triphosphate on replication DNA synthesis and unscheduled DNA synthesis in permeable mouse sarcoma cells, *Biochim. Biophys. Acta*, 606, 246, 1980.
34. Pedrali-Noy, G. and Spadari, S., Aphidicolin allows a rapid and simple evaluation of DNA repair synthesis is damaged human cells, *Mutat. Res.*, 70, 389, 1980.
35. Giulotto, E., and Mondello, C., Aphidicolin does not inhibit the repair synthesis of mitotic chromosomes, *Biochem. Biophys. Res. Commun.*, 99, 1287, 1981.
36. Hardt, N., Pedrali-Noy, G., Focher, F., and Spadari, S., Aphidicolin does not inhibit DNA repair replication in HeLa cells: an autoradiographic study, *Biochem. J.*, 199, 453, 1981.
37. Hanaoka, F., Kato, H., Ikegami, S., Ohashi, M., and Yamada, M., Aphidicolin does inhibit DNA repair replication in HeLa cells, *Biochem. Biophys. Res. Commun.*, 87, 575, 1979.
38. Kato, H., Hanaoka, F., Tetsuji, N., Yamada, K., and Yamada, M.-A., DNA repair synthesis in HeLa cell lysate, *Biochim. Biophys. Acta*, 606, 47, 1980.
39. Berger, N. A., Kurohara, K. K., Petzold, S. J., and Sikorsky, G. W., Aphidicolin inhibits eukaryotic DNA replication and repair — implications for involvement of DNA polymerase-α in both processes, *Biochem. Biophys. Res. Commun.*, 89, 218, 1979.
40. Ciarrocchi, G., Jose, J. G., and Linn, S., Further characterization of a cell-free system for measuring replicative and repair DNA synthesis with cultured human fibroblasts and evidence for involvement of DNA polymerase-α in DNA repair, *Nucl. Acids Res.*, 7, 1205, 1979.
41. Snyder, R. D. and Regan, J. D., Aphidicolin inhibits repair of DNA in UV-irradiated human fibroblasts, *Biochem. Biophys. Res. Commun.*, 99, 1088, 1981.
42. Snyder, R. D. and Regan, J. D., Differential responses of log and stationary phase human fibroblasts to inhibition of DNA repair by aphidicolin, *Biochim. Biophys. Acta*, 697, 229, 1982.
43. Seki, S., Ohashi, M., Ogura, H., and Oda, T., Possible involvement of DNA polymerases α and β in bleomycin-induced unscheduled DNA synthesis in permeable HeLa cells, *Biochem. Biophys. Res. Commun.*, 104, 1502, 1982.
44. Collins, A. R. S., Squires, S., and Johnson, R. T., Inhibitors of repair DNA synthesis, *Nucl. Acids Res.*, 10, 1203, 1982.
45. Collins, A., DNA repair in ultraviolet-irradiated HeLa cells is disrupted by aphidicolin. The inhibition of repair does not imply the absence of repair synthesis, *Biochim. Biophys. Acta*, 741, 341, 1983.
46. Van Zeeland, A. A., Bussman, C. J. N., Degrassi, F., Filon, A. R., van Kesteren-van Leeuven, A. C., Palitti, F., and Natarajan, A. T., Effects of aphidicolin on repair replication and induced chromosomal aberrations in mammalian cells, *Mutat. Res.*, 92, 379, 1982.
47. Waters, R., Aphidicolin: an inhibitor of DNA repair in human fibroblasts, *Carcinogenesis*, 2, 795, 1981.
48. Waters, R., Crocombe, K., and Mirzayans, R., The inhibition of DNA repair by aphidicolin or cytosine arabinose in X-irradiated normal and xeroderma pigmentosum fibroblasts, *Mutat. Res.*, 94, 229, 1981.
49. Bermudez, E., The inhibition of DNA repair in primary rat hepatocyte cultures by aphidicolin. Evidence for the involvement of alpha polymerase in the repair process, *Biochem. Biophys. Res. Commun.*, 109, 275, 1982.
50. Zoncheddu, A. and Accomando, R., Characteristics of the DNA polymerase β-directed unscheduled DNA synthesis in isolated nuclei from adult rat liver, *Int. J. Biochem.*, 16, 391, 1984.
51. Yamada, K., Kato, H., Hanaoka, F., and Yamada, M.-A., Unscheduled DNA synthesis induced by N-methyl-N′-nitro-N-nitrosoguanidine in HeLa cells, *Gann*, 73, 63, 1981.
52. Hanaoka, F., Yamada, K., Kato, H., and Yamada, M.-A., Enchancement of ultraviolet-induced unscheduled DNA synthesis in HeLa cell lysate by aphidicolin *in vivo*, *Biochem. Biophys. Res. Commun.*, 107, 1, 1982.
53. Miller, M. R. and Chinault, N. D., Evidence that DNA polymerase α and β participate differentially in DNA repair synthesis induced by different agents, *J. Biol. Chem.*, 257, 46, 1982.
54. Miller, M. R. and Liu, L. H., Participation of different DNA polymerases in mammalian DNA repair synthesis is not related to "patch size", *Biochem. Biophys. Res. Commun.*, 108, 1676, 1982.

55. Miller, M. R. and Chinault, D. N., The roles of DNA polymerases α, β, and γ in DNA repair synthesis induced in hamster and human cells by different DNA damaging agents, *J. Biol. Chem.*, 257, 10204, 1982.
56. Dresler, S. L., Roberts, J. D., and Lieberman, M. W., Characterization of deoxyribonucleic acid repair synthesis in permeable human fibroblasts, *Biochemistry*, 21, 2557, 1982.
57. Dresler, S. L. and Lieberman, M. W., DNA polymerase function in repair synthesis in human fibroblasts, in *ADP Ribosylation, DNA Repair, and Cancer*, Miwa, M., Ed., Japan Science Society Press, Tokyo, 1983, 253.
58. Dresler, S. L. and Lieberman, M. W., Identification of DNA polymerases involved in DNA excision repair in dipolid human fibroblasts, *J. Biol. Chem.*, 258, 9990, 1983.
59. Dresler, S. L., Comparative enzymology of UV-induced DNA repair synthesis and semiconservative DNA replication in permeable diploid human fibroblasts, *J. Biol. Chem.*, 259, 13947, 1984.
60. Bender, M. A. and Preston, R. J., Role of base damage in aberration formation: interaction of aphidicolin and X-rays, *Prog. Mutat. Res.*, 4, 37, 1981.
61. Cleaver, J. E., Structure of repaired sites in human DNA synthesized in the presence of inhibitors of DNA polymerases alpha and beta in human fibroblasts, *Biochim. Biophys. Acta*, 739, 301, 1983.
62. Smith, C. A. and Okumoto, D. S., Nature of DNA repair synthesis resistant to inhibitors of polymerase-α in human cells, *Biochemistry*, 23, 1383, 1984.
63. Nishiyama, Y., Yoshida, S., and Maeno, K., Involvement of DNA polymerase-α in host cell reactivation of UV-irradiated *Herpes simplex* virus, *J. Virol.*, 49, 598, 1984.
64. Nishiyama, Y., Tsurumi, T., Aoki, H., and Maeno, K., Identification of DNA polymerase(s) involved in the repair of viral and cellular DNA in *Herpes simplex* type 2-infected cells, *Virology*, 129, 524, 1983.
65. Cleaver, J. E., Sensitivity of excision repair in normal human, xeroderma pigmentosum and Cokayne's syndrome fibroblasts to inhibition by cytosine arabinose, *J. Cell. Physiol.*, 108, 163, 1981.
66. Morita, T., Tsutsui, Y., Nishiyama, Y., Nakamura, H., and Yoshida, S., Effects of DNA polymerase inhibitors on replicative and repair DNA synthesis in ultraviolet-irradiated HeLa cells, *Int. J. Radiat. Biol.*, 42, 47, 1982.
67. Mattern, M. R., Paone, R. F., and Day, R. S., III, Eukaryotic DNA repair is blocked at different steps by inhibitors of DNA topoisomerases and of DNA polymerases α and β, *Biochim. Biophys. Acta*, 697, 6, 1982.
68. Tyrell, R. M., Specific toxicity of aphidicolin to ultraviolet-irradiated excision proficient human skin fibroblasts, *Carcinogenesis*, 4, 327, 1983.
69. Dube, D. K., Seal, G., and Loeb, L. A., Differential heat sensitivity of mammalian DNA polymerases, *Biochem. Biophys. Res. Commun.*, 76, 483, 1977.
70. Dube, D. K., Kunkel, T. A., Seal, G., and Loeb, L. A., Distinctive properties of mammalian DNA polymerases, *Biochim. Biophys. Acta*, 561, 369, 1979.
71. Denman, D. L., Spiro, I. J., and Dewey, W. C., Effects of hyperthermia on DNA polymerase alpha and beta: relationship to thermal cell killing and radiosensitization, *Natl. Cancer Inst. Monogr.*, 61, 37, 1982.
72. Spiro, J. J., Denman, D. L., and Dewey, W. C., Effect of hyperthermia on CHO DNA polymerases α and β, *Radiat. Res.*, 89, 134, 1982.
73. Miverchi, N. F. and Dewey, W. C., Effect of glycerol and low pH on heat induced cell killing and loss of DNA polymerase activities in Chinese hamster ovary cells, *Radiat. Res.*, 99, 352, 1984.
74. Castellot, J. J., Jr., Miller, M. R., Lehtomaki, D. M., and Pardee, A. B., Comparison of DNA replication and repair enzymology using permeabilized baby hamster kidney cells, *J. Biol. Chem.*, 254, 6904, 1979.
75. Eichler, D. C., Wang, T. S.-F., Clayton, D. A., and Korn, D., *In vitro* replication of mitochondrial DNA. Elongation of the endogenous primer sequence in D loop mitochondrial DNA by human DNA polymerase-β, *J. Biol. Chem.*, 252, 7888, 1977.
76. Wang, T. S.-F. and Korn, D., Reactivity of KB cell deoxyribonucleic acid, *Biochemistry*, 19, 1782, 1980.
77. Mosbaugh, D. W. and Linn, S., Excision repair and DNA synthesis with a combination of HeLa DNA polymerase-β and DNase V, *J. Biol. Chem.*, 258, 108, 1983.
78. Wang, T. S.-F. and Korn, D., Specificity of the catalytic interaction of human DNA polymerase β with nucleic acid substrates, *Biochemistry*, 21, 1597, 1982.
79. Mosbaugh, D. W. and Linn, S., Gap filling DNA synthesis by HeLa DNA polymerase-α in an *in vitro* base excision DNA repair scheme, *J. Biol. Chem.*, 259, 10247, 1984.
80. Siedlecki, J. A., Szysko, J., Pietrzykowska, I, and Smudzka, B., Evidence implying DNA polymerase β function in excision repair, *Nucl. Acids Res.*, 8, 361, 1980.
81. Nowak, R., Zarebska, Z., and Smudzka, B., The DNA polymerase β reaction with ultraviolet-irradiated DNA incised by correnuclease, *Biochim. Biophys. Acta*, 609, 246, 1980.

82. Soltyk, A., Siedlecki, J. A., Pietrzykowska, I., and Smudzka, B., Reactions of calf thymus DNA polymerases α and β with native DNA damaged by thymine starvation or by methyl methanesulfonate treatment of *Escherichia coli* cells, *FEBS Lett.*, 125, 227, 1981.
83. Philippe, M., Wang, T. S.-F., Hanawalt, P. C., and Korn, D., DNA synthesis on UV irradiated model templates using human DNA polymerases α and β: Primer slippage to account for evident transdimer continuity in product, *Biochemie,* 64, 783, 1982.
84. Mosbaugh, D. W. and Meyer, R. R., Interaction of mammalian deoxyribonuclease V, a double-strand 3′ → 5′ and 5′ → 8′ exonuclease with deoxyribonucleic acid polymerase β from Novikoff hepatoma, *J. Biol. Chem.*, 255, 10239, 1980.
85. Meyer, R. R. and Mosbaugh, D. W., Characterization of mammalian DNase V, a double-strand bidirectional exonuclease which interacts with DNA polymerase-β in the Novikoff hepatoma, in *Mechanistic Studies of DNA Replication and Genetic Recombination,* Alberts, B., Ed., Academic Press, New York, 1980, 639.
86. Thomas, D. C. and Meyer, R. R., Deoxyribonucleic acid dependent adenosine triphosphatase from the Novikoff hepatoma. Characterization of a homogeneous adenosine triphosphatase that stimulates DNA polymerase β, *Biochemistry,* 21, 5060, 1982.
87. Bose, K., Karran, P., and Strauss, B., Repair of depurinated DNA *in vitro* by enzymes purified from human lymphoblasts, *Proc. Natl. Acad. Sci. U.S.A.*, 75, 794, 1978.
88. Evans, D. H. and Linn, S., Excision repair of pyrimidine dimers from simian virus to minichromosomes *in vitro, J. Biol. Chem.*, 259, 10252, 1984.
89. Yamada, K., Hanoka, F., and Yamada, M.-A., Effects of aphidicolin and/or 2′,3′-dideoxythymidine on DNA repair induced in HeLa cells by four types of DNA-damaging agents, *J. Biol. Chem.*, 260, 10412, 1985.

Chapter 8

FIDELTY OF DNA SYNTHESIS

Nothing is more damaging to a new truth than an old error.

J. W. von Goethe
Proverbs in Prose

I. INTRODUCTORY REMARKS

In order to perpetuate species homogeneity, DNA replication must be an exceptionally accurate process. Yet, a small number of errors must be permitted to promote species evolution. Eigen et al.[1] have argued that organisms with small genomes have evolved DNA-replicating systems with inherently high error rates, while animal cells with large genomes replicate their DNA with high fidelity. A mutation rate greater than one per genome per cell generation might be incompatible with species stability. Measurements of spontaneous mutation frequencies in animal cells suggest that the average frequency of errors in DNA replication is in the range of 10^{-9} to 10^{-11} errors per base-pair replicated.[2] This value is the net result of errors made during DNA replication and repair minus the correction of these mistakes by mechanisms such as mismatch repair. In each of these processes, DNA polymerases and associated activities play a major role. Thus the fidelity of catalysis by DNA polymerases is likely to be a major determinant in the overall accuracy of DNA replication. However, it is also likely that there are many other components of the DNA replication and repair systems in animal cells that confer accuracy and have yet to be identified.

Several laboratories have been able to copy DNA in the test tube using purified bacterial components and achieve an accuracy similar to that exhibited by bacteria in vivo.[3-5] The major contributors to this high fidelity are the prokaryotic DNA polymerases themselves. These enzymes can copy DNA templates with an accuracy as great as 10^{-7}. This accuracy is achieved by at least two sequential steps: the first involves selection of the complementary deoxynucleoside triphosphates by the DNA polymerase, and the second involves proofreading by the associated $3' \rightarrow 5'$ exonuclease, i.e., the preferential hydrolysis of newly incorporated noncomplementary nucleotides. In contrast, most purified animal cell DNA polymerases are highly error-prone and after purification lack any associated proofreading exonucleolytic activity (see Chapter 2). As a result the major contributor to fidelity by these polymerases is likely to occur at the level of nucleotide insertion. The error rates of these purified animal DNA polymerases vary from 1/5000 to 1/40,000.[6] (Error rates are expressed as the ratio of noncomplementary to total nucleotides incorporated.) When one compares these numbers to the spontaneous mutation rates in animal cells, one must conclude that our gap in knowledge may be as great as six orders of magnitude.

In this chapter we will consider the methods that are used for analyzing the fidelity of DNA synthesis in vitro and the results that have been obtained in investigations with purified DNA polymerases from animal cells. Even though these enzymes are error-prone, their accuracy is greater than that likely to arise from difference in energies between complementary and noncomplementary base pairings. Accordingly, a number of schemes have been suggested by which these enzymes enhance accuracy. Presently available data strongly suggest that other cellular components are likely to increase accuracy by working in concert with DNA polymerases. As a DNA replicating apparatus is being assembled, it will become increasingly important to measure the contribution of each of these components to the overall accuracy of DNA replication.

II. METHODS TO MEASURE THE FIDELTY OF DNA SYNTHESIS

Exceptionally sensitive methods are required for studying the accuracy of DNA replication since one ultimately desires to measure events that occur at a frequency as low as 10^{-9} to 10^{-11}. So far, these methods have been limited to measuring the incorporation of radioactive substrates that are noncomplementary to the template or to measuring mutagenesis in vitro using biologically active DNA templates.

A. Studies with Synthetic Polynucleotide Templates

The first, simplest, and most extensively used method to quantitate the fidelity of DNA polymerases involves the copying of a synthetic polynucleotide template composed of only one or two nucleotide species.[7,8] An error is defined as the incorporation of any nucleotide that is noncomplementary to the base in the template; the error rate is thus the ratio of noncomplementary to total nucleotides incorporated. High reproducibility can be obtained by the use of a double label assay, in which one noncomplementary nucleotide is labeled with a radioisotope at very high specific activity and a complementary nucleotide is labeled with a different radioisotope at very low specific activity.[8] For fidelity measurements to be reliable, the template that is copied in vitro should be free of any contaminating molecules. The alternating copolymer poly d(A-T) is particularly advantageous in this respect.[9] This template can be extensively and faithfully synthesized starting with a very small amount of primer-template by *Escherichia coli* DNA polymerase I. Furthermore, poly d(A-T) enables the use of dGTP as the noncomplementary substrate and thus avoids the use of dCTP which can spontaneously deaminate to dUTP, which mimics the complementary nucleotide, dTTP. In contrast to poly d(A-T), most other homopolymer templates are products of the random polymerization of deoxynucleoside triphosphates with deoxynucleotidyl terminal transferase.[10] Such randomly generated polymers would logically contain any contaminating deoxynucleoside triphosphates in proportion to their presence in the reaction mixture used for polymer synthesis.

A large number of controls are required for interpreting results of fidelity assays using polynucleotide templates. The most important controls are those directed at showing that the noncomplementary nucleotide is actually incorporated in phosphodiester linkage into a polynucleotide product that is coded for by the template. This can be accomplished by nearest neighbor analysis in which the α-^{32}P from the noncomplementary nucleotide is transferred to the adjacent complementary nucleotide. A major consideration is that any contaminating DNA that may be present in the reaction mixture might serve as an alternate template and direct the incorporation of substrates that are noncomplementary to the added polynucleotide. As a result of this problem, even under the most carefully controlled experimental conditions, assays with polynucleotide templates can be used only with extensively purified DNA polymerases. For practical purposes, the purity of the template, substrates, and enzymes limits the sensitivity of these assays to approximately 10^{-5}. Further, the effect on fidelity of polymerase-associated cofactors, which may be removed during enzyme purification, cannot be assessed unless they can be added in pure form. On the other hand, these assays have been particularly useful in studying agents that enhance misincorporation.[11-13] Fry et al.[14] designed a procedure for measuring fidelity in copying a poly d(A-T) template in the presence of large amounts of contaminating DNA. The experimental scheme involves the selective suppression of the template activity of contaminating DNA and then subjecting the product of the reaction to a series of restriction enzymes that cleaves any DNA synthesized on the contaminating DNA template, but fails to cleave the added synthetic template or the product of the reactions directed by this template. Using this method, the authors were able to measure the fidelity of a polymerase activity found in preparations of chromatin containing DNA as a major constituent.

The accuracy of synthesis with synthetic polynucleotide templates may not be the same as that which occurs with DNA. With synthetic polynucleotide templates, there is slippage of the primer along the template during synthesis,[15] and this slippage could be associated with increased misincorporations. Moreover, with alternating polynucleotides, the misincorporated nucleotides can more easily loop out of the DNA helix. The disadvantages of synthetic polymer templates as well as the need for greater sensitivity served as an impetus for the development of methods to measure fidelity of polymerase catalyzed copying of natural DNA templates. Nevertheless, studies with polynucleotides have remained particularly advantageous in experiments involving incorporation of analogs, mechanisms of mutagenesis, and nucleotide turnover.

B. Reversion Assays with Mutant DNA Templates

Since natural DNA contains all four bases, measurements of fidelity cannot be simply obtained by quantitating the incorporation of a radioactive "noncomplementary nucleoside triphosphate". Instead, a genetic approach has been designed.[16] A single-stranded ϕX174 DNA template, containing an amber mutation resulting from a single base change is copied in vitro by any purified DNA polymerase. After copying proceeds past the *amber* site, the partially double-stranded DNA product is transfected into *E. coli* spheroplasts, which complete replication and produce progeny phage. The error rate of the polymerase is determined from the reversion rate of the progeny phage after plating on bacteria that are permissive or nonpermissive for the amber mutation.

Because of the low background reversion frequency of *am*3 ϕX174 DNA, one can detect misincorporations at a frequency approaching 10^{-6} in reactions with all four deoxynucleoside triphosphates at equal concentration.[17] By extrapolating from the results of experiments in which the concentration of the noncomplementary nucleotide is progressively increased (pool bias), error rates as low as 10^{-8} can be ascertained.[18] Since the assay is based on measurements of biological activity, the presence of any contaminating DNA does not affect the results. In order to calculate the error rate of a polymerase from the reversion frequency of newly produced phage, two important controls are required. One must have an independent measurement of the percent of the template molecules that are copied past the amber site, and second, the percent of revertant DNA molecules that are competent to produce mutant phage in the transfection assay must be determined.[16,19] The latter has been referred to as penetrance.[16] The amber 3 locus is in a stretch of the ϕX DNA sequence that codes for the products of two required genes. The DNA sequences for the two genes are on two overlapping reading frames separated by a single nucleotide. As a result, only a limited number of substitutions at this amber site produce active gene products. In fact, all viable substitutions occur at position 587, and involve replacement of dT in the newly synthesized viral minus strand with dA, dG, or dC. By sequencing the DNA in the progeny phage, one can obtain the nature of the substitution.

In the original ϕX fidelity assay,[16] synthesis was initiated by the use of a hybridized restriction fragment whose 3′-OH terminus was 83 nucleotides from the amber site. This resulted in two important limitations with animal DNA polymerases. First, these enzymes copy long single-stranded stretches of DNA with relative inefficiency,[20] necessitating the use of large amounts of polymerase and, for reasons unknown, high concentrations of deoxynucleoside triphosphates. Second, most preparations of animal cell DNA polymerases are contaminated by small amounts of endonucleases that can cleave the ϕX template, rendering it biologically inactive. To overcome these limitations, two groups of investigators substituted for the restriction fragment a synthetic oligonucleotide whose 3′-OH terminus was only a few nucleotides distant from the

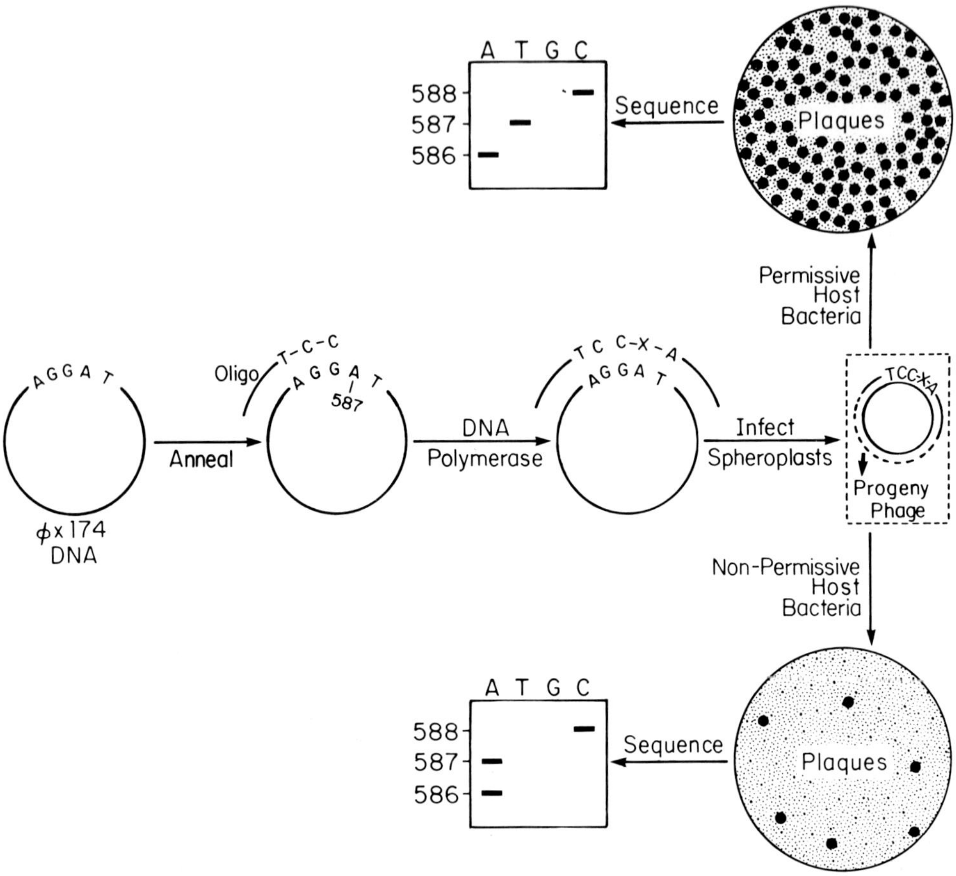

FIGURE 1. Assay for fidelity of DNA synthesis using oligonucleotide primed ϕX174 DNA. The template, *am*3 ϕX174 DNA, contains a mutation at position 587. The primer is an oligonucleotide, 15 nucleotides in length with a 3′-OH terminus, one nucleotide upstream from position 587. Synthesis of DNA in vitro is carried out with an animal DNA polymerase. Thereafter, the partially copied DNA is transfected into *E. coli* spheroplasts. Progeny phage or infective centers are scored in parallel on permissive and nonpermissive bacterial strains. The nature of the substitution can be determined by growing the revertant phage and sequencing its DNA.

amber site.[21,22] This modified ϕX fidelity assay is diagrammed in Figure 1. The use of an oligonucleotide mitigates against the effects of nucleases, since only brief periods of incubation are required for synthesis to proceed past the amber site. With the use of such oligonucleotide primers, it has been possible to measure misincorporations in unfractionated cellular homogenates which contain considerable nucleolytic activity.[22]

Essentially all of the ϕX genome is required for the production of viable phage. Thus reversion assays with ϕX DNA are for the most part limited to the detection of single base sustitutions. In principle, any biologically active DNA template could be used in place of ϕX174 DNA as long as one has a method for measuring revertants at a termination codon. Fersht and collaborators[21,23] have extended the ϕX assay to utilize amber 16 ϕX174 DNA in which substitution at multiple positions within the amber locus lead to viable revertants. They have developed criteria for classification of the types of substiutions based on plaque morphology and growth at different temperatures. Two other methods have been described for measuring fidelity in vitro using mutant ϕX DNA templates. The one developed by Liu et al.[24] requires the entire T_4 DNA replication complex containing seven purified proteins and thus is not amenable to studies

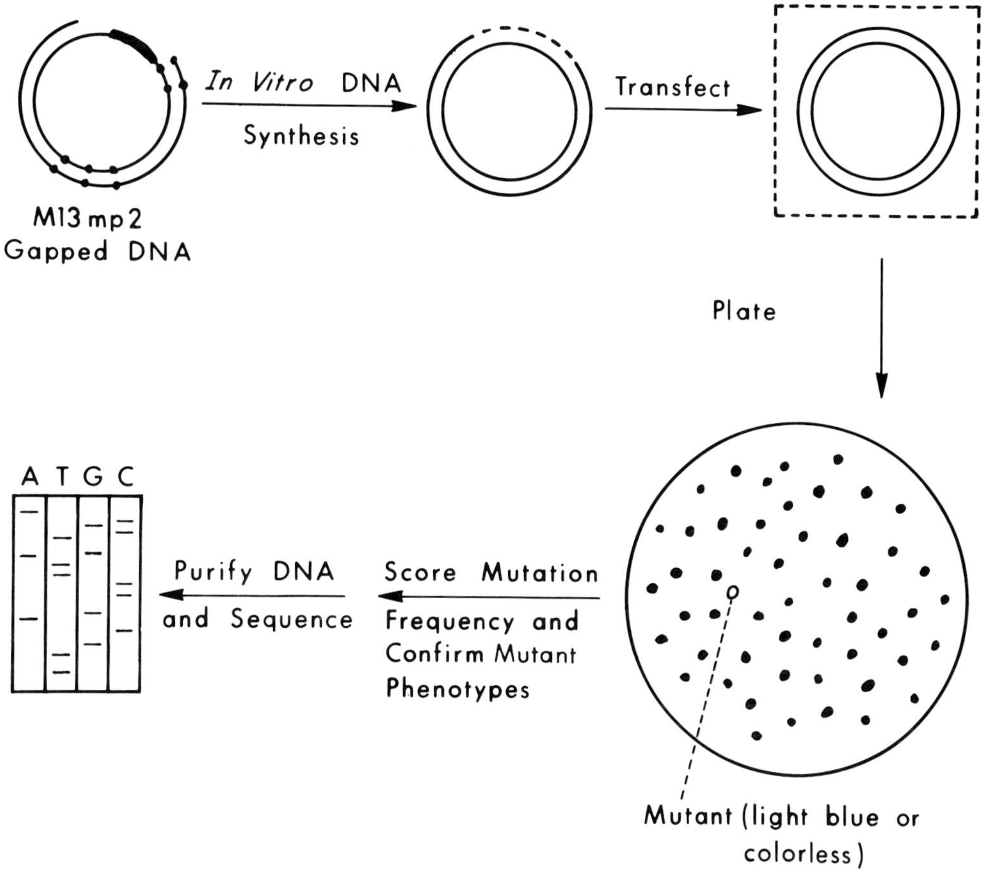

FIGURE 2. Scheme for measurement of mutation spectrum on M13mp2 recombinant DNA. The gapped DNA that contains the segment of the *lac* Z gene (thickened line) is copied in vitro with any desired animal cell DNA polymerase. The copied DNA is transfected into an *E. coli* host, which contains a deletion of the *lac* Z operon and an episome that carries the remaining coding sequence of the *lac* Z gene. Mutagenesis in vitro renders the phage-carried segment of *lac* Z nonfunctional and thus no β-galactosidase will be formed by complementation with the episomal fragment. Mutations resulting in loss of β-galactoside are unable to hydrolyze an indicator dye. DNA is isolated from individual clones and sequenced to identify the mutational spectra in the *lac* Z segment. (Derived from Kunkel, T. A., *J. Biol Chem.*, in press. With permission.)

with animal DNA polymerases. The one described by Fersht[4] utilizes *E. coli* holoenzyme III and four other purified bacterial proteins and thus cannot also be used for animal DNA polymerases.

Recently, Kunkel et al.[25] has established a forward mutagenesis system that is capable of detecting, at multiple sites, a wide spectrum of mutagenic changes, including frameshifts, deletions, duplications, and complex errors as well as different base substitutions. In this assay, a single-stranded DNA region of the hybrid bacteriophage M13mp2, containing a small segment of the *E. coli lac* operon, is copied in vitro by DNA polymerase (Figure 2). The copied DNA is transfected into an *E. coli* host that contains a deletion of the *lac* operon, but also harbors an episome that contains the remaining coding sequence of the *lac* Z gene. Functional β-galactosidase is produced by complementation of the two protein products, one of which is coded for by the in vitro sythesized DNA. Decreased complementation resulting from a mutagenic event within the gapped region can be detected as decreased hydrolysis of an indicator dye. The mutational spectrum is determined by sequencing the DNA obtained from individual β-galactosidase-defective clones. This assay is inherently not as sensitive as a reversion assay in which one scores only for changes at a single template locus. However, the M13mp2 assay is particularly advantageous for measurements of mutational spectrum and nearest neighbor effects on misincorporation, since one can score for most mutational changes within a 250-base target. Since the results are dependent on the DNA sequence of each mutant, assays for forward mutations are labor intensive, but have yielded important results.

III. FIDELITY OF DNA POLYMERASE-α

In this section we will consider the results of studies that have measured the fidelity of the catalytic component of DNA polymerase-α core enzyme. Whenever tested, both homogenous and partially purified preparations of nearly every DNA polymerase-α studied have been found to be free of any associated 3'→5' exonucleolytic activity. Moreover, Chang and Bollum[26] demonstrated that purified DNA polymerase-α from calf thymus fails to generate complementary deoxynucleoside monophosphates during catalysis, suggesting that exonucleolytic proofreading is not likely to occur. This observation has been repeated many times with preparations of DNA polymerase-α from a variety of animal sources (see Chapter 3, Section IV.A). Thus errors observed in copying both synthetic and natural DNA templates are likely to reflect mistakes made by the enzyme at the insertion step. The lack of monophosphate generation does eliminate many but not all possible mechanisms for kinetic proofreading.

A. Studies With Synthetic Polynucleotides

A large number of studies have been carried out on the fidelity by which purified preparations of DNA polymerase-α copy polynucleotide templates. Selected results are compiled in Table 1. With poly d(A-T) as a template and dGTP as the noncomplementary nucleotide, the frequency of misincorporation varies from 1 in 4000 to 1 in 20,000. In parallel experiments, the error rate of *E. coli* DNA polymerase I had been reported to be as low as 1/70,000, and an even lower error rate was obtained for T4 DNA polymerase.[20,27] In selected situations, lower error rates have been reported for polymerase-α with other synthetic templates, but these have usually only been obtained after interpolating from the results of pool bias experiments.[28,29] In these interpolations it is assumed that the error rate is proportional to the ratio of incorrect to correct nucleotides in the reaction mixture. This is justified by limited studies with DNA polymerase-α which indicates a rough linear dependence of error rate on nucleotide pool bias.[27] However, Krauss and Linn[30] failed to observe competition between the complementary and noncomplementary nucleotides using poly (dA)·poly (dT) as template, while — under the same conditions — competition was observed with polymerase-β and -γ. With poly d(A-T) as template, it has been possible to study the position of the incorporated noncomplementary nucleotides by nearest-neighbor analysis.[31] With human placenta DNA polymerase-α, at least 95% of residues of dGMP was found to be incorporated into poly d(A-T) in phosphodiester linkage in place of the complementary purine, dAMP.[27]

DNA polymerases of the α type commonly exhibit molecular heterogeneity and are found in different subspecies after purification (Chapter 3, Section II). Conceivably, one of these might be exceptionally accurate, in accord with a putative role in DNA replication. Brodsius et al.[32] have extensively purified three forms of DNA polymerase-α and measured their fidelity with poly d(A-T) as a template. Their results indicate a 15-fold difference in accuracy among the different forms. Interestingly, the smallest species — and perhaps that one which had been most extensively degraded by cellular proteases — exhibits the highest rate of misincorporation. Loeb et al.[33,34] fractionated

DNA polymerase-α by a product of Holmes et al.[35,36] and observed no differences in the fidelity between two of the forms, forms C and D. These results are not necessarily contradictory, since there is insufficient information available to determine the correspondence between the different forms in these two different purification procedures.

B. Studies with Natural DNA Template

So far, studies on the fidelity of DNA polymerase-α with natural DNA templates have been limited to the use of either *amber* 3 or *amber* 16 ϕX174 DNA. In an in-depth analysis, Kunkel and Loeb[6] copied *am*3 ϕX174 DNA with preparations of DNA polymerase-α from different cell types. The newly synthesized partially double stranded DNA was isolated by sedimentation on a neutral sucrose gradient and used to measure the frequency of revertants produced as a result of copying by the polymerase. Using this procedure, one does not need to estimate the percent of molecules copied since only copied molecules are used in the transfection assay. The results indicate that at equal nucleotide concentrations, DNA polymerase-α inserts one noncomplementary nucleotide opposite deoxyadenosine at position 587 for each 30,000 molecules copied. It is important to remember that this number refers to all three base substitutions, some of which presumably occur at a much lower frequency. Also, this assay does not score for additions, deletions or complex arrangements. Enhancement in reversion frequency is obtained by increasing the concentration of either dATP, dCTP, or dGTP in the reaction mixture, indicating that all three substitutions do indeed occur opposite position 587 in reactions catalyzed by DNA polymerase-α. So far, measurements of DNA sequence have not been reported on a sufficient number of revertants produced by copying mutant ϕX DNAs with DNA polymerase-α to establish the frequencies of the three different types of substitutions.

Grosse et al.[21] have analyzed the frequency and type of mispairings resulting from the copying of *amber* 16 ϕX174 DNA by the 9-S DNA polymerase from calf thymus. In these studies, use was made of the endogenous primase activity that is associated with the 9-S polymerase by adding to the reaction all four ribonucleoside triphosphates. By varying pool bias and scoring for plaque morphology and temperature sensitivity, these investigators were able to estimate the types of substitutions occuring at each of the three positions in the *amber* 16 locus. From their results, they calculate that the frequency of unbiased nucleotide: template mispairings for dGTP:T, dGTP:G, and dATP:G were 1/13,500, 1/22,700, and 1/35,700, respectively. With this annotation, the dNTP refers to the incorporated base. In these studies the negative results are also important: mispairings involving dCTP:A, dGTP:A, dCTP:T, and dTTP:T were below the level of detection, $<5 \times 10^{-6}$. The data also demonstrate that the frequency of mispairing is likely to be determined by which base is in the template. dATP:G mispairs were nearly an order of magnitude higher than dGTP:A mispairs. However, in such comparisons one cannot entirely eliminate the effects of neighboring nucleotides and the secondary structure of the template.

The results so far considered on the fidelity of polymerase-α indicate that this class of DNA polymerases is highly error-prone. These results are in contrast to the recent report of Kaguni et al.[37] in which it was observed that the primase-polymerase complex from *Drosophila* is as accurate as *E. coli* DNA polymerase III holoenzyme complex in copying *am*3 ϕX174 DNA. It should be noted that the *Drosophila* polymerase lacks an associated 3′→5′ activity and does not exhibit significant pyrophosphate exchange, which conceivably could provide a second mode for proofreading. The results with *Drosophila* polymerase assert that DNA polymerase-α by itself, in the form of an isolated oligomeric enzyme, is capable of carrying out catalysis with very high accuracy. This can occur without any associated exonucleolytic proofreading activity, implying that the polymerase can exhibit very high selectivity at the nucleotide insertion step.

The evidence presented for this very exceptionally high fidelity is based on measurement of reversion frequencies with nucleoside triphosphates at equal concentrations and under conditions in which the concentrations of each of the noncomplementary nucleotides is increased by 100-fold.[37] Repetitive measurements of the reversion frequencies of DNA copied using equal dNTP pools with *Drosophila* polymerase-α, Pol III holoenzyme + single-stranded binding protein, and Pol III* indicate that with each the frequency of revertants obtained after copying past the amber site is twofold greater than the background reversion frequency exhibited by uncopied DNA. The implication is that, since all these enzymes have a similar accuracy, the *Drosophila* polymerase-α could be as accurate as Pol III, which has been reported by Fersht[4] and by Loeb et al.[5] to have an error rate approaching 10^{-7}. However, in these latter studies, revertants above background were not detected in reactions carried out with nucleotide substrates at equal concentrations or in reactions in which the concentration of the noncomplementary nucleotide was increased by as much as 50-fold.

A second set of measurements are those carried out with biased nucleotide pools. Under conditions in which the concentrations of the noncomplementary nucleotide dATP was increased 100-fold over that of the complementary nucleotide, dTTP, the reversion frequency of the copied DNA was increased 42-fold for *Drosophila* polymerase-α and 100-fold for Pol III holoenzyme. As noted by the authors,[37] these results are not unambiguous, since dATP is the next nucleotide, and when present at high concentrations, would abnegate any proofreading activity by the Pol III holoenzyme complex.[38] Thus experiments at high dATP probe only the insertional activities of both enzymes since proofreading is not operative with either enzyme. In fact, these results with dATP bias would suggest the opposite: the *Drosophila* polymerase is not very accurate. However, it should also be noted that no increase in revertants was observed in the dCTP pool bias experiments where misinsertion by an error prone polymerase should enhance mutagenesis. In conclusion, the report on the exceptionally accurate catalysis by *Drosophila* polymerase-α[37] is tantalizing in suggesting that an oligomeric polymerase exhibits high fidelity. A comparison between the fidelity of the isolated catalytic subunit of *Drosophila* DNA polymerase-α and the oligomeric enzyme would be particularly instructive. Such analysis might indicate that catalysis by the multimeric polymerase is more accurate than that of the catalytic subunit and serve to define the effects of each of the polymerase subunits on fidelity.

With the exception of studies with synthetic polynucleotide templates,[39] DNA polymerase-α has been reported to have a higher fidelity than polymerase-β. Even though polymerase-α fails to excise noncomplementary nucleotides at the 3′-primer terminus and does not catalyze the generation of deoxynucleoside monophosphates,[40] limited amounts of hydrolysis of noncomplementary nucleotides would escape detection, yet would have profound effects on the fidelity of DNA polymerization. A number of criteria have been established for evaluating the contribution of proofreading to the overall fidelity of DNA synthesis in the copying of natural DNA templates. Considering that proofreading involves the selective hydrolysis of a terminally incorrect base, one can intuitively reason that misincorporations would occur more frequently as a result of increased rate of polymerization. In studies with bacterial DNA replicating complexes, Fersht and Knill-Jones[4,23] observed an enhancement in mutagenesis in vitro when the concentration of the nucleotide complementary to the next nucleotide was increased in reaction mixture. In studies with Pol I and *am*3 ϕX DNA, it was demonstrated that increasing the concentration of the next complementary nucleotide, dATP, resulted in a greater than linear increase in the reversion frequency.[41] Moreover, DNA sequence analysis of the revertants demonstrated that the most frequent substitution was dCMP. In contrast, with calf thymus polymerase-α, Abbotts and Loeb[33] reported that the reversion frequency in copying reactions carried out with elevated levels of all three incorrect dNTPs is no greater than the additive effects of the individual biased dNTPs. This additivity implies no next nucleotide effect. Under conditions in which the addition of deoxynucleoside monophosphates increases mutagenesis with *E. coli* Pol I, presumably by inhibiting the 3′→5′-exonuclease, there is no enhancement in mutagenicity of calf thymus DNA polymerase-α.[33] Similarly, the addition of pyrophosphate failed to enhance mutagenesis by polymerase-α under conditions in which the error rate of Pol I is increased fourfold. In fact, with polymerase-α, the opposite is observed; there is a fivefold enhancement in fidelity, which could be accounted for by a model proposed by Doubleday et al.[42] for rejection of incorrect triphosphates through pyrophosphate exchange or pyrophosphorolysis.

Studies on the effects of α-thiol derivatives of dNTPs on the fidelity of polymerase-α have yielded unexpected results.[33] These derivatives contain a sulfur atom in place of an oxygen on the alpha phosphorus.[43] They are incorporated into DNA with normal base pairing properties[43] and—once incorporated—they are not excised by the exonucleases associated with *E coli* Pol I.[44] Thus the observed mutagenic effects of these thiol nucleotides in copying reactions with Pol I has been attributed to the abolition of proofreading.[44] These thiol nucleotides do not affect the fidelity of synthesis with polymerase-β or avian myeloblastosis reverse transcriptase[44] enzymes lacking a proofreading exonucleolytic activity.[20] Also, with calf thymus polymerase-α, substitution of dATP or dGTP by its thiol derivative does not result in a significant increase in reversion frequency.[33] However, when dCTP is replaced by αS-dCTP, reversion frequency increased by an order of magnitude. This affect appears to be specific for polymerase-α; it is not observed with DNA polymerase-β or with AMV DNA polymerase. It is observed with both forms of C and D as well as putative holoenzyme purified from calf thymus by the method of Hübscher et al.[45] A direct interpretation of these results is that the αS-dCTP fits into the pyrimidine binding site of polymerase-α in such a way that it is recognized as dTTP and is then misinserted into the amber site of ϕX DNA. This implies that at least, the α phosphate group is involved in the recognition of the correct substrate by DNA polymerase-α.

C. Associated Proteins and Other Factors

Given that the catalytic core component of polymerase-α is highly error prone in vitro and that replication is an exceptionally accurate process in vivo, the goal of future studies will be to determine what cellular factors work in concert with polymerase-α to enhance the accuracy of DNA replication. So far, two approaches have been used to tackle this problem, and studies with these have been limited. The first has been to add purified components known to be associated with DNA replicating complexes and to measure their effects on fidelity. The second has been to assess the accuracy of very crude preparations of DNA polymerase-α, which presumably contain an undegraded holoenzyme to see if one can detect higher fidelity.

In addition to the possibility of proofreading and mismatch repair in animal cells (which we will consider separately) other factors can operate in concert with polymerase-α and directly increase the fidelity of DNA synthesis. Putative fidelity enhancing factors could operate by increasing the processivity of the polymerase, increasing the rigidity or helical structure of the template, altering the rate of polymerization, or changing the conformation of the polymerase.[20] An excellent candidate for a factor with some of these effects is single-stranded DNA binding protein. This protein has been shown to bind to single-stranded regions of DNA ahead of the advancing replicating fork. A role for SSB in modulating the accuracy of DNA replication in prokaryotes is suggested by studies,[46-48] which demonstrate that mutations in the gene for binding protein alter the overall mutation frequency of bacteriophage T4. Moreover, the addition of T4 binding protein to T4 DNA polymerase results in a 30 to 80%

reduction in turnover of noncomplementary nucleotides during the copying of poly d(A-T).[49] Kunkel et al.[50] demonstrated that the addition of *E. coli* single-stranded DNA binding protein increased the accuracy of calf thymus DNA polymerase-α in copying poly d(A-T) and ϕX DNA by 13- and 3.2-fold, respectively. These results were confirmed by Brosius et al.[32] who reported that *E. coli* single-stranded binding protein increased the accuracy of both the 7- and 5.7-S forms of calf thymus polymerase-α with poly d(A-T) as a template. Even though the enhancement in accuracy by SSB appears reproducible, so far the effects have been small compared to the accuracy requirements for in vivo replication of DNA. Studies have not yet been reported in homologous systems in which the binding protein and polymerase are derived from the same cells. In homologous systems, the magnitude of the increase in fidelity may be much greater since the stimulation of bacterial DNA polymerase activity by SSB has been reported to be maximum only with homologous proteins.[51,52]

In general, very few substances have been shown to enhance the fidelity of catalysis by purified DNA polymerases, including polymerase-α. However, these studies have been limited to adding substances to reactions catalyzed by the catalytic core of component of polymerase-α. In copying ϕX DNA with the 9-S form of calf thymus polymerase-α, Grosse et al.[21] observed the fidelity was not greatly different in reactions initiated with a deoxyoligonucleotide primer compared to those utilizing the endogenous primase associated with the polymerase. Included in these experiments was the addition of 100 μM amounts of all four ribonucleoside triphosphates, which also must not have affected fidelity. Furthermore, since the accuracy achieved in these experiments was similar to that previously reported by Kunkel and Loeb[6] using only the catalytic core component of calf thymus polymerase-α, the comparison must also imply that the presence of primase does not enhance fidelity. This comparison is with reservations because of the studies with *Drosophila* oligomeric α-polymerase, which possesses exceptionally high fidelity.[37] In other studies, it was reported that neither the presence of rATP (in amounts up to 1 mM nor Ap_4A affects the fidelity of the 5.7-S polymerase from calf thymus when assayed with poly d(A-T) as a template.[32] Polyamines appear to enhance the accuracy of calf thymus polymerase-α, but the enhancement is limited to about twofold. Interestingly, Mikhailov and Androsova[53] recently reported that spermidine decreases the affinity of loach DNA polymerase-α for DNA. However, in spite of this effect, it also increases the fidelity of catalysis with poly(dA).oligo(dT) by 3.8-fold. In summary, the limited number of additions experiments that have been reported have failed to demonstrate the presence of any factor that enhances the fidelity of DNA polymerase-α by more than tenfold.

In principle, with the use of an oligonucleotide initiator that is positioned adjacent to the amber site on ϕX174 DNA, it should be possible to measure fidelity in very crude preparations of DNA polymerase-α prior to separation of critical components during purification.[34] This type of assay has been successfully used to document misincorporation by cellular homogenates,[22] but has not yet been exploited for studying fidelity of putative replicating complexes. The major difficulty is the error-proneness of DNA polymerase-β, which, when present, would be responsible for most misincorporations with a natural DNA template (*vide infra*). Another difficulty involves the use of salt extraction procedure for separation of DNA polymerase-α. The presence of 0.5 M monovalent salts (those commonly used) may dissociate some components of the putative replicative complex. Nevertheless, studies with crude complexes have not been adequately utilized for analyzing the fidelity of DNA synthesis in eukaryotic cells.

IV. FIDELITY OF DNA POLYMERASE-β

In principle, studies on the fidelity of DNA polymerase-β might be most easy to

Table 1
FIDELITY OF PURIFIED ANIMAL DNA POLYMERASES WITH SYNTHETIC POLYNUCLEOTIDE TEMPLATES

Enzyme source	Template	Error rate	Ref.
DNA polymerase-α			
Calf thymus	poly[d(A-T)]	1/4,500	50
Rat liver	poly[d(A-T)]	1/16,500	61
Human placenta	poly[d(A-T)]	1/12,000	27
Human fibroblasts	poly[d(A-T)]	1/15,400[a]	62
Human lymphocyte	poly[d(A-T)]	1/5,330	39
HeLa cell	poly[d(A-T)]	1/4,830	39
Regenerating rat liver	poly[d(A-T)]	1/100,000[a]	28
Subspecies			
Calf thymus 9 S	poly[d(A-T)]	1/160,000[a]	32
Calf thymus 7 S	poly[d(A-T)]	1/14,000	32
Calf thymus 5.7 S	poly[d(A-T)]	1/4,200	32
Form C (7.3 S)	poly[d(A-T)]	1/12,900	34
Form D (6.8 S)	poly[d(A-T)]	1/9,120	34
DNA polymerase-β			
Calf thymus	poly[d(A-T)]	1/11,800	39
Calf liver	poly[d(A-T)]	1/15,200	39
Rat liver	poly[d(A-T)]	1/18,900	50
Human placenta	poly[d(A-T)]	1/20,000	27
Calf thymus	poly(dA) · oligo(dT)	1/180,000[b]	28
DNA polymerase-γ			
Fetal calf liver	poly[d(A-T)]	1/7,250	20
HeLa cell	poly[d(A-T)]	1/3,120	20

Note: The data in this table are from Mg^{2+}-activated reactions using extensively purified DNA polymerases. The selection of data was heavily biased for studies that utilized poly[d(A-T)] as a template for the reasons indicated in the text. In all the tabulated studies, dGTP served as a noncomplementary substrate.

[a] Interpolated from pool bias experiments.
[b] Determined by subtracting misincorporation due to terminal transferase activity.

accomplish since the enzyme is of low molecular weight, readily purified to homogeneity, and free of any associated exonucleolytic activity. The fact that this enzyme is involved in DNA repair might suggest that the polymerase is more error-prone than DNA polymerase-α. This view is likely to be naive since there are few if any studies that define the rate of mutagenesis during repair synthesis of DNA, and there is no reason *a priori* to think that DNA repair is error-prone in eukaryotes. Also DNA polymerase-α is associated with a larger replicating complex — the components of which are likely to enhance fidelity — and therefore the core catalytic activity need not be more accurate than DNA polymerase-β.

A. Studies with Synthetic Polynucleotide Templates

In the case of DNA polymerase-β, measurements of error rates with synthetic polynucleotide templates do not correlate with those obtained using natural DNA. Seal et al.[27] and Dube et al.[39] reported that, under similar conditions, DNA polymerase-β from a variety of sources was more accurate than DNA polymerase-α in copying poly d(A-T) (Table 1). Nearest neighbor analysis indicated that 86% of the dGMP incorporated by β-polymerase was inserted as a single base substitution in place of dAMP, while

14% substituted for dTMP. In contrast, DNA polymerase-α was more stringent in substitutions — essentially all of the dGMP was incorporated in place of dAMP. The greater accuracy of DNA polymerase-β has not been confirmed with natural DNA templates and is likely to reflect a unique property of poly d(A-T). In fact, in all studies with natural DNA templates so far reported, DNA polymerase-α is more accurate than DNA polymerase-β.

Krauss and Linn[30] isolated multiple species of DNA polymerase-β from human placenta that exhibited differences in fidelity with polynucleotide templates. Interestingly, the forms that sedimented faster on glycerol gradients exhibited the higher error rates, suggesting that proteolysis might reduce fidelity in a manner inferred from the results of studies reported with DNA polymerase-α.[32,35]

In an effort to study the accuracy of chromosomal replication, Fry et al.[14] developed an assay for the fidelity of copying polynucleotides in which any copied endogenous DNA contaminant was subsequently degraded by digestion with restriction enzymes. Using chromatin as a source for DNA polymerase, they found that the predominant activity that copied poly d(A-T) was DNA polymerase-β. With this system the error rate for DNA polymerase-β from two different species of mice was approximately 1/1100.[54] The apparent discord among the various measurements of accuracy of purified and chromatin-associated DNA polymerase-β can be interpreted in several ways. It is possible that distinct forms of polymerase-β have inherently different fidelities, and the one associated with chromatin is particularly error-prone. Second, polymerase-β may be susceptible to proteolytic degradation, which renders it more error-prone, and this degradation occurs during chromatin isolation. Alternatively, the chromatin associated polymerase may interact with proteins that reduce its accuracy.

B. Studies with Natural DNA Templates

Detailed studies on the error rate of DNA polymerase-β from rat hepatoma and mouse myeloma using *am*3 ϕX174 DNA as a template were reported by Kunkel and Loeb.[6] They measured reversion frequencies after purifying the partially double-stranded product of the polymerase reaction, thus avoiding any uncertainties concerning the percent of the template molecules copied. The error rates varied from 1/3000 to 1/6600. This error rate is significantly greater than that exhibited by DNA polymerase-α in assays carried out in parallel under the same conditions with the same *amber* DNA. In reactions in which graded amounts of all three noncomplementary deoxynucleotides were increased coordinately, there was a strikingly parallel increase in reversion frequency.[55] This activity within a limited range suggests a lack of any proofreading activity, even by a kinetic mechanism that does not generate deoxynucleoside monophosphates.

The possibility of proofreading with polymerase-β has not been as extensively probed as with polymerase-α. In part, the higher error with natural DNA templates[6] renders this mechanism less likely. However, the frequency of noncomplementary single base substitutions is lower than what one would anticipate from base-pairing and is thus likely to reflect a contribution of polymerase-β to the accuracy of base selection. As mentioned above, the linearity of enhanced misincorporation as a function of increased concentration of noncomplementary nucleotides argues against a next nucleotide effect, which is a hallmark of exonucleolytic proofreading. Moreover, the substitution of α-thiol dNTPs for the corresponding incorrect substrates does not enhance mutagenesis even with α-thiol dCTP.

The introduction of the M13mp2 system by Kunkel[25] has made it possible to examine all classes of mutations that result from copying a natural DNA template by a purified DNA polymerase. As a first application of this system, the mutational spectrum resulting from copying natural DNA by β-polymerase was established.[25] To do so, the

DNA sequence of some nucleotides was determined for 460 independent mutants. The following conclusions can be drawn. First, preparations of DNA β-polymerases from two different species exhibited a remarkably similar mutational spectrum. Second, the frequency of single base substitution averaged over many sites and all 12 possible mispairs is 1/5000, a value essentially identical to that obtained in the ϕX fidelity system.[6] Third, the most frequently detected mispairings were transitions as would be predicted from theoretical considerations.[20] Fourth, an unexpectedly large number of pyrimidine:pyrimidine mispairs were also observed. Fifth, a considerable number of frameshift mutations were obtained, and these were primarily deletions. Last, a large number of deletions were also produced by polymerase-β. The mutational spectra observed in studies with polymerase-β are decidedly different than those observed with DNA polymerases-α and -β.[132] This fact considered in context with the remarkable similarity in the spectra of mutations resultant from copying with two different polymerases of the β-type argues that these enzymes must profoundly influence the selection of which base is inserted at each site during DNA synthesis. If selection were dictated by Watson-Crick base pairings and base stacking interactions alone, one would not expect to see characteristic differences between the different types of polymerases.

C. Associated Factors

Only a limited number of studies has been conducted on the effects of associated factors on the fidelity of DNA polymerase-β in a manner quantitatively similar to that of polymerase-α.[50] Single-stranded binding protein increases the fidelity of catalysis with polymerase-β. The enhancement in fidelity with poly d(A-T) and natural DNA templates is 9.4- and 2.8-fold, respectively.[55] DNAse V, a bidirectional exonuclease exhibiting both a 3′→5′ and a 5′→3′ activity was isolated during the purification of polymerase-β (Chapter 4, Section III.A). Addition of this exonuclease to reaction catalyzed by polymerase-β resulted in only a fourfold enhancement in accuracy.[56] Conceivably this effect could be much greater in vivo or it is possible that separation of the two activities, polymerase-β and DNAase V, during purification results in irreversible changes that lessen subsequent cooperation.

V. FIDELITY OF DNA POLYMERASE-γ AND -δ

It is unfortunate that studies on the fidelity of polymerase-γ have not been more extensive. Based on the small size of the mitochondrial genome in animal cells, the large number of mitochondria in each cell, and the tenfold higher rate of changes in mitochondrial DNA sequence during evolution compared to that of nuclear DNA,[57] one might predict that polymerase-γ is highly error prone. However, in an extensive analysis of the DNA sequence of fragments of human mitochondrial DNA cloned in recombinant plasmids, only one error in 100,000 nucleotides was observed.[58,59] This sequence preservation could be indicative of an unexpectedly high accuracy for mitochondrial DNA replication.

So far, only scattered measurements have been obtained on the fidelity of purified preparation of polymerase-γ. With poly d(A-T) as a template, misincorporation of dGMP was observed at a frequency of 1/7000 and 1/3100 for enzymes from calf liver and HeLa cells, respectively (Table 1). These results are in reasonable accord with studies in the ϕX system in which an error rate of 1/6000 to 1/8000 was established (Table 2). However, these are scattered observations, and more detailed studies are required before firm conclusions can be drawn about the accuracy of this enzyme.

That DNA polymerase-δ exists as an independent entity is predicated on its tight association with a 3′→5′-exonucleolytic activity.[60] This activity has been shown to hydrolyze noncomplementary nucleotides at the 3′-primer terminus. However, this

Table 2
FIDELITY OF PURIFIED DNA POLYMERASES WITH NATURAL DNA TEMPLATES

Enzyme source	Template	Noncompl. nucleotide temp. nucl.	Error rate	Ref.
DNA polymerase-α				
Calf thymus	*am*3 ϕX	C, A, or G	1/24,000—31,000	6
Mouse myeloma	*am*3 ϕX	C, A, G	1/47,500	6
Calf thymus 9 S	*am*16 ϕX		1/10,000—1/100,000	21
DNA polymerase-β				
Rat hepatoma	*am*3 ϕX	C, A, or G	1/6,600	6
Mouse myeloma	*am*3 ϕX	C, A, or G	1/3,000—1/4,600	6
HeLa cell	M13mp2	All four	1/5,000	25
Chicken embryo	M13mp2	All four	1/5,000	25
DNA polymerase-γ				
HeLa cell	*am*3 ϕX	C, A, or G	1/6,600—1/8,100	6
E. coli polymerases				
Pol I	*am*3 ϕX	C	1/680,000	19
Pol III holoenzyme	*am*3 ϕX	C	$<1/10^7$	5
T4 DNA polymerase	*am*3 ϕX	C	$<1/10^7$	63

Note: All error rates that are $<1/10^5$ were interpolated from pool bias experiments as described in each of the references. The designation C, A, G indicates that any of dCMP, dAMP, and dGMP, respectively, could be misincorporated. In the M13mp2 assay, the error rate is that averaged over some 90 positions in the *lac* Z gene.

does not constitute adequate evidence for proofreading without demonstrating that this exonuclease mediates enhanced fidelity. Methods are at hand — i.e., the effects of the next complementary nucleotide, monophosphate, pyrophosphate, and thiol nucleotides analogs on fidelity — to establish the accuracy of polymerase-δ and the contribution of its associated exonuclease to this accuracy. This is particularly important since it is possible that this activity is present ubiquitously in animal cells and has a role in DNA replication.

VI. MECHANISM FOR ENHANCED FIDELITY BY ANIMAL DNA POLYMERASES

Considerable evidence has been accummulated to indicate that most purified animal DNA polymerases lack an associated exonucleolytic activity. The major exception is DNA polymerase-δ, and even in this case we lack studies to show that this exonuclease functions to enhance accuracy. Given this situation, any increase in fidelity greater than that indicated by differences in free energies between complementary and noncomplementary base pairings must be mediated by a base selective function of animal DNA polymerase. The difference in free energy between correct and incorrect Watson-Crick base-pairings at equilibrium is not precisely known. However, based on a variety of physical studies, this difference is usually estimated to be one to three Kcal/mol,[20] which would predict that catalysis in the absence of an enzyme would produce one error for every 100 nucleotides polymerized. This estimate has received considerable support from studies of Orgel and associates[64] on the nonenzymatic template-directed polymerization of oligonucleotides, which indeed demonstrates an error rate of 1 in 200. The fact that the error rate of DNA polymerase-α can be as low as 1/40,000 implies that this polymerase has a role in base selection, as does polymerase-β with its higher error rate of 1/5000. The fact that animal DNA polymerases exhibit different error rates, lower than that of base pairing, argues against a passive polymerase model[65] in its strictest sense. In considering possible mechanisms for enzyme-mediated increases in the fidelity of DNA synthesis by animal DNA polymerases, we will focus only on those that do not involve proofreading or the generation of deoxynucleoside monophosphates.

A. The K_m Discrimination Model

It can be argued that the simplest model for enhanced fidelity without proofreading involves an enzyme-mediated increase in the difference in free energy, ΔG, between correct and incorrect base-pairings. In other words, ΔG in the presence of polymerase is greater than in its absence. This concept is central to the Km discrimination model of Goodman and associates.[65,66] Such an increase in ΔG could result from a hydrophobic environment at the substrate binding site on the polymerase that excludes the water.

In this model, it is assumed that the "on rates" for binding at the growing point are diffusion limited and are the same for correct and incorrect substrates. Discrimination is based on differences in "off rates". The key prediction is that the error rate is proportional to differences in the K_m between correct and incorrect nucleotides. With KB cell DNA polymerase-α, Watanabe and Goodman[66] tested this concept. They observed that the K_m for the incorporation of the incorrect nucleotide, dCTP, opposite 2-aminopurine in the template was 20- to 40-fold higher than that of the correct nucleotide, dTTP. Similarly, the K_m of 2-aminopurine triphosphate is sixfold greater than dATP with thymidine in the template.[66] In both cases, the differences in K_ms are proportional to the misincorporation rates of the respective substrates. Even though these studies are in support of K_m discrimination mechanism for enhanced fidelity by DNA polymerase-α, they consider base pairing only by an analog which — by definition — does not have the same geometry and enzyme contact point as normal nucleotide substrates. Thus these studies may not probe the mechanism of base selection that is operative with normal nucleotide substrates. The significance of these observations would be considerably strengthened by detailed studies using normal substrates: the predicted difference in K_m for DNA polymerase-α would be three to five orders of magnitude (the reciprocal of the misinsertion frequency). Moreover, this differential should be different for DNA polymerases-α and -β. By the use of DNA sequence gels, it should be possible to quantitate the incorporation of a single nucleotide by DNA polymerase.

B. Conformation Model

There are three strong but phenomenological arguments that the mechanism for enhanced base selection by animal DNA polymerases involves a change in enzyme conformation at each nucleotide addition step. First, based on studies with *E. coli* polymerase I, it is assumed that DNA polymerases have one site for the binding of all four deoxynucleoside triphosphates.[67] Only in the case of polymerase-α has there been evidence for more than one site (Chapter 3). If there are less than four nucleotide binding sites on a DNA polymerase, these enzymes must have a mechanism to accommodate nucleotides or base-pairs of diverse structure. Second, animal cell DNA polymerases of the different classes all lacking any associated exonucleolytic activity have widely differing error rates when copying the same DNA template.[6,20] This implies that these enzymes have different capacities for base-selection. Third, NMR studies with Pol I indicate that DNA polymerase, in the absence of template, holds its substrates in a β-like conformation.[68,69] This suggests that the enzyme has multiple contact points for interaction with each substrate. Definitive evidence for a conformation model for base-selection by animal cell DNA polymerases is lacking and may have to await cloning of these polymerases followed by introducing mutations at particular sites. However, a literate mechanism would predict a difference in the V_{max} for the correct and incorrect nucleotides that would be proportional to the error rate of different polymerases.[66] These measurements could be obtained by using a single synthetic oligonucleotide hybridized to a ϕX template[22] and measuring misincorporation of the first nucleotide inserted as a function of the concentration of complementary and noncomplementary nucleotides. Also, a literate mechanism is likely to involve multiple contact points between the polymerase and each of the four different nucleotide substrates. A detailed study of nucleotide analogs with methyl groups at different positions should demonstrate that the methyl groups at nonbase-pairing positions alter fidelity. The demonstration that misincorporation of α-thiol dCTP occurs at a much higher frequency than dCTP with polymerase-α and not with polymerase-β[33] provides a clue that polymerase-α coordinates with multiple sites on each deoxynucleoside triphosphate.

C. Energy Relay

In order to account for the fidelity of eukaryotic DNA polymerases and other polymerases lacking a proofreading exonuclease, Hopfield[70] proposed an ingenious mechanism in which the energy released by phosphate bond cleavage is applied by the polymerase to proofread the insertion of the following nucleotide. The most straightforward prediction of this hypothesis is that the incorporation of the first nucleotide should be more error prone than the incorporation of subsequent nucleotides. Two laboratories tested this hypothesis. Grosse et al.[21] compared the frequency of misincorporation in reactions primed with a synthetic oligonucleotide that was one nucleotide from an amber site on ϕX DNA to that obtained with a restriction fragment that primes further upstream from the amber site. No difference in reversion frequency was observed. However, these authors failed to observe competition between complementary and noncomplementary nucleotides using the synthetic oligomer, implying that there still might be something special about the first nucleotide to be incorporated that was not understood. Abbotts and Loeb[71] compared reversion frequencies obtained by copying *am*3 ϕX174 DNA with DNA polymerases-α and -β using synthetic oligonucleotides that were one, three, and ten nucleotides from the amber site. No differences in reversion frequencies were obtained as a function of primer position. Moreover, the change in reversion frequency as a result of changes in pool bias were the same with each of the primers, including the one that was only one nucleotide away. Thus these authors failed to observe any difference between incorporation of the first nucleotide and incorporation of subsequent nucleotides. These combined studies argue against an energy relay mechanism for the enhancement of base-insertion by animal DNA polymerases. However, they do not rule out the possibility that auxillary proteins or interactions with an unincorporated high energy compound might impart the ability to a DNA polymerase to utilize an energy relay mechanism for increasing accuracy.

An energy relay mechanism for increasing fidelity could be central to the use of RNA primers for the initiation of DNA replication in vivo. By the concerted action of primase and DNA polymerase, the insertion of the final ribonucleotide onto an RNA primer could be the source of energy to ensure that the first deoxyribonucleotide was incorporated by the polymerase with high fidelity. Since the RNA primers are eraseable, it would not be necessary to evolve a mechanism for guaranteeing high fidelity for the incorporation of the first ribonucleotide.

D. Pyrophosphate-Mediated Proofreading

Hopfield[72] and Ninio and Bernardi[73] initially suggested that proofreading might occur prior to exonucleolytic excision by the release of pyrophosphate and the deoxynucleoside monophosphate from an enzyme-nucleoside triphosphate intermediate. In such a kinetic proofreading scheme, the addition of pyrophosphate could increase misincorporation by uncoupling base selection from proofreading.[74] In contrast, Doubleday et al.[42] postulated a different effect of added pyrophosphate, which may pertain to DNA polymerase-α. In their scheme, the incorrect substrate is rejected by pyrophosphorolysis or by pyrophosphate exchange without the generation of deoxynucleoside monophosphates. A prediction of this mechanism is that the addition of pyrophosphate should enhance the fidelity of catalysis by α-polymerase. Results they report with synthetic polynucleotides are in accord with this prediction. Moreover, studies with calf thymus polymerase-α copying ϕX DNA demonstrate a fourfold enhancement in fidelity by the addition of 7 mM pyrophosphate.[33] This is in contrast to the decreased fidelity observed with *E. coli* polymerase I,[76] as would be predicted on the basis of kinetic proofreading.[72,73] One must conclude that pyrophosphate concentration may modulate fidelity in eukaryotic cells, but it is unlikely to be a major factor in establishing the overall accuracy of DNA replication. However, it can always be argued that the effects of pyrophosphate might be greater with a multiprotein complex that is involved in DNA replication.

VII. MECHANISMS FOR ACCURACY IN ANIMAL CELLS

From the results presented, it is unlikely that the fidelity of catalysis by DNA polymerases (as determined after enzyme purification) is adequate to account for the accuracy of DNA replication in vivo. One is forced to postulate powerful fidelity enhancing factors that work in concert with DNA polymerases or to postulate the correction of errors after polymerization. The latter mechanism would be analogous to mismatch correction in bacteria.[75,77]

A. Proofreading in Animal Cells

It is reasonable to postulate that proofreading could be a universal method to increase the fidelity of DNA synthesis. When one considers the aesthetic simplicity of a two-step mechanism to achieve accuracy, it is difficult to visualize how such a mechanism could be deleted during the course of evolution. With respect to animal DNA polymerase, one could simply argue that the exonucleolytic function resides on a separate protein component that is lost during the course of purification. In support of this concept are the many reports on the presence of exonucleolytic activities associated with a variety of DNA polymerases isolated predominantly from simple eukaryotes (see Chapters 3 and 9). Hence, with the *Ustilago* polymerase, the terminal mismatch is hydrolyzed before synthesis begins, which suggests a role for the exonuclease in proofreading.[78] Furthermore, DNA polymerase-δ is essentially identified on the basis of an associated exonuclease, and polymerase-β has been reported to be loosely associated with a specific exonuclease, DNase V.[79] However, we lack fidelity measurements with any of these exonuclease-polymerase complexes that demonstrate an unusually high accuracy or that implicate proofreading by their associated exonuclease. That a proofreading exonuclease is weakly associated with DNA polymerase-α is suggested by Radman et al.[80] who reported a progressive loss in fidelity during the purification of DNA polymerase-α, and that this loss corresponded with removal of an exonuclease hydrolyzing terminally mismatched nucleotides. That proofreading occurs in eukaryotic cells is suggested by the data of Akiyoshi,[81] who reported that the conversion of dGTP to dGMP in isolated Yoshida sarcoma nuclei was dependent on DNA synthesis. The dependency on DNA synthesis is suggested by the requirement for dATP and the inhibition by NEM, aphidicolin, spermine, and KCl. However, as the author notes, since nuclei contain a dGTPase activity one cannot rule out activation of a DNA-dependent dGTPase. Thus there is no substantial evidence for proofreading in eukaryotic cells. In fact, there is evidence to the contrary. The frequency of incorporation of 2-aminopurine triphosphate vs. dATP by isolated preparation of polymerase-α is the same as that found in nuclei, suggesting that there is no added proofreading activity in nuclei.[82] This seems to be a reasonable test for proofreading, since prokaryotic DNA polymerases recognize 2-aminopurine nucleotides as noncomplementary bases. Also, the contribution of proofreading to fidelity by prokaryotic DNA polymerases has been estimated at 10- to 300-fold.[41,44,48] However, this contribution may be prohibitively high in cost of metabolic energy, since it requires hydrolysis of some 10% of the nucleotides polymerized.[83] An even higher level of proofreading is theoretically possible, but it would require a significant fraction of the cell's total energy resources. Thus, if proofreading occurs in animal cells, other mechanisms may also be required to account for the high accuracy of DNA replication.

B. Mismatch Correction in Animal Cells

The lack of a proofreading exonuclease in purified eukaryotic DNA polymerases and the potentially prohibitive energy cost of proofreading highlights the possibility of a postsynthetic mismatch correction system being a major contributor to fidelity in eukaryotic cells. This system would presumably depend on excision of a mismatched base by a specific endonuclease, the formation of a gap by exonucleolytic hydrolysis, and resynthesis by DNA polymerase. The fidelity of such a system can be exceptionally high. The accuracy of the corrective resynthesis would be equivalent to the accuracy of the polymerase used for resynthesis multiplied by the number of nucleotides replaced. Repetitive cycles of mismatch repair could provide exceptionally high accuracy, if the number of nucleotides replaced per corrective event were limited. In bacteria, evidence indicates that this pathway can contribute as much as three orders of magnitude to the accuracy of DNA replication.[84] Recognition of the newly synthesized strand is dependent on a delay in methylation of adenine residues in specific GATC sequences.[84] Since the presence of 6-methyl adenosine has not been reported in eukaryotic DNA, it is likely that, if mismatch correction occurs, different signals ensure that excision occurs only in the daughter strand. Modrich and associates[85,86] have designed an in vitro system that mimics the requirements for mismatch repair in bacteria. One surprising finding is that the resynthesized gap is a few thousand nucleotides in length. This observation causes one to rethink the possibilities for mismatch repair in eukaryotic cells. The lack of 6-methyladenine in eukaryotic DNA so far has been determined only on the basis of chemical analysis — and the sensitivity of detection indicates that it constitutes less than 0.1% of the nucleotide residues. Based on the size of the resynthesized DNA in bacteria, it could be argued that the amount of 6-methyladenine that could have escaped detection in eukaryotic DNA might be sufficient to guarantee strand recognition. More sensitive methods of analysis are now available to establish the presence of altered bases in eukaryotic DNA.[87] However, any mismatch repair system that resynthesizes a few thousand nucleotides per repair event is not energy cost-efficient. Thus a much smaller patch size and a different mechanism for strand recognition would be required for mismatch repair to be a major factor in enhancing the fidelity of DNA replication in animal cells. In eukaryotic cells, other signals for strand recognition should be considered. These might include a subset of methylcytidine residues in DNA, or strand recognition even could be based on the association of histones or other chromosomal proteins with parental DNA.[20]

VIII. MUTAGENESIS BY DNA POLYMERASES

The high error rates of eukaryotic DNA polymerases suggests, by itself, that minor perturbation in the mechanism of fidelity at the level of incorporation by DNA polymerases may be a major contributor to mutagenesis. This argument predicts that small fluctuations in the frequency of misinsertion by DNA polymerases would produce large numbers of errors, which could overwhelm the corrective capacities in cells. Thus identification of mutagens that alter the fidelity of catalysis by DNA polymerase may simultaneously identify the molecular basis for mutagenesis by these agents.

A. Nucleotide Pools

The results presented in this chapter have repetitively indicated, with only a few exceptions, that the error rates of eukaryotic DNA polymerases are directly proportional to the ratio of incorrect to correct nucleotide substrates in the reaction mixture. Even with an added proofreading step by prokaryotic DNA polymerases or by a putative associated exonuclease in eukaryotes, this proportionality is still valid for average frequencies of misincorporation over a large template span. Thus changes in deoxynucleotide pools in vivo might enhance misincorporation frequency by DNA polymerases and lead to mutagenesis. Since in animal cells, the amount of deoxynucleoside triphosphate present at any given time is sufficient for only 15 to 30 sec of DNA synthesis,[88] and since imbalances in nucleotide ratio could be mutagenic,[89] intuitively, one might expect cells to stringently regulate nucleoside triphosphate concentrations. However, if such controls exist they are not absolute, since the addition of exogenous nucleosides to cell cultures alters intracellular nucleotide pools.[90,91] Minimization of fluctuations in nucleotide concentrations at the replication fork could be achieved by replication complexes, which contain proteins that are involved in the control of nucleotide concentration, particularly if these proteins channel nucleotide substrates directly onto the DNA polymerase.[92,93]

A number of observations in the literature suggest that eukaryotic cells with altered nucleotide metabolism exhibit an increased rate of spontaneous mutagenesis. Meuth et al.[89,94] isolated three mammalian cell lines with increased intracellular dCTP concentrations that exhibited an increased frequency of spontaneous mutagenesis. Weinberg et al.[95] demonstrated that the alterations in nucleotide pool sizes in mutagenic T-lymphosarcoma cell lines resulted from alterations in ribonucleotide reductase and deoxycytidylate deaminase. Most pertinently, normalization of the dNTP pools in deaminase-deficient cells by adding exogenous nucleotides suppressed their mutator phenotype. Other biological observations, including bromodeoxyuridine mutagenesis,[96,97] deoxynucleoside toxicity and mutagenesis,[98] and potentiation of mutagenesis by alkylating agents[92] have been attributed to selective alterations in nucleotide pools.

Recent reports have demonstrated that exposure of animal[99,100] and bacterial[101] cells to mutagens results in rapid changes in deoxynucleoside triphosphate pools. Das et al.[99] demonstrated that exposure of Chinese hamster V79 cells to UV and other mutagens results within 2 to 4 hr in a four- and eightfold increase in the intracellular concentrations of dTTP and dATP, respectively. Newman and Miller[100] demonstrated similar increases in dTTP concentration after UV treatment of Chinese hamster ovary cells, as well as a decline in dCTP. Alterations in nucleotide pools may provide an explanation for nontargeted mutagenesis — that is, changes in DNA sequence not in immediate proximity to damaged nucleotides in DNA. Thus, if damage to DNA results in alterations in nucleotide pools, the biases in nucleotide concentration at the replication fork could cause random misincorporation by DNA polymerase during resynthesis of the gap produced after excision of the damaged region.

B. Incorporation of Nucleotide Analogs

Only scattered observations have appeared on the incorporation of deoxynucleoside triphosphate analogs by animal DNA polymerases. These include the incorporation of nucleoside triphosphate such as aminopurine,[66] bromodeoxyuridine, 6-methyladenine, and phosphorothioates.[44] Eadie et al.[102] have proposed that mutagen alteration of deoxynucleoside triphosphates in cells is a significant factor in mutagenesis by alkylating agents and related compounds. A careful study on the rates of incorporation of deoxynucleoside triphosphate analogs by DNA animal polymerases and the potential of the incorporated analog for mutagenesis is required in order to evaluate this hypothesis.

C. Effect of Mutagenic Metal Ions

A number of metals, many of which are ubiquitous in our environment, have been shown to be mutagens and potent carcinogens.[103] Sirover and Loeb[104] reported a positive correlation between the metal ions that decrease the fidelity of DNA synthesis in vitro and those that are reported to be mutagens and carcinogens in vivo. These studies were initially carried out with the DNA polymerase from avian myeloblastosis virus. However, many of these metal ions have been shown to diminish the fidelity of synthesis using DNA polymerase-α and -β.[27,62] Metal ions that alter the fidelity of DNA synthesis appear to fall into two classes. The first are those that can substitute for Mg^{2+} and serve as an added metal activator.[105] Among these are Mn^{2+} and Co^{2+}, which can serve as metal activators for DNA polymerases-α and -β from human placenta.[27] Substitution of Mn^{2+} for Mg^{2+} has been shown to enhance misincorporation by DNA polymerase-α from human placenta,[27] calf thymus,[28,106] and human fibroblasts,[30] as well as by DNA polymerase-β[27] and DNA polymerase-γ from human placenta.[30] It has been assumed that substitution occurs at the active site. However, recent experiments that compare binding of Mn^{2+} with changes in fidelity by DNA polymerases indicate that this may not be the case. The change in mutagenesis at low Mn^{2+} concentrations corresponds to change in the binding of the template, and not to the enzyme.[106] A second group of metals that diminish the fidelity of DNA synthesis do not serve as activators of DNA polymerase. It is assumed that many of these metal ions alter the fidelity of DNA synthesis by binding to the DNA template. However, there is evidence for one exception: Be^{2+}, a metal with an unusually low exchange rate, forms a stable complex with avian myeloblastosis DNA polymerase, and this complex is error prone.[107]

D. Template Modifications

Covalent modification of DNA templates has always been considered a major pathway for mutagenesis by chemical carcinogens. It is assumed that DNA polymerases bypass unrepaired nucleotide adducts in the DNA template and in so doing, incorporate deoxynucleotides that are noncomplementary to the template nucleotides. Because of the relative ease of chemically synthesizing ribonucleotide analogs, the effects of template base modifications have been most extensively studied with synthetic polyribonucleotides and RNA polymerase.[108,109] So far, systematic studies are lacking with DNA polymerases from animal cells. Such studies are needed to assess the mutagenic potential of modified DNA in animal cells. The most exacting studies will be those involving site-specific mutagenesis in which one inserts a chemically synthesized analog at a specified position on a template and determines the frequency and type of nucleotides incorporated opposite the template modification.

Of particular interest are recent studies on the copying of DNA containing apurinic sites. These sites can arise spontaneously due to the inherent instability of glycosylic bond in DNA[110] or by removal of chemical altered bases by cellular glycosylases.[111] Mutagenesis via apurinic sites appears to proceed by the preferential insertion of deoxyadenosine as a single base substitution.[112,113] Under conditions in which *E. coli* DNA polymerase I can by-pass less than 10% of the apurinic sites, DNA polymerases-α and -β can copy past 60 and 40% of the sites, respectively.[113] In transfection experiments with *am3* ϕX DNA containing apurinic sites, mutagenesis is observed only after the induction of the SOS response,[114,115] and this response could be mediated by an alteration in DNA polymerases.[116,117] The ability of animal cell DNA polymerases to readily copy past apurinic sites on DNA[113] suggests that mutagenesis by this mechanism may not require alterations in the DNA replicating complexes in animal cells.[118]

E. Alterations in DNA Polymerases

The fact that animal cell DNA polymerases have different error rates brings into focus the possibility that substitution of one DNA polymerase for another may have important mutagenic consequences for the cell. The similarity of the requirements for catalysis of DNA synthesis in vitro by different animal DNA polymerases supports such a possibility. It has been observed that during the copying of *am3* ϕX174 DNA by polymerase-α, the addition of DNA polymerase-β increases the frequency of misincorporation.[119] If this occurs in vivo, it provides an interesting mechanism for mutagenesis in animal cells. For example, DNA polymerases from RNA tumor viruses are error prone,[120] and these might substitute for a cellular DNA polymerase and copy the host cell DNA at a high error rate. Thus the viral DNA polymerase could be mutagenic in eukaryotic cells.

Errors in the synthesis of DNA polymerase have been postulated to be significant in the pathogenesis of tumor progression[121] and aging.[122,123] The intriguing possibility that the fidelity of DNA polymerase-α is diminished during aging is considered in Chapter 9. With respect to tumor progression, the presence of an error-prone DNA polymerase was first reported in lymphocytes from patients with leukemia.[124] However, the presence of terminal transferase in these cells has precluded rigorous interpretations. Chan and Becker[125] noted the appearance of an abberant form of DNA polymerase-α during the induction of liver cancer in rats exposed to the hepatocarcinogen, *N*-2-acetylaminofluorone. The fidelity of the aberrant form was several fold less in copying synthetic polynucleotides than DNA polymerase-α purified from controls. With the introduction of fidelity assays using natural DNA templates and the rapid purification of DNA polymerase-α by immunoaffinity columns, it is particularly desirable to repeat these important studies.

An even more speculative possibility is modulation of the fidelity of DNA polymerases by postsynthetic processes such as phosphorylation. Aberrant phosphorylation of proteins during viral carcinogenesis has been extensively demonstrated. In fact, the sarc gene and a number of related oncogenes are protein kinases.[126] The phosphorylated products can be localized in nuclei. Conceivably, DNA polymerase may be one of the targets for phosphorylation by protein kinases. If this phosphorylation of a species of DNA polymerase results in increased infidelity, it could account for random and progressive accumulation of mutants during tumor progression.[127,128]

F. SOS Repair in Animal Cells

Mutagenesis by many agents in bacteria is dependent on a series of coordinately regulated genes referred to as the SOS system,[129] named after the universal distress signal. The system can be induced by blockage of DNA synthesis as a result of damage to DNA by chemicals that form bulky adducts. In bacteria, the induction of the SOS response proceeds by the inactivation of a common repressor, the *lex* A protein, resulting in the increased expression of some 17 genes, causing a pleotropic response, including mutagenesis.[116] There is evidence that mutagenesis during the SOS response results from alteration in DNA polymerase;[116,117] the altered replicating complex is able to bypass bulky adducts or apurinic sites.[118,119] However, there is controversy over whether or not a similar system is present in animal cells.[130,131] Currently, one piece of strong evidence for the SOS response in animal cells is increased reversion of temperature sensitive mutants of SV40 to wild-type phenotype when the UV irradiated virus is grown in UV irradiated monkey cells.[131] The production of viral revertants is likely to involve DNA polymerase-α. However, it is not known whether or not this process is mediated by the induction of an error prone DNA polymerase.

IX. UPDATE

A. Mutational Spectra

The sequencing of forward mutation produced in M13mp2 DNA by eukaryotic DNA polymerases has provided the first insights into the spectrum of mutations resulting from catalysis by purified eukaryotic DNA polymerases.[25] The first studies established the mutational spectra of DNA polymerase-β 25 and these studies have now been extended to DNA polymerases-α and γ.[133]

1. DNA Polymerase-α

Pol-α, purified from five different sources, produced forward mutations in M13mp2 DNA at a frequency of 0.8 to 1.6% per single round of gap-filling DNA synthesis. This frequency is lower than that reported for DNA polymerase-β.[25] Even though the majority of mutation was single-base substitutions, single-base frameshifts and large deletions were also observed.

2. DNA Polymerase-γ

Pol-γ was unexpectedly more accurate than pol-α, producing mutants at a frequency of 0.3 to 0.5%. With polymerase-γ, 90% of the mutants resulted from single-base substitutions. This finding is in accord with the high frequency of single-base differences in mitochondrial DNA between individuals.[134]

3. Relation of Fidelity to Processivity

In this system, the relative accuracy of the three eukaryotic DNA polymerases tested is $\gamma > \alpha > \beta$ for base substitutions, frameshifts, and deletions. This order parallels the relative processivity of these enzymes. A detailed analysis of the effects of changing processivity and fidelity should substantiate this relationship and may provide mechanistic insights into base selection by DNA polymerases.

REFERENCES

1. Eigen, M., Gardiner, W., Schuster, P., and Winkler-Oswatitsch, R., Origin of genetic information, *Sci. Am.*, 244, 88, 1981.
2. Drake, J. W., *The Molecular Basis of Mutation*, Holden Dag, San Francisco, 1970.
3. Hibner, U. and Alberts, B. M., Fidelity of DNA replication catalysed *in vitro* on a natural DNA template by the T4 bacteriophage multi-enzyme complex, *Nature (London)*, 285, 300, 1980.
4. Fersht, A. R., Fidelity of replication of phage ΦX174 DNA by DNA polymerase III holoenzyme: spontaneous mutation by misincorporation, *Proc. Natl. Acad. Sci. U.S.A.*, 76, 4946, 1979.
5. Loeb, L. A., Kunkel, T. A., and Schaaper, R. M., Fidelity of copying natural DNA templates, in *Mechanistic Studies in DNA Replication and Genetic Recombination*, Vol. 19, ICN-UCLA Symp. Mol. Cell. Biol., Academic Press, New York, 1980, 735.
6. Kunkel, T. A. and Loeb, L. A., Fidelity of mammalian DNA polymerases, *Science*, 213, 765, 1981.
7. Hall, Z. W. and Lehman, I. R., An *in vitro* transversion by a mutationally altered T4-induced DNA polymerase, *J. Mol. Biol.*, 36, 321, 1968.
8. Battula, N. and Loeb, L. A., The infidelity of avian myeloblastosis virus deoxyribonucleic acid polymerase in polynucleotide replication, *J. Biol. Chem.*, 249, 4086, 1974.

9. Agarwal, S. S., Dube, D. K., and Loeb, L. A., On the fidelity of DNA replication. VII. Accuracy of *E. coli* DNA polymerase I, *J. Biol. Chem.*, 254, 101, 1979.
10. Bollum, F. J., Terminal deoxynucleotidyl transferase: biological studies, *Adv. Enzymol.*, 47, 347, 1978.
11. Saffhill, R., The effect of ionising radiation and chemical methylation upon the activity and accuracy of *E. coli* DNA polymerase I, *Biochem. Biophys. Res. Commun.*, 61, 752, 1974.
12. Sirover, M. A. and Loeb, L. A., Infidelity of DNA synthesis *in vitro:* screening for potential metal mutagens or carcinogens, *Science*, 194, 1434, 1976.
13. Singer, B. and Kusmierek, J. T., Chemical mutagenesis, *Annu. Rev. Biochem.*, 52, 655, 1982.
14. Fry, M., Shearman, C. W., Martin, G. M., and Loeb, L. A., On the fidelity of DNA synthesis directed by chromatin-associated DNA polymerase-β, *Biochemistry*, 19, 5939, 1980.
15. Chang, L. M. S. Cassani, G. R., and Bollum, F. J., Deoxynucleotide-polymerizing enzymes of calf thymus gland. VII. Replication of homopolymers, *J. Biol. Chem.*, 247, 7718, 1972.
16. Weymouth, L. A. and Loeb, L. A., Mutagenesis during *in vitro* DNA synthesis, *Proc. Natl. Acad. Sci. U.S.A.*, 75, 1924, 1978.
17. Kunkel, T. A., James, E. A., and Loeb, L. A., The use of ΦX174 *amber* mutations to study DNA repair, in *DNA Repair: A Laboratory Manual of Research Procedures*, Marcel Dekker, New York, 1983, 223.
18. Kunkel, T. A. and Loeb, L. A., On the fidelity of DNA replication. X. Effect of divalent metal ion activators and deoxyribonucleoside triphosphate pools on *in vitro* mutagenesis, *J. Biol. Chem.*, 254, 5718, 1979.
19. Kunkel, T. A. and Loeb, L. A., On the fidelity of DNA replication: the accuracy of *Escherichia coli* DNA polymerase I in copying natural DNA *in vitro*, *J. Biol. Chem.*, 255, 9961, 1980.
20. Loeb, L. A. and Kunkel, T. A., Fidelity of DNA synthesis, *Annu. Rev. Biochem.*, 52, 429, 1982.
21. Grosse, F., Krauss, G., Knill-Jones, J. W., and Fersht, A. R., Accuracy of DNA polymerase-α in copying natural DNA, *EMBO J.*, 2, 1515, 1983.
22. Abbotts, J. and Loeb, L. A., On the fidelity of DNA replication: Use of synthetic oligonucleotide-initiated reactions, *Biochim. Biophys. Acta*, in press.
23. Fersht, A. R. and Knill-Jones, J. W., Fidelity of replication of bacteriophage ΦX174 DNA *in vitro* and *in vivo*, *J. Mol. Biol.*, 165, 655, 1983.
24. Liu, C. C., Burke, R. L., Hibner, U., Barry, J., and Alberts, B. M., Probing DNA replication mechanisms with the T4 bacteriophage *in vitro* systems, *Cold Spring Harbor Symp. Quant, Biol.*, 43, 469, 1979.
25. Kunkel, T. A., The mutational specificity of DNA polymerase-β during *in vitro* DNA synthesis: production of frameshift, hase substitution and deletion mutations, *J. Biol. Chem.*, in press.
26. Chang. L. M. S. and Bollum, F. J., A comparison of associated enzyme activities in various DNA polymerases, *J. Biol. Chem.*, 248, 3398, 1973.
27. Seal, G., Shearman, C. W., and Loeb, L. A., On the fidelity of DNA replication: studies with human placenta DNA polymerases, *J. Biol. Chem.*, 254, 5229, 1979.
28. Chang, L. M. S., Low molecular weight deoxyribonucleic acid polymerase from calf thymus chromatin. II. Initiation and fidelity of homopolymer replication, *J. Biol. Chem.*, 248, 6983, 1973.
29. Salisbury, J. G., O'Connor, P. J., and Saffhill, R., Molecular size and fidelity of DNA polymerase-α from the regenerating liver of the rat, *Biochim. Biophys. Acta*, 517, 181, 1978.
30. Krauss, S. W. and Linn, S., Fidelity of fractionated deoxyribonucleic acid polymerases from human placenta, *Biochemistry*, 19, 220, 1980.
31. Kornberg, A., *DNA Replication*, W. H. Freeman, San Francisco, 1980.
32. Brosius, S., Grosse, F., and Krauss, G., Subspecies of DNA polymerase-α from calf thymus with different fidelity in copying synthetic template-primers, *Nucl. Acids Res.*, 11, 193, 1983.
33. Abbotts, J. and Loeb, L. A., DNA polymerase-α and models for proofreading, *Nucl. Acids Res.*, in press.
34. Loeb, L. A., Abbotts, J., Liu, P. K., and Silber, J. R., Fidelity of DNA synthesis in eucaryotic cells, in *Cellular Responses to DNA Damage*, UCLA Symp. Molec. Cell Biol., Friedberg, E. C. and Bridges, B. A., Eds., Alan R. Liss, New York, 1983, 471.
35. Holmes, A. W., Hesselwood, I. P., and Johnston, I. R., The occurrence of multiple activities in the high-molecular-weight DNA polymerase fraction of mammalian tissues, *Eur. J. Biochem.*, 43, 487, 1974.
36. Holmes, A. M. and Johnston, I. R., DNA polymerases of eukaryotes, *FEBS Lett.*, 60, 233, 1975.
37. Kaguni, L. S., DiFrancesco, R. A., and Lehman, I. R., The DNA polymerase-primase from *Drosophila melanogaster* embryos. Rate and fidelity of polymerization on single-stranded DNA templates, *J. Biol. Chem.*, 259, 9314, 1984.
38. Fersht, A. R. and Knill-Jones, J. W., DNA polymerase accuracy and spontaneous mutation rates: frequencies of purine·purine, purine·pyrimidine and pyrimidine·pyrimidine mismatches during DNA replication, *Proc. Natl. Acad. Sci. U.S.A.*, 78, 4251, 1981.

39. Dube, D. K., Kunkel, T. A., Seal, G., and Loeb, L. A., Distinctive properties of mammalian DNA polymerases, *Biochim. Biophys. Acta*, 561, 369, 1979.
40. Fisher, P. A., Wang, T. S-F., and Korn, D., Enzymological characterization of DNA polymerase-α, *J. Biol. Chem.*, 254, 6128, 1979.
41. Kunkel, T. A., Schaaper, R. M., Beckman, R. A., and Loeb, L. A., On the fidelity of DNA replication. XII. Effect of the next nucleotide on proofreading, *J. Biol. Chem.*, 256, 9883, 1981.
42. Doubleday, O. P., Lecomte, P. J., and Radman, M., A mechanism for nucleotide selection associated with the pyrophosphate exchange activity of DNA polymerase-α, in *Cellular Responses to DNA Damage*, Friedberg, E. C. and Bridges, B. A., Eds., Alan R. Liss, 1983, 489.
43. Eckstein, F., Nucleotide analogues for the study of enzyme mechanisms, *Trends Biochem. Sci.*, June, 157, 1980.
44. Kunkel, T. A., Eckstein, F., Mildvan, A. S., Koplitz, R. M., and Loeb, L. A., Deoxynucleoside [1-thio]triphosphates prevent proofreading during *in vitro* DNA synthesis, *Proc. Natl. Acad. Sci. U.S.A.*, 78, 6734, 1981.
45. Hübscher, V., Gerschwiler, P., and McMawter, G. K., A mammalian DNA polymerase-α holoenzyme functioning on defined *in vitro*-like templates, *EMBO J.*, 1, 1513, 1982.
46. Bernstein, C., Bernstein, H., Mufti, S., and Strom, B., Stimulation of mutation in phage T4 by lesions in gene 32 and by thymidine imbalance, *Mutat. Res.*, 16, 113, 1972.
47. Koch, R. E., McGaw, M. K., and Drake, J. W., Mutator mutations in bacteriophage T4 gene 32 (on an unwinding protein), *J. Virol.*, 19, 490, 1976.
48. Watanabe, S. M. and Goodman, M. F., Mutator and antimutator phenotypes of suppressed amber mutants in genes 32, 41, 44, 45, and 62 in bacteriophage T4, *J. Virol.*, 25, 73, 1978.
49. Gillin, F. D. and Nossal, N. G., Control of mutation frequency by bacteriophage T4 DNA polymerase II. Accuracy of nucleotide selection by the L88 mutator, CB120 anti mutator, and wild type phage T4 DNA polymerases, *J. Biol. Chem.*, 251, 5225, 1976.
50. Kunkel, T. A., Meyer, R., and Loeb, L. A., Single-strand binding protein enhances the fidelity of DNA synthesis *in vitro*, *Proc. Natl. Acad. Sci. U.S.A.*, 76, 6331, 1979.
51. Champoux, J. J., Proteins that affect DNA conformation, *Annu. Rev. Biochem.*, 47, 449, 1978.
52. Banks, G. R. and Spanos, A., The isolation and properties of a DNA-unwinding protein from *Ustilago mavdis*, *J. Mol. Biol.*, 93, 63, 1975.
53. Mikhailov, V. S. and Androsova, I. M., Effect of spermine on interaction of DNA polymerase-α from the loach *(Misgurnus fossilis)* eggs with DNA, *Biochim. Biophys. Acta*, 783, 6, 1984.
54. Fry, M., Loeb, L. A., and Martin, G. M., On the activity and fidelity of chromatin-associated hepatic DNA polymerase-β in aging murine species of different life spans, *J. Cell. Physiol.*, 106, 435, 1981.
55. Kunkel, T. A., Silber, J. R., and Loeb, L. A., The mutagenic effect of deoxynucleotide substrate imbalances during DNA synthesis with mammalian DNA polymerases, *Mutat. Res.*, 94, 413, 1982.
56. Probst, G. S., Stalker, D. M., Mosbaugh, D. W., and Meyer, R. R., Stimulation of DNA polymerase by factors isolated from Novikoff hepatoma, *Proc. Natl. Acad. Sci. U.S.A.*, 72, 1171, 1975.
57. Brown, W. M., Polymorphism in mitochondrial DNA of humans as revealed by restriction endonuclease analysis, *Proc. Natl. Acad. Sci. U.S.A.*, 77, 3605, 1980.
58. Monnat, R. J., Jr. and Loeb, L. A., Nucleotide sequence preservation of human mitochondrial DNA, *Proc. Natl. Acad. Sci. U.S.A.*, in press.
59. Monnat, R. J., Maxwell, C. L., and Loeb, L. A., Nucleotide sequence preservation of human leukemic mitochondrial DNA, *Cancer Res.*, in press.
60. Byrnes, J. J., Downey, K. M., Blank, V. L., and So, A. G., A new mammalian DNA polymerases with 3′ to 5′ exonuclease activity: DNA polymerase-δ, *Biochemistry*, 15, 2817, 1976.
61. Chan, J. Y. H. and Becker, F. F., Decreased fidelity of DNA polymerase activity during *N*-2-fluorenylacetamide hepatocarcinogenesis, *Proc. Natl. Acad. Sci. U.S.A.*, 76, 814, 1979.
62. Murray, V. and Holliday, R., Increased error frequency of DNA polymerases from senescent human fibroblasts, *J. Biol. Chem.*, 145, 55, 1981.
63. Kunkel, T. A., Loeb, L. A., and Goodman, M. F., On the fidelity of DNA replication: the accuracy of T4 DNA polymerases in copying ϕX174 DNA *in vitro*, *J. Biol. Chem.*, 259, 1539, 1984.
64. Lohrmann, R. and Orgel, L. E., Efficient catalysis of polycytidylic acid-directed oligoguanylate formation by Ph^{2+}, *J. Mol. Biol.*, 142, 555, 1980.
65. Goodman, M. F., Hopkins, R., and Gore, W. C., 2-aminopurine-induced mutagenesis in T4 bacteriophage: a model relating mutation frequency to 2-aminopurine incorporation in DNA, *Proc. Natl. Acad. Sci. U.S.A.*, 74, 4806, 1977.
66. Watanabe, S. M. and Goodman, M. F., Kinetic measurement of 2-aminopurine-cytosine and 2-aminopurine-thymidine base pairs as a test of DNA polymerase fidelity mechanisms, *Proc. Natl. Acad. Sci. U.S.A.*, 79, 6429, 1982.
67. Englund, P. T., Huberman, J. A., Jovin, T. M., and Kornberg, A., Enzymatic synthesis of deoxyribonucleic acid. XXX. Binding of triphosphates to deoxyribonucleic acid polymerase, *J. Biol. Chem.*, 244, 3038, 1969.

68. Sloan, D. L., Loeb, L. A., Mildvan, A. S., and Feldmann, R. J., Conformation of deoxynucleoside triphosphate substrates on DNA polymerase I from *E. coli* as determined by nuclear magnetic relaxation, *J. Biol. Chem.*, 250, 8913, 1975.
69. Ferrin, L. J., Beckman, R. A., Loeb, L. A., and Mildvan, A. S., Kinetic and magnetic resonance studies of the interaction of Mn^{2+}, substrates and templates with DNA polymerases, in *Manganese in Metabolism and Enzyme Function,* Wedler, F. C. and Schramm, V. L., Eds., Academic Press, New York, in press.
70. Hopfield, J. J., The energy relay: a proofreading scheme based on dynamic cooperativity and lacking all characteristic symptoms of kinetic proofreading in DNA replication and protein synthesis, *Proc. Natl. Acad. Sci. U.S.A.*, 77, 5248, 1980.
71. Abbotts, J. and Loeb, L. A., On the fidelity of DNA replication: lack of primer position effect on the fidelity of mammalian DNA polymerases, *J. Biol. Chem.*, 259, 6712, 1984.
72. Hopfield, J. J., Kinetic proofreading: a new mechanism for reducing errors in biosynthetic processes requiring high specificity, *Proc. Natl. Acad. Sci. U.S.A.*, 71, 4135, 1974.
73. Bernardi, F. and Ninio, J., The accuracy of DNA replication, *Biochimie,* 60, 762, 1978.
74. Loeb, L. A., Dube, D. K., Beckman, R. A., and Gopinathan, K. P., On the fidelity of DNA replication: nucleoside monophosphate generation during polymerization, *J. Biol. Chem.*, 256, 3978, 1981.
75. Wagner, R., Jr. and Meselson, M., Repair tracts in mismatched DNA heteroduplexes, *Proc. Natl. Acad. Sci. U.S.A.*, 73, 4135, 1976.
76. Kunkel, T. A., Beckman, R. A., and Loeb, L. A., unpublished results.
77. Glickman, B. W. and Radman, M., *Escherichia coli* mutator mutants deficient in methylation-instructed DNA mismatch correction, *Proc. Natl. Acad. Sci. U.S.A.*, 77, 1063, 1980.
78. Yarranton, G. Y. and Banks, G. R., A DNA polymerase from *Ustilago maydis.* Evidence of proofreading by the associated 3′ → 5′ dexoyribonuclease activity, *Eur. J. Biochem.*, 77, 521, 1977.
79. Mosbaugh, D. W. and Meyer, R. R., Interaction of mammalian deoxyribonuclease V, a double strand 3′ → 5′ and 5′ → 3′ exonuclease with deoxyribonucleic acid polymerase-β from Novikoff hepatoma, *J. Biol. Chem.*, 255, 10239, 1980.
80. Radman, M., Spadari, S., and Villani, G., Mutagenesis and cell transformation by ultraviolet irradiation: many hypotheses, few results, *Cancer Inst. Monogr.*, 50, 121, 1978.
81. Akiyoshi, H., Conversion of dNTP to dNMP dependent of DNA synthesis in isolated Yoshida sarcoma nuclei, *Biochim. Biophys. Acta,* 696, 332, 1982.
82. Wang, M-L., Stellwagen, R. H., and Goodman, M. F., Evidence for the absence of DNA proofreading in HeLa cell nuclei, *J. Biol. Chem.*, 256, 7097, 1981.
83. Fersht, A. R., Knill-Jones, J. W., and Tsui, W.-C., Kinetic basis of spontaneous mutation. Misinsertion frequencies and cost of proofreading by DNA polymerases of *Escherichia coli, J. Mol. Biol.*, 156, 37, 1982.
84. Glickman, B., van den Elsen, P., and Radman, M., Induced mutagenesis in *dam*⁻ mutants of *Escherichia coli*: a role for 6-methyladenine residues in mutation avoidance, *Molec. Gen. Genet.*, 163, 307, 1978.
85. Lu, A. L., Clark, S., and Modrich, P., Methyl-directed repair of DNA base-pair mismatches *in vitro, Proc. Natl. Acad. Sci. U.S.A.*, 80, 4639, 1983.
86. Lu, A. L., Welsh, K., Clark, S., Su, S. S., and Modrich, P., Repair of DNA base-pair mismatches in extracts of *Escherichia coli, Cold Spring Harbor Symp. Quant. Biol.*, in press.
87. Randerath, K., Reddy, M. V., and Gupta, R. C., ^{32}P-labeling test for DNA damage, *Proc. Natl. Acad. Sci. U.S.A.*, 78, 6126, 1981.
88. Nicander, B. and Reichard, P., Dynamics of pyrimidine deoxynucleoside triphosphate pools in relationship to DNA synthesis in 3T6 mouse fibroblasts, *Proc. Natl. Acad. Sci. U.S.A.*, 80, 1347, 1983.
89. Meuth, M., L'Heureux-Huard, and Trudel, M., Characterization of a mutator gene in Chinese Hamster ovary cells, *Proc. Natl. Acad. Sci. U.S.A.*, 76, 6505, 1979.
90. Tattersall, M. H. N., Ganeshaguru, K., and Hoffbrand, A. V., The effect of external deoxyribonucleosides on deoxyribonucleoside triphosphate concentrations in human lymphocytes, *Biochem. Pharmacol.*, 24, 1495, 1975.
91. Peterson, A. R., Landolph, J. R., Peterson, H., and Heidelberger, C., Mutagenesis in Chinese Hamster cells is facilitated by thymidine and deoxycitidine, *Nature (London),* 276, 508, 1978.
92. Mathews, C. K., North, T. W., and Reddy, G. P. V., Multienzyme complexes in DNA precursor biosynthesis, in *Advances in Enzyme Regulation,* Vol. 17, Weber, G., Ed., Pergamon Press, 1979, 133—156.
93. Reddy, G. P. V. and Pardee, A. B., Multienzyme complex for metabolic channeling in mammalian DNA replication, *Proc. Natl. Acad. Sci. U.S.A.*, 77, 3312, 1980.
94. Meuth, M., Trudel, M., and Siminovitch, L., Selection of Chinese hamster cells auxotrophic for thymidine by 1-β-D-arabino furanosyl cytosine, *Somat. Cell Genet.*, 5, 303, 1979.

95. Weinberg, G., Ullman, B., and Martin, D. W., Mutator phenotypes in mammalian cell mutants with distinct biochemical defects and abnormal deoxyribonucleoside triphosphate pools, *Proc. Natl. Acad. Sci. U.S.A.*, 78, 2447, 1981.
96. Ashman, C. R. and Davidson, R. L., Bromodeoxyuridine mutagenesis in mammalian cells is related to deoxyribonucleotide pool imbalances, *Molecular Cell Biol.*, p. 254, 1981.
97. Davidson, R. L. and Kaufman, E. R., Bromodeoxyuridine mutagenesis in mammalian cells is stimulated by thymidine and suppressed by deoxycytidine, *Nature (London)*, 276, 722, 1978.
98. Bjursall, G. and Reichard, P., Effects of thymidine on deoxyribonucleoside triphosphate pools and deoxyribonucleic acid synthesis in Chinese hamster ovary cells, *J. Biol. Chem.*, 248, 3904, 1973.
99. Das, S. K., Benditt, E. P., and Loeb, L. A., Rapid changes in deoxynucleoside triphosphate pools in mammalian cells treated with mutagens, *Biochem. Biophys. Res. Commun.*, 114, 458, 1983.
100. Newman, C. N. and Miller, J. H., Kinetics of UV-induced changes in deoxynucleoside triphosphate pools in Chinese hamster ovary cells and their effect on measurements of DNA synthesis, *Biochem. Biophys. Res. Commun.*, 116, 1064, 1983.
101. Das, S. K. and Loeb, L. A., UV irradiation alters deoxynucleoside triphosphate pools in *Escherichia coli*, *Mutat. Res.*, 131, 97, 1984.
102. Eadie, J. S., Conrad, M., Toorchen, D., and Topal, M. D., Mechanisms of mutagenesis by O^6-methylguanine, *Nature (London)*, 308, 201, 1984.
103. Sunderman, F. W., Jr., Carcinogenic effects of metals, *Fed. Proc. Am. Soc. Exp. Biol.*, 37, 40, 1978.
104. Sirover, M. A. and Loeb, L. A., Infidelity of DNA synthesis *in vitro:* screening for potential metal mutagens or carcinogens, *Science*, 194, 1434, 1976.
105. Sirover, M. A. and Loeb, L. A., On the fidelity of DNA replication. Effect of metals on synthesis with avian myeloblastosis DNA polymerase, *J. Biol. Chem.*, 252, 3605, 1977.
106. Beckman, R. A., Mildvan, A. S., and Loeb, L. A., On the fidelity of DNA replication: manganese mutagenesis *in vitro*, in preparation.
107. Sirover, M. A. and Loeb, L. A., Metal-induced infidelity during DNA synthesis, *Proc. Natl. Acad. Sci. U.S.A.*, 73, 2331, 1976.
108. Singer, B. and Grunberger, D., *Molecular Biology of Mutagens and Carcinogens*, Plenum Press, New York, 1983.
109. Singer, B. and Kroger, M., Participation of modified nucleosides in translation and transcription, *Prog. Nucl. Acids Res. Mol. Biol.*, 23, 151, 1979.
110. Lindahl, T. and Nyberg, B., Rate of depurination of native deoxyribonucleic acid, *Biochemistry*, 11, 3610, 1972.
111. Lindahl, T., DNA repair enzymes, *Annu. Rev. Biochem.*, 51, 61, 1982.
112. Schaaper, R. M., Kunkel, T. A., and Loeb, L. A., Infidelity of DNA synthesis associated with bypass of apurinic sites, *Proc. Natl. Acad. Sci. U.S.A.*, 80, 487, 1983.
113. Kunkel, T. A., Schaaper, R. M., and Loeb, L. A., Depurination-induced infidelity of DNA synthesis with purified DNA replication proteins, *in vitro*, *Biochemistry*, 22, 2378, 1983.
114. Schaaper, R. M. and Loeb, L. A., Depurination causes mutations in SOS-induced cells, *Proc. Natl. Acad. Sci. U.S.A.*, 78, 1773, 1981.
115. Kunkel, T. A., The mutational specificity of depurination, *Proc. Natl. Acad. Sci. U.S.A.*, 81, 1494, 1984.
116. Walker, G. C., Mutagenesis and inducible responses to deoxyribonucleic acid damage in *Escherichia coli*, *Microbiol. Rev.*, 48, 60, 1984.
117. Echols, H., Lu, C., and Bergers, P. M. J., Mutator strains of *Escherichia coli* and *dnaQ*, with defective exonucleolytic editing by DNA polymerase III holoenzyme, *Proc. Natl. Acad. Sci. U.S.A.*, 80, 2189, 1983.
118. Loeb, L. A., Apurinic sites as mutagenic intermediates, *Cell*, in press.
119. Kunkel, T. A. and Loeb, L. A., unpublished results.
120. Gopinathan, K. P., Weymouth, L. A., Kunkel, T. A., and Loeb, L. A., Mutagenesis *in vitro* by DNA polymerase from an RNA tumor virus, *Nature (London)*, 278, 857, 1979.
121. Loeb, L. A., Springgate, C. F., and Battula, N., Errors in DNA replication as a basis of malignant changes, *Cancer Res.*, 34, 2311, 1974.
122. Linn, S., Kairis, M., and Holiday, R., Decreased fidelity of DNA polymerase activity isolated from aging human fibroblasts, *Proc. Natl. Acad. Sci. U.S.A.*, 73, 2818, 1976.
123. Holliday, R. and Tarrant, G. M., Altered enzymes in aging human fibroblasts, *Nature (London)*, 238, 26, 1972.
124. Springgate, C. F. and Loeb, L. A., Mutagenic DNA polymerase in human leukemic cells, *Proc. Natl. Acad. Sci. U.S.A.*, 70, 245, 1973.
125. Chan, J. Y. H. and Becker, F. F., Decreased fidelity of DNA polymerase-α activity during *N*-2-flourenylacetamide hepatocarcinogenesis, *Proc. Natl. Acad. Sci. U.S.A.*, 76, 814, 1979.
126. Collett, M. S., Purchio, A. F., and Erikson, R. L., Avian sarcoma virus transforming protein, $p60^{src}$ shows protein kinase activity specific for tyrosine, *Nature (London)*, 285, 167, 1980.

127. Nowell, P. C., The clonal evolution of tumor cell populations, *Science,* 194, 23, 1976.
128. Cifone, M. A. and Fidler, I. J., Increasing metastatic potential is associated with increasing genetic instability of clones isolated from murine neoplasms, *Proc. Natl. Acad. Sci. U.S.A.,* 78, 6949, 1981.
129. Witkin, E. M., Ultraviolet mutagenesis and inducible DNA repair in *Escherichia coli, Bacteriol. Rev.,* 40, 869, 1976.
130. Rossman, T. G. and Klein, C. B., The mammalian SOS system: a case of misplaced analogies, *Cancer Invest.,* in press.
131. Sarasin, A., The SOS response in mammalian cells, *Cancer Invest.,* in press.
132. Kunkel, T. A., personal communication.
133. Kunkel, T. A., The mutational specificity of DNA polymerases-α and -γ during in vitro DNA synthesis, *J. Biol. Chem.,* in press.
134. Cann, R. L., Brown, W. M., and Wilson, A. C., Polymorphic sites and the mechanism of evolution in human mitochondrial DNA, *Genetics,* 106, 479, 1984.

Chapter 9

BIOLOGY OF ANIMAL CELL DNA POLYMERASES

I. EVOLUTION OF DNA POLYMERASES

Only a few systematic studies on the phylogeny of eukaryotic cell DNA polymerases have been carried out, and our knowledge of the evolution of these enzymes is, therefore, fragmentary. Phylogenetic data on eukaryotic DNA polymerases are in large part limited at present to identification of the types of polymerases that are present in organisms of different phyla and classes. Only a few studies have been conducted on the extent of structural homology between DNA polymerases of different eukaryotes. It is clear that with the recent advent of activity gel electrophoresis, selective inhibitors of the various polymerase types and particularly, monospecific antibodies against polymerases of the α and β categories, more comprehensive investigations on the evolution of DNA polymerases become feasible.

Polymerases of the α class have been identified throughout the animal and plant kingdoms. However, the degree of evolutionary conservation of polymerase α is not known. Polymerase-β is present in all multicellular organisms, and its structure is highly conserved among vertebrates. Enzymes of the β-class have been detected in some but not all protozoans and perhaps in some fungi but not in plants. Polymerase-γ has been found in all the examined vertebrates whereas its presence in lower eukaryotes is less clear. In this section, some of the available data on the phylogeny of DNA polymerases of the α, β, and γ categories are presented. These data relate in the main to the identification of the types of DNA polymerases in various phyla, classes, and species. Detailed studies on the evolution of these enzymes must await the sequencing of their cloned genes and/or the enzymic proteins.

A. DNA Polymerase-α

DNA polymerase-α, loosely defined as a DNA polymerase with size exceeding 100 kdaltons that is inhibited by thiol group reactive agents, is ubiquitous in protozoa, fungi, plant cells, invertebrates, and animal cells. Yet, detailed and systematic structural analysis of the α-type enzymes are still missing, and thus a general picture on the evolutionary conservation or divergence of α-polymerase is not available.

In surveys of the catalytic subunit of replicative DNA polymerases of bacteria and eukaryotes, Hubscher et al.[1] and Scovassi et al.[2] identified on activity gel electrophoregrams >70-kdalton DNA polymerizing polypeptides from extracts of both lower and higher eukaryotes. That the size and structure of the catalytic subunit of α-polymerase is evolutionarily conserved among mammals has been indicated by Chang et al.[3] These authors used a polyclonal monospecific antibody against calf thymus α-polymerase to demonstrate the presence of antigenically cross-reacting 155-kdalton catalytic subunit of α-polymerase in several mammalian species. It should be noted, however, that although common antigenic determinants indeed appear to exist in α-polymerases from different species, divergence also occurs as revealed by the highly discriminatory monoclonal antibodies against α-polymerase. In some instances anti-α-polymerases monoclonal antibodies reveal interspecies immunological cross-reactivity,[4-6] whereas in others antibodies uniquely interact with species-specific determinants.[4,5]

Analysis of α-polymerases from diverse species reveals similarities between polymerases of mammals and lower vetebrates such as birds and and amphibians, and invertebrates such as insects. Hence, similarities in native molecular size, subunit structure, association with DNA primase and sensitivity to selective inhibitors were noted for α-

polymerases from chick cells,[7,8] *Xenopus laevis*[9,10] and *Drosophila melanogaster* embryos,[11,12] as well as for various mammalian species.[13-15] That a considerable degree of evolutionary conservation exists in the catalytic subunit of α-polymerases from phyla as divergent as vertebrates and anthropods was shown by Méchali and de Recondo.[16] These authors have reported similarity in the tryptic peptide patterns of homogenous catalytic polypeptides of α-polymerases from rat and *D. melanogaster,* in addition to close resemblance in the subunit composition of the native enzymes. Invertebrates, such as echinodermata, were also found to possess an α-type DNA polymerase, although the extent of its structural homology with vertebrate polymerase-α is unknown.[17,18]

In low eukaryotes such as fungi, algae, and protozoa, divergence from the canonical polymerase-α defining criteria becomes more evident (for general reviews on polymerases of plants and lower eukaryotes cf. References 19 to 21). In the fungus *Saccharomyces cervisiae* two DNA polymerases designated A(I) and B(II) have been identified.[22] Both polymerases are large (150 to 170 kdaltons) and are inhibited by *N*-ethylmaleimide and aphidicolin. However, polymerase B is associated with a 3′→5′ exonuclease activity, whereas the antigenically distinct polymerase A is devoid of such an activity.[22] That yeast polymerase A is very similar to α-polymerase is evident from its size, response to inhibitors, ability to extend ribonucleotidyl primers, and tight association with DNA primase, as well as by the subunit structure of the homogenous enzyme.[23,24] In the fungus *Neurospora crassa* properties of a 147-kdalton enzyme appear to generally conform with those of an α-polymerase.[25] However, in the basomycete *Ustilago maydis,* an N-ethylmaleimide sensitive 180 to 200 kdalton DNA polymerase is associated with a 3′→5′ exonuclease, copies poly(rA) templates at an exceptionally high efficiency, and require 120 mKCℓ for maximum activity.[26,27] In the unicellular algae *Chlorella,* two DNA polymerases designated I and II have been identified recently.[28] While neither polymerase has been identified unequivocally as an α-type enzyme, polymerase I had more properties in common with animal cell polymerase-α than did polymerase II, which was somewhat akin to γ-polymerase. Detailed studies on the DNA polymerases of the free-living *Tetrahymena pyriformis* were conducted recently by Sakai and Watanabe.[29-31] One of the two polymerases which were isolated from that unicellular eukaryote was surprisingly similar to vertebrate α-polymerase in that the activity of the nearly homogeneous enzyme was inhibited by *N*-ethylmaleimide, aphidicolin, and ara CTP and that the enzyme was composed of two subunits of 135 and 66 kdaltons.[29,31]

In summary, a replicative enzyme whose properties generally conform with those of the α category of the animal cell DNA polymerase is ubiquitous in all eukaryotes. Furthermore, it has been argued that the size of the catalytic core of the replicative polymerase is maintained in bacteria, plants, amd eukaryotes.[1,2,32,33] Yet, any judgment on the evolutionary conservation of polymerases of the α type must await a systematic structural analysis of the enzyme in a large number of diverse species, classes, and phyla.

B. DNA Polymerase-β

In a pioneering phylogenetic investigation, Chang[34] surveyed a spectrum of species for activity of polymerase-β. Defining polymerase-β as a DNA polymerizing activity with size <50 kdaltons, which is resistant to *N*-ethylmaleimide, that enzyme was discerned in organisms ranging from mammalia to metazoans. The β-type enzyme was not detected, however, in insect embryos, higher plants, fungi, free-living protozoans, and bacteria.[34,35] The stage in evolution at which the β-polymerase appeared and the degree of its evolutionary preservation thereafter is not clear. First, the ubiquity of the β-polymerase gene(s) in all multicellular organisms is disputable since it appears that insect embryos are devoid of the enzyme (see Section IV for update), whereas its presence in adult insects is still a controversial question (cf. Section II of this chapter). Second, although protozoans generally do not possess a β-type DNA polymerase, there have been cases in which unicellular eukaryotes were found to contain β-class polymerases. Hence, Chang et al.[36] reported that extracts of the unicellular parasite *Trypanosoma brucei* contained both α- and β-polymerases that were immunologically distinguishable from host animal DNA polymerases. In a recent study, Sakai and Watanabe[30] disputed earlier claims for the existence of an *N*-ethylmaleimide resistant DNA polymerase in *Tetrahymena pyriformis.* Yet, in addition to an α-like DNA polymerase in that protozoan, it also contained an *N*-ethylmaleimide sensitive, 70-kdalton polymerase, which was, however, inhibited by d_2TTP but not by aphidicolin and ara CTP. It was argued that despite its sensitivity to thiol-group blocker and its size, most of the properties of the 70-kdalton enzyme were akin to those of β-polymerase. Although a β-class enzyme is undetectable in fungi such as yeast and *Ustilago,* recent reports suggest that DNA polymerase that displays partial resemblance to β-polymerase may be present in some fungi species. Thus the slime mold *Dictyostelium discoideum* was found to contain a β-type polymerase whose 70-kdalton size exceeded that of polymerase-β.[37] A 40-kdalton, *N*-ethylmaleimide resistant DNA polymerase was isolated from *N. crassa* and by some physical and catalytic criteria was more similar to γ-polymerase than to polymerase-β.[25] The claim for the presence of β-type polymerase in fungi should, therefore, be critically examined in order to elucidate the evolutionary stage at which this enzyme appeared as a novel distinct entity.

The precise step in evolution at which polymerase-β appeared notwithstanding, it seems that this enzyme evolutionarily highly conserved in multicellular eukaryotes. Chang et al.[3] showed that anti-calf thymus DNA polymerase-β antibody interacted specifically in a sensitive double-antibody precipitation test with β-polymerase from a bird species, an amphibian, crustacean, and a free-living protozoan. Furthermore, *in situ* detection of the β-enzyme by gel electrophoresis revealed that in all phyla tested, it was of similar size; ∼45 kdaltons.[3] Tanabe et al.[38] found that DNA polymerases from several mammalian species and from chick embryo displayed considerable similarity in their tryptic peptide patterns. That study was recently expanded to show that highly purified β-polymerases from mouse, chick and salmon cells are all composed of a single 40-kdalton polypeptide.[39] Furthermore, two-dimensional peptide mapping of the enzymes indicated that a part of the polymerase structure remained highly conserved in the course of evolution. It has been argued that the persistence of conserved structural features of β-polymerase through eons of evolutionary time indicates that a "constant region" of the enzymic protein is vital for its activity and possibly for survival.[39]

C. DNA Polymerase-γ

The phylogeny of mitochondrial DNA polymerase-γ was initially studied by Scovassi et al.[40] The only detectable polymerase in isolated mitochondria of diverse vertebrates was of the γ-class, and similarity in antigenicity and peptide map pattern was demonstrated for polymerase-γ from mouse and chick cells.[40] That polymerase-γ also appears in the invertebrate echinoderms was recently reported by Haraguchi and Nagano[18] who isolated a 7-S γ-type DNA polymerase from oocytes of the starfish *Asterina pectinifera.* The presence of γ-polymerase in fungi and plants is less clear. It has been reported that although isolated mitochondria of a fungus and plant species appear to contain γ-polymerase, they also exhibited α-polymerase activity.[41] Whether this was due to contamination of the mitochondria with polymerase-α or to unique properties of the γ enzyme in lower eukaryotes is not clear.

II. DNA POLYMERASES IN DEVELOPMENT

A. Differentiation

A limited number of investigations have focused on changes in polymerase types and their activities in the course of embryonal development and differentiation. Hence, few conclusions can be drawn on the place of DNA polymerases in differentiation. Perhaps the single commonly recurring observation in diverse differentiating systems is the progressive decline in α-polymerase activity as the proliferative activity of the differentiating cells diminishes. Other observations suggest that in some systems α-polymerase may be retained by nondividing cells, perhaps as a DNA repair enzyme. It conversely appears that during DNA endoreduplication in rat embryogenesis, DNA polymerase-β may play a central role in replicative synthesis of DNA. Further, although DNA polymerase-β is undetectable in embryonic insect tissue, it was traced in adult cells, tantalizingly suggesting that expression of genetic information for that enzyme might occur during insect development. Some differentiating cell systems in which DNA polymerases were investigated are surveyed in this section.

Spermatogenesis — Perhaps the most extensively documented case of change in DNA polymerase activities in differentiation is that of spermatogenesis in mouse or rat. Spermatogenesis takes place in the seminiferous tubules of the testes. Progressive differentiation during spermatogenesis is associated with a decline in the proliferative activity of spermatogonial cells. Testicular tissue contains activities of both DNA polymerases-α and -β.[42,43] A six- to sevenfold decrease in the level of polymerase-α and two- to threefold increase in the activity of β-polymerase were monitored during spermatogonial maturation. It has been suggested that levels of α- and β-polymerases correlate, respectively, with phases of DNA replication and of postrecombination repair.[42] Grippo et al.[44] provided evidence in support of this view by showing that both α- and β-polymerases, resolved by velocity sedimentation, were present in premeiotic cells. However, in meiotic and postmeiotic cells, the β enzyme was the only detected DNA polymerase. The activity of β-polymerase reached a maximum during the pachytene stage of meiosis, presumably because that enzyme was required for repair synthesis following chromosomal crossing over.[44] Other investigators have reported that although activity of testicular polymerase-α declines as DNA replication ceases, it remains detectable in spermatogonia, spermatocytes, and spermatides.[43,46-47] Direct *in situ* assay of the activities of α- and β-polymerase followed by autoradiography of radioactively labeled precursors incorporated into DNA revealed that only in very late stages of spermatogenesis do both the activities of α- and β-polymerases decline.[47] *In situ* detection of α- and β-polymerases in chick testis using fluorescently labeled monospecific antibodies against α- and β-polymerases revealed that DNA polymerase-α was traceable only in spermatogonia/spermatocytes, whereas β-polymerases could be detected in all testicular cells.[48] It has been suggested that preservation of α-polymerase in spermatogonia, spermatocytes, and spermatide that do not replicate their genome is required due to a role played by α-polymerase in DNA repair.[47]

Neurons — Hübscher and associates[49,50] determined activities of α-, β-, and γ-polymerases during progressive development of neurons in rat brain. The level of α-polymerase activity correlated with cell proliferation in the neuronal tissue. As cells differentiated and their rate of division declined, the activity of α-polymerase decreased. By contrast, activities of DNA polymerase-β and -γ remained relatively constant during the course of cell differentiation. Nondividing, fully differentiated neurons were devoid of detectable polymerase-α; only the activities of polymerase-β and -γ were preserved in these postmitotic cells.[49,50] Similar observations were made in neural retina from chick: activity of α-polymerase decreased progressively between days 7 and 11 of embryonal development, and this decrease paralleled the decline in mitotic activity.[51]

It was noted, however, that a constant residual activity of α-polymerase remained even after hatching. Level of DNA polymerase-β in neural retina cells was between 7 and 14 days of embryonal development and then decreased until hatching.

Muscle cells — Studies on DNA polymerases in differentiating cardiac muscle of the rat[52] revealed that the activity of α-polymerase decreased 80-fold between days 1 and 17 after birth, paralleling the progressively diminishing rate of DNA synthesis. By contrast, levels of β-polymerase remained constant throughout neonatal development of the cardiac muscle tissue.

Blood cells — Bollum and colleagues[53] noted that an α type polymerase activity is elevated in immature, rapidly dividing, human leukemic hematopoietic cells and is lower in mature quiescent cells. By contrast, the level of β-polymerase did not correlate with proliferative activity of the cells. In a subsequent work,[54] similar observations were made for the differentiating mouse erythropoietic cells. Other authors[55,56] similarly reported that levels of both α- and β-polymerases were high in dividing avian erythroblasts. However, as the division rate of maturing cells diminished, the activity of polymerase-α decreased rapidly, whereas that of β-polymerase remained constant and declined only in the terminal stages of differentiation.

Trophoblasts — Kalf and associates[57-59] discovered distinctive features of DNA polymerase activity and function in the developing polyploid rat giant trophoblasts. During the process of embryo implantation, rat trophoectoderm is transformed into trophoblast cells, which progressively divide at a slower rate without curtailing DNA replication. As a result, giant polyploid cells are formed through a process of genomic endoreduplication, which occurs without detectable mitoses between days 12 and 16 of pregnancy. It was initially found that the activity of β-polymerase remained unaltered in trophoblasts prior to and during endoreduplication. By contrast, the level of polymerase-γ decreased twofold between days 11 and 12 of gestation, whereas that of α-polymerase dropped tenfold at that stage.[57] Furthermore, mixing experiments in which *N*-ethylmaleimide-treated day 11 cell extract was added to day 12 extracts suggested that the sharp decline in polymerase-α activity between days 11 and 12 was due to loss of inhibitor-insensitive stimulatory factor rather than to disappearance of the enzyme itself. This was subsequently shown when an 85-kdalton *N*-ethylmaleimide-resistant protein which specifically stimulated α-polymerase was isolated from dividing trophoblasts.[58] The stimulatory protein was found to be present in trophoblasts at the early phase of gestation, but it was not detected in day 12 trophoblasts that do not divide but carry out endoreduplication of DNA. If α-polymerase is functionally ineffective during endoreduplication, which polymerase is then responsible for DNA replication in the nondividing giant trophoblasts? Siegel and Kalf[59] monitored effects of a large variety of selective inhibitors of DNA polymerases on the in vitro DNA endoreduplication in isolated 13 day trophoblast nuclei, which were found to contain only DNA polymerases-α and -β. The sensitivity of nuclear endoreduplication to d_2TTP and to potassium phosphate was identical to that of isolated β-polymerase. By contrast, in vitro endoreduplication was relatively resistant to *N*-ethylmaleimide as was β-polymerase, and totally insensitive to aphidicolin like polymerase-β and unlike α-polymerase. The conclusion is surprising: β-polymerase appears to be the major enzyme that is responsible for DNA replication without cell division in rat trophoblasts.[59]

B. Embryogenesis and Development

Only a few systems have been studied for changes in DNA polymerase activities during oogenesis, egg fertilization, embryogenesis, and development. One observation, which required additional substantiation, is that in insects β-polymerase is absent in the embryo (see Section IV for update) but might appear in the adult. Another, unrelated line of data suggests that in some organisms DNA polymerases are stockpiled in excessive amounts during oogenesis in preparation for their subsequent use in the course of the very rapid DNA synthesis during early embryogenesis.

Insect development — The types of DNA polymerases present in insects at different stages of embryonal development is controversial. In an initial phylogenetic survey, Chang[36] failed to detect β-polymerase in insect embryos. The prevalence of polymerase-β in adult insects is still in question. Furia et al.[60] isolated β-like polymerase from adult *D. melanogaster* flies. Scovassi et al.[2] detected on activity gels a band corresponding to β-polymerase in cockroach testes extract, whereas no such activity was identified in crude extracts of Mediterranean fly embryos. It has been reported recently that the only detectable DNA polymerase in *D. melanogaster* throughout embryonal development and in the adult fly was of the α-type and that its activity correlated with DNA replication activity.[61] Although β-polymerase was not detected in some reports using adult insects, the consensual reports on its absence from embryonal tissues and some observations on its presence in adult cells are intriguing. The possibility remains that β-polymerase synthesis is activated at a late stage of embryonal development, whereas at earlier stages, α-polymerase is the predominant insect cell DNA polymerase (see Section IV for update).

Oogenesis and embryogenesis of sea urchin — Following fertilization of sea urchin eggs, very rapid DNA replication and cell division ensue (cf. Chapter 2). Stockpiling of DNA polymerase-α in the unfertilized eggs in preparation for rapid DNA synthesis during early embryogenesis was indicated in early reports by Fansler and Loeb.[62] It has been observed that a high level of oocyte total DNA polymerase activity, which presumably represents α-polymerase, remained roughly constant after fertilization and during early embryogenesis. Furthermore, it appeared that the polymerase was accumulated in the cytoplasm of the unfertilized egg and was translocated to the nucleus during each cell division in early embryo.[62] Shioda et al.[63] noted that levels of α-polymerase remained constant during early embryogenesis of the sea urchin and that only at the gastrulation stage did it increase twofold due to the appearance of a new form of the enzyme. Further evidence indicated that the α enzyme was stored in the rough endoplasmic reticulum of the unfertilized egg, and it migrated to the nucleus toward early embryogenesis.[64]

Oogenesis and embryogenesis in *Xenopus laevis* — Massive amounts of different DNAs are accumulated during various stages of oogenesis and embryogenesis of the frog *X. laevis.* At different stages of oogenesis, large amounts of ribosomal and mitochondrial DNAs are presynthesized for subsequent use during embryogenesis. Following fertilization and commencement of embryonal development, the major biosynthetic activity is for the synthesis of chromosomal DNA at a rate sufficient to accumulate roughly 10^5 equivalents of DNA during the first 24 hr after fertilization.[65] Recently, Benbow and colleagues[66] examined the levels of the various classes of DNA polymerases at different stages during oogenesis and embryogenesis in *X. laevis.* Oocytes, eggs, and embryos contain a mitochondrial DNA polymerase-γ, polymerase-β, and two forms of α-polymerase denoted α_1 and α_2, which are, respectively, associated with and devoid of DNA primase.[9,10] It has been found that in early oocytes, polymerase-γ is the predominant DNA polymerase. During oocyte development, the activities of polymerases-α_1 and -α_2 increase more than 100-fold, β-polymerase by 15-fold, and γ-polymerase by 8-fold. Oocyte maturation and ovulation involves an additional twofold increase in the amount of polymerases of all classes so that the mature oocyte contains five orders of magnitude more DNA polymerase activity per cell than the adult somatic cell.[66] No further increase in the level of all types of polymerases occurs after fertilization and cleavage — a period of maximum DNA replication. It has been concluded, therefore, that synthesis of DNA polymerases is not directly coupled to DNA duplication activity during embryogenesis but, rather, these enzymes are stockpiled in the course of oogenesis to be used after fertilization.[66] These observations are in line with the earlier reports on sea urchin embryogenesis.[62-64]

Embryogenesis in other organisms — It was reported that DNA polymerases-α_1 and -α_2 and -β remained at a constant level throughout early embryogenesis in the axolotl, *Ambystoma mexicanum*.[67] It was noted, however, that at the tailbud stage, the activity of axolotl polymerase-β substantially increased.[67] Levels of DNA polymerase-α, -β, and -γ were recently compared in extracts of mature eggs and liver and testes cells of the teleost fish *Misgurnus fossilis*.[68] Whereas amounts of polymerase-γ were similar in the three cell types, polymerase-α levels were high in eggs and undetectable in mature hepatocytes. Conversely, the levels of β-polymerase in mature eggs were two orders of magnitude lower than in liver and testes cells. Hence, unlike α-polymerase and in contrast to β-polymerase in eggs of echinoderms and amphibiae, fish polymerase-β appears not to be accumulated during oocyte maturation. It was also found that in parallel with previous observations in other organisms, activities of β- and γ-polymerases did not change during early embryogenesis of the fish.[68] Hence, these enzymes appear to be stockpiled during oogenesis in amounts that suffice for the ensuing embryonal development.

III. DNA POLYMERASES IN AGING

Two aspects of DNA polymerases activity have been investigated in relation to aging. First, the question of linkage between diminishing cell proliferation during aging and DNA polymerase activities has been addressed. Is the reduction in the proliferative potential of aging cells accompanied by or possibly caused by, characteristic changes in the level or properties of DNA polymerases? Second, DNA polymerases from young and old cells have been compared in respect to their fidelity in polymerizing DNA. Some theories of aging argue that a decrease in the fidelity of DNA, RNA, and/or protein synthesis during aging leads to an accumulation of somatic mutations that are increasingly detrimental to the functionality and survival of the organism.[69,70] Detection of a consistent decrease in the accuracy of DNA synthesis catalyzed by polymerases from aging cells would, therefore, provide critical supporting evidence for these "error catastrophe" theories.

Two general types of experimental systems have been used to study DNA polymerases during aging. First, polymerases isolated from young and old organisms or from species with different lifespans have been compared for their activities and accuracy. Second, diploid animal cells possess a limited proliferative potential in culture, and after a period of division at an exponential rate, cell growth progressively slows and cells enter into a phase of in vitro "senescence".[71,72] Hence, cell aging in culture was examined in respect to changes in activity, physical and catalytic properties, and fidelity of DNA polymerases. In this section we survey results that in toto provide contradictory evidence on the activity and fidelity of DNA polymerases during aging.

A. Diminishing Cell Proliferation During Aging and Activities of DNA Polymerases

Measurements of levels of DNA polymerase-α in early passage ("young") and late passage ("old") cultured diploid fibroblastic cells have yielded variable results. Linn et al.[73] reported that the level of α-polymerase in extracts of senescent cultured human MRC-5 cells was very low compared with young cells and barely detectable eight passages before cessation of growth. Murray[74] and Murray and Holliday[75] reported that the specific activity of α-polymerase, expressed as units enzymic activity per milligram of cellular protein, decreased only two- to fourfold in late passage MRC-5 cells. Krauss-Wald and Linn[76] reported that the level of total extractable polymerase-α decreased with increasing in vitro "age" of IMR-90 cultured human diploid fibroblasts.

Weisman-Shomer et al.[77] showed, however, that the level of chromatin-bound DNA polymerase-α expressed as enzyme activity per chromatin DNA remained almost unchanged in the course of in vitro aging of mouse fibroblasts. Some of these discrepant results can be explained by the use of different procedures for cell extract preparation, solubilization of polymerase-α, and assay conditions for the determination of polymerase activity. An additional explanation was provided recently by Pendergrass et al.[78] These authors observed that the specific activity of α-polymerase expressed as enzyme units per milligram of cellular protein decreased with increasing cell age in culture for human diploid fibroblasts. However, the size and protein content of cells increase at late passages. Hence, normalization of the level of polymerase-α to be expressed as enzyme activity per cell, revealed that the cellular content of α-polymerase, which is almost synonymous with expression of enzyme activity per cell DNA,[77] was closely similar for early and late passage cells. Furthermore, similar increases in α-polymerase activity per cell were observed after exposure of young and old cultured fibroblasts to mitogens. Yet, due to the increased protein content of the senescent fibroblasts, polymerase-α was operatively "diluted" in the enlarged cell mass of late passage cells.[78]

It has also been found that the cellular content of α-polymerase in several strains of diploid and heteroploid cells was directly related to their ability to reinitiate DNA synthesis upon their fusion with senescent cultured fibroblasts.[79] It was argued that "threshold" levels of α-polymerase are required to induce DNA replication in late passage fibroblasts — only cells that were able to supply adequate levels of α-polymerase could induce DNA replication in the fused senescent cell. In contrast to the decreased cell content of α-polymerase, several authors have reported activity of β-polymerase remained relatively unaltered in the course of in vitro cell aging.[74,76,77] It was also reported that the specific activity of polymerase-γ remained constant throughout the in vitro life span of MRC-5 human fibroblasts.[74]

Declining activity of α-polymerase was recently observed in the regenerating liver of aged mice.[80] The decrease in the level of chromatin-bound hepatic α-polymerase in old mice was not due to redistribution of the enzyme within the cell, and it reflected decreased cell proliferation in the regenerative liver of aging mice. It is not clear, however, whether the decline in α-polymerase activity results from the decrease in proportion of dividing cells in old mouse cell liver or rather, diminished activity of α-polymerase is a cause for reduced cell proliferation. In contrast to the declining activity of α-polymerase in aging mouse hepatocytes, activity of β-polymerase was found to remain constant during aging in both intact[81] and regenerating[80] liver.

Several authors have reported changes in the physical and catalytic properties of DNA polymerase of the α-type isolated from senescent cultured cells. Fry and Weisman-Shomer[82] observed that several physical properties of two molecular forms of chick fibroblast DNA polymerase-α were altered in senescent cells. Most notably, a fraction of the enzyme molecules population was rendered more heat labile in late passage cells. The augmented heat sensitivity could reflect changes in the primary structure of the polymerase or posttranslational modifications of the enzyme. Alternatively, altered heat lability could result from increased partial proteolysis of polymerase isolated from old cells. Indeed, Murray[74] reported that two forms of α-polymerase were discerned in extracts of MRC-5 human fibroblasts, one of which had a sedimentation coefficient of 7 S and the other, which appeared to be a product of partial proteolysis, sedimented at 5 S. The proportion of the partially degraded enzyme increased from 7% in young cells to 29% in senescent fibroblasts. Whether that increase was due to augmented proteolytic activity in old cells or to higher sensitivity of senescent cell polymerase to proteolytic cleavage is not clear. Krauss-Wald and Linn[76] resolved by DEAE cellulose column chromatography three molecular forms of α-polymerase from early passage IMR-90 cultured human fibroblasts. The activity of two of the forms declined preferentially in late passage cells, and a fourth new form appeared in the senescent cultures. Because of the exceptional susceptibility of DNA polymerase-α to partial proteolysis, changes in molecular form patterns and even augmentation in heat lability as a function of age are difficult to interpret.

B. Fidelity of DNA Polymerases in Aging Cells

If enhanced mutagenesis is a hallmark of cellular senescence, fidelity of DNA replication and repair should be lower in aged cells than in young ones. Hence, several authors undertook to compare the accuracy of DNA synthesis catalyzed by polymerases from young and old cells.

With one exception,[82] all comparative studies on fidelity of DNA polymerases from early passage and late passage cultured cells indicated that the accuracy of DNA polymerases from senescent cells is lower relative to that of young cells.[73-76] DNA polymerase-α partially purified from MRC-5 cells exhibited error rates in copying poly[d(AT)] that were greater for α polymerase from late passage fibroblasts than for young cells.[73,75] Figure 1 depicts the progressive decline in accuracy of DNA polymerase-α during in vitro aging of MRC-5 cells. In a subsequent work, polymerases-α, -β, and -γ from senescent cultured human fibroblasts were each found to copy synthetic primer-templates less accurately than polymerases from early passage cells.[76]

In contrast to the consistent decrease in fidelity of DNA polymerases from cells aged in vitro, all measurements on DNA polymerases in cells from aging organisms indicate that their accuracy does not change with age. Agarwal et al.[83] measured the fidelity of copying of synthetic DNA templates by DNA polymerases, which were isolated from lymphocytes of normal young and old human individuals. No age-related change in the fidelity of lymphocyte polymerase-α was detected. Developing a method for the measurement of accuracy of copying synthetic polynucleotides by chromatin-bound β-polymerase,[84] Fry et al.[81] found that the fidelity of chromatin-associated hepatic β-polymerase did not change during the aging of two mouse species, *Mus musculus* and *Peromyscus leucopus*. Moreover, the twofold difference in the life span of these two murine species was not reflected in the inherent fidelities of their respective polymerases-β, which were similar.[81] Silber et al.[85] reported recently that both α- and β-polymerases, which were partially purified from regenerative mouse livers of young and old mice, exhibited indistinguishable accuracy in copying in vitro natural ϕX174 phage DNA template (see Figure 2). Furthermore, the polymerases from young and old mice displayed similar heat inactivation curves and possessed indistinguishable chromatographic behavior. Based on these results, it was argued that the altered fidelity and physical properties of polymerases from aging cultured cells may reflect changes that characterize cell growth in vitro rather than aging.[85,86] Supporting this claim is the observation that reduced fidelity of DNA polymerases-α, -β and -γ was observed not only for late passage human cells but also for young cells grown to postconfluency.[76] Further, a proteolytically degraded 5-S form of α-polymerase, whose level increased in senescent cultures of human MRC-5 cells, had a lower fidelity than the higher size 7-S enzyme.[74]

In summary, although more work is required to determine whether age-related changes in fidelity of DNA polymerases reflect physiological phenomena in cultured cells, there is no evidence at present for such alterations in aging organisms. The suggested dichotomy between decreased fidelity of polymerases observed with cultured cells aging in vitro and lack of enhanced inaccuracy of DNA synthesis in aging animals, must be more firmly established. Yet, current results suggest that the two situations are not synonymous and it might be that increased error proneness of DNA replication in cells aging in vitro may not reflect the in vivo situation. Alternatively, it may be nec-

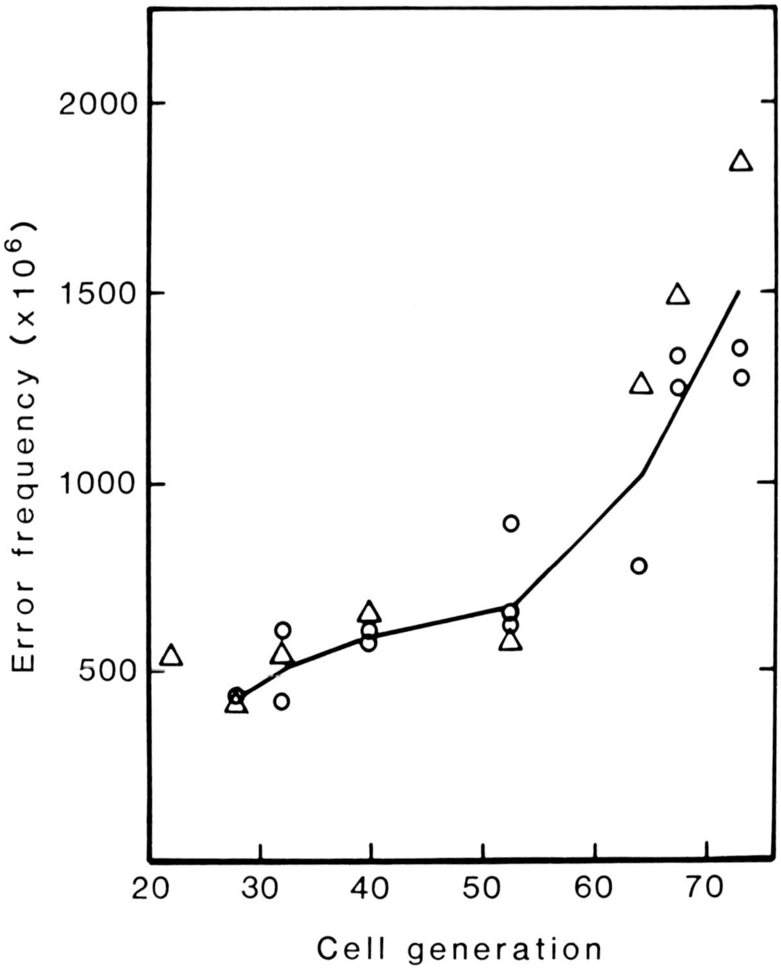

FIGURE 1. Increased error frequency of DNA polymerase-α from MRC-5 cells during cell aging in culture. Polymerase-α was partially purified from extracts of MRC-5 cells collected at progressive cell generations. Fidelity of DNA synthesis was monitored by measuring the rate of misincorporation of dGMP with poly[d(A-T)] template using magnesium as a metal activator. (Adapted from Murray, V. and Holliday, R., *J. Mol. Biol.*, 146, 55, 1981. With permission.)

essary to define particular populations of cells that are of crucial importance in determining organismal age and in which accumulation of errors in DNA replication and protein synthesis may be of primary importance.[86] And last, the possibility that proteins auxiliary to DNA polymerases change during cell aging is still unexplored and requires investigation.

IV. UPDATE

A. Evolution of DNA Polymerase-β

Although several authors reported that a polymerase of the β-type could not be detected in embryos of the insect *D. melanogaster,*[11,34,61] a polymerase that is distinguishable from polymerase-β and that displays some attributes of a β-class enzyme has been purified recently from this organism.[90] The β-like activity is typically sensitive to d_2TTP inhibition and relatively resistant to aphidicolin and MalNet. However, in contrast to the 40-kdalton size of the mammalian polymerase-β, tne *Drosophila* enzyme

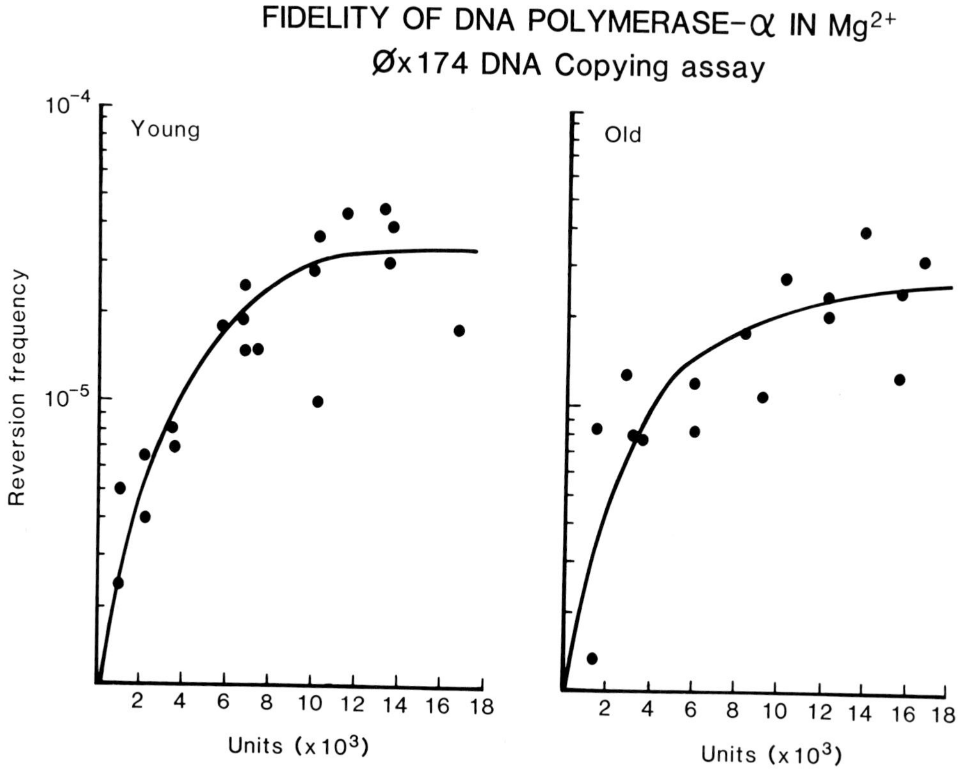

FIGURE 2. Unaltered fidelity of hepatic DNA polymerase-α from young and old mouse. DNA polymerase-α was partially purified from regenerating liver of 6- and 28-month-old *Mus musculus*. Fidelity of the isolated enzymes was measured by monitoring the reversion frequency of in vitro copying of ΦX-174 *am 3* DNA (cf. Chapter 8 for methodology). Template copying was conducted in presence of magnesium ions and an excess of incorrect deoxynucleoside triphosphates. Experimental procedures are detailed in Reference 85.

had an apparent mass of 110 kdaltons.[90] It appears, therefore, that in addition to polymerase-α, some insects may already have a β-class polymerase.

B. DNA Polymerase Activities in Aging Cells

Correlation between the declining proliferative activity of senescing cultured human fibroblasts and the intracelular level of their DNA polymerases has been reexamined.[87] As cell growth rate declines with age in culture, activities of both polymerases-α and -β decease. However, when enzyme activities are normalized for the fraction of remaining dividing cells in the senescent cultures, only the level of polymerase-α that was assayed in either permeabilized or lyzed fibroblasts declined whereas the activity of polymerase-β per dividing cell remained unaltered.[87] This observation is in line with previous reports on the selective decrease with age in the activity of polymerase-α from human fibroblasts,[76,79] quail oviduct,[88] mouse bone marrow,[89] and mouse regenerating liver.[80] Interestingly, however, the activity of polymerase-α that is associated with the nuclear matrix of senescing fibroblasts was reported to remain unchanged with age.[87] This observation could be interpreted either as indicating that declining DNA synthesis rates are due to changes in chromatin structure rather than decreasing polymerase levels[87] or that the matrix does not represent the in vivo site of replication.

REFERENCES

1. Hübscher, U., Spanos, A., Albert, W., Grummt, F., and Banks, G. R., Evidence that a high molecular weight replicative DNA polymerase is conserved during evolution, *Proc. Natl. Acad. Sci. U.S.A.*, 78, 6771, 1981.
2. Scovassi, A. I., Torsello, S., Plevani, P., Badaracco, G. F., and Bertazzoni, U., Active polypeptide fragments common to prokaryotic, eukaryotic and mitochondrial DNA polymerases, *EMBO J.*, 1, 1161, 1982.
3. Chang, L. M. S., Plevani, P., and Bollum, F. J., Evolutionary conservation of DNA polymerase β structure, *Proc. Natl. Acad. Sci. U.S.A.*, 79, 758, 1982.
4. Tanaka, S., Hu, S.-Z., Wang, T. S.-F., and Korn, D., Preparation and preliminary characterization of monoclonal antibodies against human DNA polymerase α, *J. Biol. Chem.*, 257, 8386, 1982.
5. Masaki, S., Shiku, H., Kanenda, T., Koiwai, O., and Yoshida, S., Production and characterization of monoclonal antibody against 10 S DNA polymerase α from calf thymus, *Nucl. Acids Res.*, 10, 4703, 1982.
6. Yagura, T., Tanaka, S., Kozu, T., Seno, T., and Korn, D., Tight association of DNA primase with a subspecies of mouse DNA polymerase α, *J. Biol. Chem.*, 258, 6698, 1983.
7. Yamaguchi, M., Tanabe, K., Takahashi, T., and Matsukage, A., Chick embryo DNA polymerase α. Polypeptide components and their microheterogeneity, *J. Biol. Chem.*, 257, 4484, 1982.
8. Matsukage, A., Yamaguchi, M., Tanabe, K., Taguchi, Y. N., Nichizawa, M., and Takahashi, T., DNA chain elongation mechanism of DNA polymerase α, and β, in *New Approaches in Eukaryotic DNA Replications*, de Recondo, A. M., Ed., Plenum Press, New York, 1983, 81.
9. Shioda, M., Nelson, E. M., Bayne, M. L., and Benbow, R. M., DNA primase activity associated with DNA polymerase α from *Xenopus laevis* ovaries, *Proc. Natl. Acad. Sci. U.S.A.*, 79, 7209, 1982.
10. Nelson, E. M., Stowers, D. J., Bayne, M. L., and Benbow, R. M., Classification of DNA polymerase activities from ovaries of the frog *Xenopus laevis*, *Dev. Biol.*, 96, 11, 1983.
11. Banks, G. R., Boezi, J. A., and Lehman, J. R., A high molecular weight DNA polymerase from *Drosophila melanogaster* embryos. Purification and partial characterization, *J. Biol. Chem.*, 254, 9886, 1979.
12. Kaguni, L. S., Rossignol, J.-M., Conaway, R. C., and Lehman, I. R., Isolation of an intact DNA polymerase-primase from embryos of *Drosophila melanogaster*, *Proc. Natl. Acad. Sci. U.S.A.*, 80, 2221, 1983.
13. Méchali, M., Abadiedebat, J., and de Recondo, A. M., Eukaryotic DNA polymerase α. Structural analysis of the enzyme from regenerating rat liver, *J. Biol. Chem.*, 255, 2114, 1980.
14. Wang, T. S.-F., Hu, S. Z., and Korn, D., DNA primase from KB cells. Characterization of primase activity tightly associated with immunoaffinity-purified DNA polymerase-α, *J. Biol. Chem.*, 259, 1854, 1984.
15. Masaki, S., Tanabe, K., and Yoshida, S., Large polypeptides of 10 S DNA polymerase α from calf thymus: rapid isolation using monoclonal antibody and tryptic peptide mapping analysis, *Nucl. Acids Res.*, 12, 4455, 1984.
16. Méchali, M. and de Recondo, A. M., Eukaryotic DNA polymerase α: Tryptic map analysis and comparison between the rat liver and *Drosophila* enzymes, *Biochem. Int.*, 4, 465, 1982.
17. Fansler, B. S. and Loeb, L. A., Sea urchin nuclear DNA polymerase, *Methods Enzymol.*, 29, 53, 1974.
18. Haraguchi, T. and Nagano, H., Isolation and characterization of DNA polymerases from mature oocytes of the starfish, *Asterina pectinifera*, *J. Biochem. (Tokyo)*, 93, 687, 1983.
19. McLennan, A. G. and Keir, H. M., The deoxyribonucleic acid polymerases of non-vertebrate eukaryotes, *Biochem. Soc. Symp.*, 42, 55, 1977.
20. Sala, F., Galli, M. G., Levi, M., Burroni, D., Parisi, B., Pedrali-Noy, G., and Spadari, S., Functional roles of plant α-like and β-like DNA polymerases, *FEBS Lett.*, 124, 112, 1981.
21. Sala, F., Galli, M. G., Pedrali-Noy, G., and Spadari, S., Effect of aphidicolin on the replication of plant DNA, in *Molecular Form and Function of Plant Genomes*, Van Vloten-Doting, L., Eds., Plenum Press, New York, 1985, in press.
22. Wintersberger, E., Yeast DNA polymerases: antigenic relationship, use of RNA primer and associated exonuclease activity, *Eur. J. Biochem.*, 84, 167, 1978.
23. Badaracco, G., Capucci, L., Plevani, P., and Chang, L. M. S., Polypeptide structure of DNA polymerase I from *Saccharomyces cerevisiae*, *J. Biol. Chem.*, 258, 10720, 1983.
24. Singh, H. and Dumas, L. B., A DNA primase that copurified with the major DNA polymerase from the yeast *Saccharomyces cerevisiae*, *J. Biol. Chem.*, 259, 7936, 1984.
25. Stauder, G., Riesemann, H., Joester, W. M., and Joester, K. E., Purification and properties of a low molecular weight DNA polymerase from *Neurospora crassa*, *Biochim. Biophys. Acta*, 741, 308, 1983.

26. Banks, G. R., Holloman, W. K., Kairis, M. V., Spanos, A., and Yarranton, T., A DNA polymerase from *Ustilago maydis*. I. Purification and properties of the polymerase activity, *Eur. J. Biochem.*, 62, 131, 1976.
27. Banks, G. R. and Yarranton, G. T., A DNA polymerase from *Ustilago maydis*. II. Properties of the associated deoxyribonuclease activity, *Eur. J. Biochem.*, 62, 143, 1976.
28. Aoshima, J., Kubota, M., Nishimura, T., and Tatsuichi, I., DNA polymerases of *Chlorella*. II. Characterization to distinguish two enzymes, *J. Biochem. (Tokyo)*, 96, 461, 1984.
29. Sakai, A. and Watanabe, Y., DNA polymerases of *Tetrahymena pyriformis*. I. Characterization of two *N*-ethylmaleimide-sensitive DNA polymerases from exponentially growing cells, *J. Biochem. (Tokyo)*, 91, 845, 1982.
30. Sakai, A. and Watanabe, Y., DNA polymerases of *Tetrahymena pyriformis*. II. Does an *N*-ethylmaleimide-resistant polymerase exist in *Tetrahymena*?, *J. Biochem. (Tokyo)*, 91, 855, 1982.
31. Sakai, A. and Watanabe, Y., *Tetrahymena* α-type DNA polymerase A: purification and subunit analysis, *J. Biochem. (Tokyo)*, 92, 1241, 1982.
32. Hübscher, U., DNA polymerases in prokaryotes and eukaryotes: mode of action and biological implication, *Experientia*, 39, 1, 1983.
33. Hübscher, U., DNA polymerase holoenzymes, *Trends Biochem. Sci.*, 9, 390, 1984.
34. Chang, L. M. S., Phylogeny of DNA polymerase β, *Science*, 191, 1183, 1976.
35. Scovassi, A. I., Plevani, P., and Bertazzoni, U., Eukaryotic DNA polymerases, *Trends Biochem. Sci.*, 5, 335, 1980.
36. Chang, L. M. S., Cheriathundam, E., Mahoney, E. M., and Cerami, A., DNA polymerases in parasitic protozoans differ from host enzymes, *Science*, 208, 510, 1980.
37. Baril, E. F., Scheiner, C., and Pederson, T., A β-like DNA polymerase activity in the slime mold *Dictyostelium discoideum*, *Proc. Natl. Acad. Sci. U.S.A.*, 77, 3317, 1980.
38. Tanabe, K., Yamaguchi, M., Matsukage, A., and Takahashi, T., Structural homology of DNA polymerase β from various mammalian cells, *J. Biol. Chem.*, 256, 3098, 1981.
39. Tanabe, K., Yamaguchi, M., Matsukage, A., and Takahashi, T., Conservation of molecular structure of DNA polymerase β in vertebrates probed by tryptic peptide mapping, *J. Biochem. (Tokyo)*, 96, 365, 1984.
40. Scovassi, A. I., Wicker, R., and Bertazzoni, U., A phylogenetic study on vertebrate mitochondrial DNA polymerase, *Eur. J. Biochem.*, 100, 491, 1979.
41. Bertazzoni, U. and Scovassi, A. I., Mitochondrial DNA polymerase of eukaryotic cells, in *The Organization and Expression of the Mitochondrial Genome*, Kroon, A. M. and Saccone, C., Eds., Elsevier/North Holland Biomedical Press, Amsterdam, 1980, 287.
42. Hecht, N. B., Farrell, D., and Davidson, D., Changing DNA polymerase activities during the development of the testis in the mouse, *Dev. Biol.*, 48, 56, 1976.
43. Hecht, N. B., Farrell, D., and Williams, J. L., DNA polymerases in mouse spermatogenic cells separated by sedimentation velocity, *Biochim. Biophys. Acta*, 561, 358, 1979.
44. Grippo, P., Geremia, R., Locorotondo, G., and Monesi, V., DNA-dependent DNA polymerase species in male germ cells of the mouse, *Cell Differ.*, 7, 237, 1978.
45. Philippe, M. and Chevallier, Ph., Study of dogfish *(Scyliohinus caniculus)* deoxyribonucleic acid polymerase α and β. Extraction, characterization and changes during spermatogenesis, *Biochem. J.*, 189, 635, 1980.
46. Hecht, N. B. and Parvinen, M., DNA synthesis catalyzed by endogenous templates and DNA-dependent DNA polymerases in spermatogenic cells from rat, *Exp. Cell Res.*, 135, 103, 1981.
47. Parvinen, M. and Hecht, N. B., Identification of living spermatogenic cells of the mouse by transillumination-phase contrast microscopic technique for *"in situ"* analyses of DNA polymerase activities, *Histochemistry*, 71, 567, 1981.
48. Matsukage, A., Yamamoto, S., Yamaguchi, M., Kusakabe, M., and Takahashi, T., Immunochemical locatlization of chick DNA polymerase α and β^+, *J. Cell Physiol.*, 117, 266, 1983.
49. Hübscher, U., Kuenzle, C. C., and Spadari, S., Variation of DNA polymerase-α, -β and -γ during perinatal growth and differentiation, *Nucl. Acids Res.*, 4, 2917, 1977.
50. Hübscher, U., Kuenzle, C. C., Limacher, W., Scherrer, P., and Spadari, S., Functions of DNA polymerase α, β and γ in neurons during development, *Cold Spring Harbor Symp. Quant. Biol.*, 43, 625, 1978.
51. Carre, D. S. and Pieau, C., Variation of DNA polymerase activities in chick neuronal retina as a function of age, *Differentiation*, 15, 161, 1979.
52. Claycomb, W. C., Biochemical aspects of cardiac muscle differentiation. Deoxyribonucleic acid synthesis and nuclear and cytoplasmic deoxyribonucleic acid polymerase activity, *J. Biol. Chem.*, 250, 3229, 1975.
53. Coleman, M. S., Hutton, J. J., and Bollum, F. J., DNA polymerases in normal and leukemic human hematopoietic cells, *Blood*, 44, 19, 1974.

54. Roodman, G. D., Hutton, J. J., and Bollum, F. J., DNA polymerase, thymidine kinase and DNA synthesis in erythropoietic mouse spleen cells separated on bovine serum albumin gradients, *Biochem. Biophys. Acta,* 425, 478, 1976.
55. Longacre, S. S. and Rutter, W. J., Nucleotide polymerases in the developing avian erythrocyte, *J. Biol. Chem.,* 252, 273, 1972.
56. Wang, H. F. and Popenoe, E. A., Variation of deoxyribonucleic acid polymerase activities during avian erythropoiesis, *Biochim. Biophys. Acta,* 474, 98, 1977.
57. Kalf, G. F., Merione, R. M., Kerlavage, B. L., and Koszalka, T. R., Variation of DNA polymerase activities of rat giant trophoblast cells in mid-gestation, *Dev. Biol.,* 84, 351, 1981.
58. Kalf, G. F., Metrione, R. M., and Koszalka, T. R., A protein stimulatory factor for DNA polymerase α in rat giant trophoblast cells, *Biochem. Biophys. Res. Commun.,* 100, 566, 1981.
59. Siegel, R. L. and Kalf, G. F., DNA polymerase β involvement in DNA endoreduplication in rat giant trophoblast cells, *J. Biol. Chem.,* 257, 1785, 1982.
60. Furia, M., Polito, L. C., Locorotondo, G., and Grippo, P., Identification of a DNA polymerase-β-like form in *Drosophila melanogaster* adult flies, *Nucl. Acids Res.,* 6, 3399, 1979.
61. Dusenbery, R. L. and Smith, P. D., DNA polymerase activity in developmental stages of *Drosophila melanogaster, Dev. Genet.,* 3, 309, 1983.
62. Fansler, B. and Loeb, L. A., Sea urchin nuclear DNA polymerase. II. Changing localization during early development, *Exp. Cell Res.,* 57, 305, 1969.
63. Shioda, M., Nagano, H., and Mano, Y., Transition of DNA polymerase-α and endoplasmic reticulum during gastrulation of sea urchin, *Dev. Biol.,* 91, 111, 1982.
64. Shioda, M., Nagano, H., and Mano, Y., Association of DNA polymerase-α and -β with rough endoplasmic reticulum in sea-urchin eggs and changes in the subcellular distribution during early embryogenesis, *Eur. J. Biochem.,* 108, 345, 1980.
65. Benbow, R. M., Activation of DNA synthesis during early embryogenesis, in *Biology of Fertilization,* Metz, C. B. and Monroy, A., Eds., Academic Press, New York, 1984, 299.
66. Zierler, M. K., Marini, N. J., Stowers, D. L., and Benbow, R. M., Stockpiling of DNA polymerases during oogenesis and embryogenesis in the frog *Xenopus laevis, J. Biol. Chem.,* 260, 974, 1985.
67. Carre, D., Signoret, J., Lefresne, J., and David, J. C., Enzymes involved in DNA replication in the axolotl. I. Analysis of the forms and activities of DNA polymerase and DNA ligase during development, *Dev. Biol.,* 87, 114, 1981.
68. Mikhailov, V. S. and Gulyamov, D. B., Changes in DNA polymerase α, β, activities during early development of the teleost fish *Misgurnus fossilis* (loach), *Eur. J. Biochem.,* 135, 303, 1983.
69. Orgel, L. E., Ageing of clones of mammalian cells, *Nature (London),* 243, 441, 1973.
70. Burnet, F. M., *Intrinsic Mutagenesis: A Genetic Approach to Aging,* John Wiley & Sons, New York, 1974.
71. Hayflick, L., The cellular basis of human aging, in *Handbook of the Biology of Aging,* Finch, C. E. and Hayflick, L., Eds., Van Nostrand Reinhold, New York, 1977, 159.
72. Norwood, T. H. and Smith, J. R., The cultured fibroblast-like cell as a model for the study of aging, in *Handbook on Aging,* Finch, C. E. and Schneider, E. L., Eds., Van Nostrand Reinhold, New York, 1984, in press.
73. Linn, S., Kairis, M., and Holliday, R., Decreased fidelity of DNA polymerase activity isolated from aging human fibroblasts, *Proc. Natl. Acad. Sci., U.S.A.,* 73, 2818, 1976.
74. Murray, V., Properties of DNA polymerases from young and aging human fibroblasts, *Mech. Age. Dev.,* 16, 327, 1981.
75. Murray, V. and Holliday, R., Increased error frequency of DNA polymerases from senescent human fibroblasts, *J. Mol. Biol.,* 146, 55, 1981.
76. Krauss-Wald, S. and Linn, S., Changes in DNA polymerase α, β and γ during the replicative lifespan of cultured human fibroblasts, *Biochemistry,* 21, 1002, 1982.
77. Weisman-Shomer, P., Kaftory, A., and Fry, M., Replicative activity of isolated chromatin from proliferating and quiescent early passage and aging cultured mouse cells, *J. Cell Physiol.,* 101, 219, 1979.
78. Pendergrass, W. R., Saulewicz, A. C., Salk, D., and Norwood, T., DNA polymerase alpha is inducable in postmitotic senescent human diploid fibrobalstic cells, submitted for publication, 1984.
79. Pendergrass, W. R., Saulewicz, A. C., Burmer, G. C., Rabinovitch, P. S., Norwood, T. H., and Martin, G. M., Evidence that a critical threshold of DNA polymerase-alpha activity may be required for the initiation of DNA synthesis in mammalian cell heterokaryons, *J. Cell Physiol.,* 113, 141, 1982.
80. Fry, M., Silber, J., Loeb, L. A., and Martin, G. M., Delayed and reduced cell replication and diminishing levels of DNA polymerase-α in regenerating liver of aging mice, *J. Cell. Physiol.,* 118, 225, 1984.
81. Fry, M., Loeb, L. A., and Martin, G. M., On the activity and fidelity of chromatin-associated hepatic DNA polymerase-β in aging murine species of different life spans, *J. Cell Physiol.,* 106, 435, 1981.

82. Fry, M. and Weisman-Shomer, P., Altered nuclear deoxyribonucleic acid α-polymerase in senescent cultured chick embryo fibroblasts, *Biochemistry,* 15, 4319, 1976.
83. Agarwal, S. S., Tufner, M., and Loeb, L. A., DNA replication in human lymphocytes during aging, *J. Cell Physiol.,* 96, 235, 1978.
84. Fry, M., Shearman, C. W., Martin, G. M., and Loeb, L. A., On the accuracy of DNA synthesis directed by chromatin-associated DNA polymerase β, *Biochemistry,* 19, 5939, 1980.
85. Silber, J. R., Fry, M., Martin, G. M., and Loeb, L. A., Fidelity of DNA polymerases isolated from regenerating liver chromatin of aging *Mus musculus, J. Biol. Chem.,* 260, 1304, 1985.
86. Loeb, L. A., Fry, M., Silber, J. R., Koplitz, M., and Martin, G. M., Fidelity of DNA synthesis during aging, in *Molecular Basis of Aging,* Roy, A. K. and Chatterjee, B., Eds., Academic Press, New York, 1984, 47.
87. Collins, J. M. and Chu, A. K., Reduction of DNA synthesis in aging but still proliferating cells, *J. Cell. Physiol.,* 124, 165, 1985.
88. Muller, W. E. and Zahn, R. K., Age-dependent gene induction in quail oviduct. VII. Alteration of DNA polymerase activities in response to progesterone treatment, *Mech. Aging Dev.,* 9, 527, 1979.
89. Muller, W. E., Zahn, R. K., Genrsten, W., and Munsch, N., Age-dependent alterations of DNA synthesis. Terminal deoxynucleotidyl transferase and DNA polymerase activities in bone marrow subpopulations from mice, *Mech. Aging Dev.,* 13, 119, 1980.
90. Sakaguchi, K. and Boyd, J. B., Purification and characterization of a DNA polymerase β from *Drosophila, J. Biol. Chem.,* 260, 10406, 1985.

Chapter 10

FUTURE PERSPECTIVES

Vision is the art of seeing things invisible.

Jonathan Swift
Miscellanies, 1726

I. INTRODUCTION

The rapid progress made in recent years in unraveling the molecular details of structure and function of animal cell DNA polymerases has significantly increased our understanding of the physical, enzymological, and biological properties of these cardinal enzymes. The impetus of this work is the basic conviction that these enzymes are responsible for the replication and repair of the cell's genetic information. Yet it appears that with the acquisition of an increasing body of knowledge, a Pandora's box of challenging new problems now opened, so that we are faced with more unanswered questions now than those asked at earlier stages in the study of DNA polymerases. In this concluding chapter, some of the more tantalizing problems that presently wait investigation are delineated. We have also attempted to indicate possible experimental approaches, which may be applicable to the solution of some of the presented problems. It should be emphasized, however, that only selected problems are presented in this chapter, the choice of which was ultimately subjective. Additional, and not less intriguing, questions which are not described herein await solutions.

II. STRUCTURAL ANALYSIS OF DNA POLYMERASES

At the time of this writing, homogeneous or near homogeneous preparations of DNA polymerases-α, -β, -γ, and -δ from several cell types are available. The purification and structural analyses of the polymerases have recently been simplified with the introduction of monospecific polyclonal and monoclonal antibodies against DNA polymerases. The isolation of messenger RNA for DNA polymerase-β[1] as well as for the catalytic polypeptide of polymerase-α[2] have been reported. Yet, our knowledge of the detailed structure of animal cell DNA polymerases is still only fragmentary. First, the question of the number and structure of subunits of polymerases-α and -γ has not been resolved. Second, although purification to homogeneity of small amounts of DNA polymerases has been attained, practically nothing is known of either the primary or the three-dimensional structure of any eukaryotic DNA polymerase. No understanding of structure-function relationship within a polymerase's molecule is possible without acquiring substantial knowledge on the structure of the polymerase. Hence, future studies will have to grapple with the problem of the detailed description of the structure of DNA polymerases.

Due to its relatively simple structure and availability in pure form, DNA polymerase-β appears to be the favored first candidate for a comprehensive structural analysis. It now seems that such an analysis will best be conducted by the application of recombinant DNA techniques. Messenger RNA for β-polymerase has been isolated and translated in vitro into an enzymatically active polypeptide.[1] Hence, it appears that c-DNA to β-polymerase m-RNA will be prepared in the near future and will be used to identify, isolate, and clone the gene for this enzyme. Alternatively, determination of a partial amino acid sequence of the available homogeneous β-polymerase might enable construction of a synthetic nucleic acid probe that can be used to identify and isolate a cloned gene for the polymerase. Construction of β-polymerase c-DNA and isolation of a structural gene for the enzyme will permit determination of the base sequence of the two and will eventually indirectly yield the primary structure of the polymerase. Furthermore, the availability of cloned β-polymerase gene or c-DNA might enable production of β-polymerase polypeptide in quantities large enough for a detailed study of the three-dimensional structure of the enzyme. Application of X-ray crystallography and nuclear magnetic resonance analysis, as well as site-specific mutagenesis of the enzyme molecule will eventually lead to detailed understanding of the mechanism of catalysis for a DNA polymerase. Such studies should clarify elusive problems such as the number and nature of dNTP binding sites in the enzyme molecule, template/primer interactive site(s), and possible interaction between various sites in the polymerase molecule. It is also clear that the isolation and sequencing of gene(s) for β-polymerase might eventually yield useful information on their chromosomal localization, fine gene structure, and possible adjacent regulatory sequences.

Eliciting molecular details of the structure of polymerase-α poses a considerably more formidable problem than that of the structure of β-polymerase. First, the size of the catalytic polypeptide of α-polymerase, 180 to 200 kdaltons, appears to be one of the largest single polypeptides known for any enzyme. As described in Chapter 3, attempts to isolate an intact catalytic core of α-polymerase have met with difficulties, the major one being the extreme susceptibility of the enzyme to partial proteolytic cleavage, which generates catalytically active fragments of the polymerase. Second, unlike β-polymerase, which is a single polypeptide, α-polymerase is a multimeric enzyme, consisting of several subunits whose precise number and structure are not fully known.

A reasonable approach to the structural analysis of polymerase-α involves study of the structure of the catalytic subunit followed by investigation of accessory subunits. Here again, the most promising methodology applicable to the problem at hand is that of molecular cloning of recombinant DNA. In principle, the availability of monoclonal antibodies should have led to the isolation of clones that express recombinant α-polymerase c-DNA, but to our knowledge there are no reports yet on successful application of this approach. However, a recent study indicated that an RNA fraction that appears to contain translationally active messenger RNA for α-polymerase has been isolated.[2] Characterization and further purification of α-polymerase messenger RNA might be instrumental in constructing c-DNA, which may be used for the identification and isolation of a cloned structural gene for the catalytic core of α-polymerase. An alternative approach to the isolation of the structural gene for α-polymerase is marked by the successful transfection of cultured cells with DNA from cells that possess mutated aphidicolin-resistant DNA polymerase-α.[3] Molecular cloning of the transforming DNA and use of aphidicolin-resistance as a selectable marker for polymerase-α might soon lead to the isolation of the structural gene for this enzyme.

A possible major impediment to the cloning of the catalytic subunit of α-polymerase is posed by the exceptionally large molecular size of this polypeptide. The size of the α-polymerase gene may exceed the maximum capacity of vector phage for insertion of recombinant DNA fragments. Molecular cloning of the α-polymerase gene may be aided with the recent recognition that the gene for the catalytic core of that enzyme is linked with the X-chromosome in human cells.[4] This information might prove useful for enrichment of α-polymerase genes in DNA preparations. Localization of the α-polymerase structural gene at the X-chromosome is also a first step in the analysis of chromosomal linkage of the several subunits that comprise together an α-polymerase "holo" enzyme molecule. It will be most interesting to learn whether the various subunits are coded by genetically linked loci or by genes dispersed among different chromosomes.

Whatever approach to the molecular cloning of α-polymerase catalytic core c-DNA or structural gene will prove successful, once their nucleic acid sequence is determined, the primary structure of the enzymically active polypeptide will become available. Moreover, molecular cloning of cDNA in bacteria or of the structural gene in eukaryotic cells might enable their expression to produce milligram amounts of enzyme, which are not attainable with conventional protein purification techniques. Yet, if postsynthetic modifications are required for the in vitro activation or modulation of the activity of polymerase-α, the expression of a cloned gene for that enzyme may not suffice to produce physiologically active polymerase. Availability of α-polymerase catalytic core in large amounts will enhance the analysis of the three-dimensional structure of the enzyme and open the way for an understanding of structure-function relationship within the polymerase molecule. Further, availability of large amounts of the catalytic core of α-polymerase and its structural gene might aid considerably in elucidation of the tantalizing question of the physiological significance of proteolytic cleavage of α-polymerase *(vide infra).*

The identification, isolation, and structural analysis of the full set of accessory subunits of α-polymerase poses another series of complex problems. Functions of some subunits/auxiliary proteins of α-polymerase have been defined, and assay systems for their identification have been constructed. Among these cofactors are DNA primase, Ap_4A binding activity, and primer recognition activity. However, it appears likely that other subunits/ancillary proteins, which have not yet been defined and isolated, interact with the α-polymerase catalytic core in the cell. A promising approach to the isolation of auxiliary subunits, which had already met with initial success, is the use of monoclonal antibodies in a single or few step immunoaffinity purification scheme.[5-8] An alternative approach will become practical with the availability of milligram amounts of α-polymerase catalytic polypeptide, which might then be used to bind ancillary subunits in affinity chromatography purification schemes. Once polypeptides other than the catalytic core are isolated, the structure of the α-polymerase "holo"-enzyme might be deduced. Further, the basis for the molecular heterogeneity of α-polymerase and its physiological significance might be illuminated (*vide infra).* With the isolation of an α-polymerase holoenzyme, it is possible that new polymerase-associated functions will be revealed. Factors that enhance fidelity or processivity may be identified. Proteins which recognize specific origin of replications and/or signals for cessation of synthesis in the genome might also be distinguished.

With an admirable effort, only a single laboratory to date has succeeded in purifying DNA polymerase-γ to homogeneity.[9,10] It is not clear yet whether the suggested tetrameric structure, $(47\ \text{kdaltons})_4$ for chick and probably mouse cell γ-polymerase is the universal multimeric structure of that enzyme. The low number of γ-polymerase molecules within animal cells poses a considerable impediment in the way of purifying that enzyme to homogeneity and in obtaining large amounts of the polymerase. It appears that molecular cloning of the gene for γ-polymerase is the method of choice for the structural analysis of that polymerase. Isolation of the gene for γ-polymerase and of workable amounts of the enzymic protein will shed light on the problems of the precise multimeric structure of that enzyme, its molecular heterogeneity, and its possible association with accessory cofactors.

III. PHYSIOLOGICAL VARIATIONS IN THE LEVEL OF α-POLYMERASE, MOLECULAR MICROHETEROGENEITY, AND PROTEOLYSIS

Ample experimental evidence indicates that the activity of α-polymerase increases in cells during periods of active DNA replication and declines during quiescence (cf. Chapter 6, Section II.A.2). Recent results obtained by *in situ* immunodetection of α-polymerase reinforce the suggestion that cell entry into DNA replicative phase involves *de novo* synthesis of polymerase-α molecules. However, all these results pertain to the catalytic core of the polymerase whose level was assayed, either biochemically or immunochemically. It is completely unknown at present, however, whether all or part of the polymerase accessory proteins/subunits are coordinately synthesized with the catalytic core. Furthermore, nothing is known on the mechanism of the inactivation of the polymerase at the end of S phase. Several possibilities exist, one of which is proteolytic degradation of the polymerase. Yet it appears that the catalytic activity of α-polymerase does not disappear at the end of S phase. A subtler mechanism of activation/inactivation of the polymerasae could be proteolytic cleavage of attachment/dissociation site(s) for essential subunit(s) during cell entry into the DNA replicative period and its cessation.

A large body of evidence, reviewed throughout Chapter 3, indicates that polymerase-α appears in animal cells in multiple molecular forms. These molecular species could result from partial proteolysis and be of no physiological significance. Alternatively, they could reflect different combinations of α-polymerase catalytic core with varying numbers of distinct subunits, as is the case with forms of polymerase that are either associated with or devoid of DNA primase (cf. Chapter 3, Section IV.B.1). An attractive mechanism for generation of active and physiologically inactive α-polymerase may proceed through a limited proteolytic cleavage of the polymerase. Hence, a catalytic core may be inactive in DNA replication in vivo unless it is partially cleaved so that it can combine with requisite accessory cofactor(s). Alternatively, defined proteolytic cleavage may render the enzyme inactive in DNA replication in vivo by dissociating essential auxilary cofactor(s) from the catalytic core. The dissociated catalytic polypeptide maintains apparent polymerase activity in vitro, thus confusing the traditional biochemist. Finally, multiple molecular forms of polymerase-α may be generated by postsynthetic modifications, such as phosphorylation, ADP-ribosylation, acetylation, etc. It is clear that these, and other speculations on the physiological significance of the molecular heterogeneity of α-polymerase and the place of proteolysis in its generation, will become subject to critical examination only when a better understanding of the structure of the intact multimeric α-polymerase is attained.

IV. THE NATURE OF THE REPLICATIVE COMPLEX

It is likely that polymerase-α comprises a part of a multiprotein replicative complex (cf. Chapter 6, Section II.A.4). Yet the nature, number, and function of components of that purported supramolecular structure are still conjectural. Several basic questions in regard to the place of α-polymerase in the hypothetical replication complex are open for study.

First, amply documented evidence indicates that entry of animal cells into replicative phase involves *de novo* synthesis and increase in the cellular activity of DNA polymerase-α. A cardinal question concerns the nature of stimuli for synthesis of α-polymerase in cells that progress into the division cycle. The tight controls of cell proliferation and genomic replication suggest that synthesis of DNA polymerase-α is a stringently regulated process that probably involves multiple steps. One should hope that, with the cloning of gene(s) for α-polymerase, the regulatory sequence that are responsible for gene activation will also be identified. Future investigations might also reveal possible links between growth regulating-genes of the protooncogenic type and α-polymerase or its associated cofactors. Once polymerase-α is synthesized, what multimeric structure does the polymerase assume, and what are its enzymatic properties when it becomes complexed with other proteins in the replicative machinery? Supposedly, α-polymerase should be associated with all its accesssory subunits/auxiliary factors, which enable it to carry out rapid, regulated, and accurate replication of DNA. Which are the cofactors that associate with the polymerase in such a complex? Moreover, is there only a single type of complex? It has been suggested recently that polymerase-α combines differently with DNA primase, RNase H, DNA topoisomerase, $3' \rightarrow 5'$ exonuclease, and DNA methylase to form at least two different complexes that were purported to be involved in leading and lagging strand replication, respectively.[11] Much work is required to rigorously demonstrate specific and physiologically regulated stoichiometric association of α-polymerase with other components of the eukaryotic "replisome" particle. Further, is polymerase-α the only eukaryotic DNA polymerase that is complexed within a supramolecular replicative machinery? Inconclusive evidence reviewed in preceding chapters suggest that DNA polymerases-β and -γ may perhaps also participate in certain aspects of genomic DNA replication. Do these two types of polymerase also associate with other proteins to perform replicative functions? Last, experiments reviewed in Chapter 6, Section II.A.4, suggest that "replitase", a high-molecular-weight multienzyme complex, consists of α-polymerase and a multitude of enzymes, which conduct channeling of early precursors of deoxynucleoside triphosphates into DNA. In the face of some contrasting evidence (cf. Chapter 6, II.A.4), the presence in vivo of the replitase complex has to be critically examined.

Whatever the constitution of a replicative complex is, its possible existence poses highly intriguing questions. Hence, if a multicomponent replicative complex is convincingly shown to be present in animal cells, one would like to know what is the physiological signal for formation of the complex during the cell cycle. Is polymerase-α processed posttranslationally in order to interact with other proteins in the complex? What holds together the various components of the complex, and what is the signal for its dissolution? Answers to these questions impinge on the general problem of the stringent normal regulation of DNA replication in animal cells and alternations in this tightly controlled process during neoplastic growth.

V. FIDELITY OF DNA REPLICATION

The evidence reviewed in Chapter 8 indicates that purified eukaryotic DNA polymerases are highly error prone. The accuracy of animal cell DNA polymerases exceeds that fidelity attributable to the energy available from base pairing, thus suggesting that by some unidentified mechanism these enzymes are capable of functioning in base selection. Yet the most accurate eukaryotic DNA polymerases that were tested have error rates not lower than 10^{-5}, whereas the estimated mutation rate in animal cells is 10^{-9} to 10^{-11}. Seldom in the history of biological science has the extent of ignorance spanned as great a range as six orders of magnitude. The simplest explanation of the gap between accuracy of isolated polymerases and fidelity of DNA replication in vivo is that a separate $3' \rightarrow 5'$ exonuclease is an accessory protein that acts in concert with α-polymerase to enhance fidelity through proofreading. Since such a mechanism is so prominent in DNA polymerization in bacteria, it is difficult to conceive that it is completely discarded in the course of evolution. A second possibility is that DNA binding proteins could enhance precision of DNA replication at the base insertion step by fixing a favorable conformation of the template strand. Alternatively, mismatch correction, i.e., the excision of a noncomplementary base after its polymerization, is a potentially powerful mechanism for attaining accuracy.

Assays can now be designed to search for proofreading exonucleases, template-interactive, factors and mismatch correcting factor(s). The putative replicative complexes appear to constitute a fertile ground to reap this harvest. In an alternative brute force approach, one can proceed by screening fractionated cell proteins for factors that enhance the accuracy of DNA polymerase-α. In fact, we should be receptive for the possibility that such searches will reveal fundamental differences between mechanisms by which fidelity of DNA replication is attained by prokaryotes and eukaryotes.

VI. ROLES OF DNA POLYMERASES IN DNA REPAIR SYNTHESIS AND THEIR POSSIBLE INCLUSION IN MULTIPROTEIN REPAIR COMPLEXES

Both DNA polymerase-α and -β appear to be involved in different as well as common aspects of repair synthesis of DNA (see Chapter 7). Yet the question of the precise role of each class of polymerase in repair synthesis has not yet been resolved. Factors such as the nature of damage to DNA, dose of damaging agent, cell type, and growth conditions all seem to be significant affectors in the mobilization of DNA polymerases-α or -β for repair synthesis. Much work is required for definition of the rules that govern the participation of polymerases of different types in DNA repair. Utilization of monoclonal antibodies against polymerases-α and -β for the specific arrest of defined modes of repair synthesis in isolated subcellular systems might prove useful for the resolution of the problem of roles of polymerases in repair synthesis.

Some evidence indicates that response of α-polymerase to selective inhibitors differs in the DNA replication and repair modes. It is possible that different molecular species of α-polymerase take part in repair and replication synthesis. In such a case, what is the subunit construction of a repairing α-polymerase? Alternatively, α-polymerase might be complexed with repair proteins in a singular supramolecular repair machinery, and this association affects the enzyme's properties and function in vivo. Use of anti-α polymerase monoclonal antibodies in an immunoaffinity purification scheme with extracts of cells that perform repair synthesis might lead to isolation of "repairing" α-polymerase subspecies and/or α-polymerase constaining repair complex.

Evidence surveyed in Chapter 7 indicates that DNase V and a DNA dependent ATPase III interact specifically with DNA polymerase-β. Moreover, some data suggest that β-polymerase might be present in the cell in a supramolecular structure that contains other repair-related proteins. Several questions come to mind. Is such a purported repair complex a constitutive cellular component or does it assemble in response to massive damage to DNA? If the latter is correct, what is the signal for assembly and what leads to dissolution of such a supramolecular structure? Is β-polymerase the only polymerase in the hypothetical complex or does it interact with polymerases-α, -γ or -δ in it? Which are the other components of such a complex? Is for instance, a fidelity enhancing cofactor included in it? Last, is there only a single type of repair complex or are there several different complexes assembled for different modes of repair DNA synthesis? Answering all or some of these questions will require development of procedures for the isolation of intact repair complexes and establishment of assay systems for the assessment of repair activity in subcellular systems.

Another group of questions relates to the possible existence of error-prone repair pathways in animal cells and to the role of DNA polymerase(s) in such a pathway. In bacteria there are at least two sets of proteins that are induced in response to DNA damage, and it is still a matter of conjecture whether such proteins appear in damaged animal cells. If such inducible proteins exist, are they involved in the bypass of bulky DNA adducts at the expense of introducing somatic mutations? Most importantly, does such a system contain an error-prone DNA polymerase or polymerase-containing unfaithful repair complex? Use of viral probes, which contain site specific mutations and their repair by host animal cells, may serve as an assay system for the detection of an error-prone repair system and for its analysis.

VII. FUNCTION OF DNA POLYMERASE-γ IN THE CELL NUCLEUS

A considerable portion of the cellular activity of DNA polymerase-γ is confined to the cell nucleus. Although it is quite clear that polymerase-γ is the sole enzyme responsible for synthesis of mitochondrial DNA, the function of that polymerase in nuclei is still a complete mystery. The uniquely high processivity of polymerase-γ as well as its very high affinity for deoxynucleoside triphosphate substrates are singular among DNA polymerases of animal cells. Whether these characteristics are related to the in vivo role of γ-polymerase in the cell nucleus remains a subject for future research. Even with respect to the function of γ-polymerase in mitochondria, it is not clear whether— in addition to its purported role in replication of mitochondrial DNA—this polymerase also takes part in repair of the mitochondrial genome if it occurs.

VIII. DISTRIBUTION, PROPERTIES, AND FUNCTION OF DNA POLYMERASE-δ

To date, DNA polymerase-δ has been isolated only from rabbit bone marrow and calf thymus, and difference in structure were noted between the δ-enzymes from these two sources. Several questions come to mind in regard to this intriguing 3′ → 5′ exonuclease-containing enzyme. First, is DNA polymerase-δ restricted to selected tissues or is it as pervasive as DNA polymerases of the α, β, and γ categories? If it is limited to specific tissues, what is the unique role of δ-polymerase in these selected tissues, and why are other cells devoid of that enzyme? Since in several respects polymerase-δ resembles α-polymerase, it will be interesting to learn what relation, if any, exists between the two enzymes. Monoclonal antibodies against α-polymerase should be useful in identifying structural features common to the two polymerases. Raising monospecific antibodies against δ-polymerase might reciprocally complement such investigation and in addition, specific immunoglobulins may be useful in future investigations on the cellular localization and role of δ-polymerase. Last, does the intimate association of DNA polymerase-δ with 3′ → 5′ exonuclease bestow on it a proofreading option and thus render it more accurate than other types of animal DNA polymerases? Assay systems to examine this question are available, and it will be most interesting to assess the fidelity of δ-polymerases, which may resemble more the accuracies of bacterial rather than those of animal cell DNA polymerases.

IX. FUNCTION OF DNA POLYMERASES IN DNA RECOMBINATION

After patiently proceeding through the preceding chapters and reaching this late point of the discussion, the reader must have noticed that the problem of participation of DNA polymerases in recombination has not been addressed. Genetic recombination involves the formation of base pairing between homologous or largely nonhomologous strands of two different segments of DNA and the exchange of genetic material between the paired strands. It is clear that, in analogy with the well-studied recombination in bacteria and phages, animal cells also conduct similar processes. This is mainly inferred from the generation of recombinant viruses after infection of cells with physically and genetically distinguishable virus strains. Very little is known, however, about the mechanism of recombination, intermediates formed, and enzymology of the process in animal cells. One major reason for this ignorance was the absence of defined DNA probes that recombine within animal cells and permit analysis of the molecular details of the recombination events. Only recently, an SV40-pBR322 hybrid plasmid has been constructed by Subramani and Berg[12] to be used for the detection and quantitation of homologous and nonhomologous recombination events in cultured monkey cells. This approach allows analysis of recombination events in mammalian cells and allows analysis of recombination at the DNA sequence level. It is reasonable to propose that, with the availability of these and similar DNA probes, the study of the enzymology of recombination in animal cells is now feasible. One would inquire which class(es) of polymerase take part in recombination, what are the mechanisms of polymerase action in recombination events, and what other proteins and cofactors participate in these processes. Since DNA recombination is at the root of genetic stability and variance, the elucidation of the role of DNA polymerases in this process is of cardinal importance.

X. CONCLUSION

The questions raised in this chapter are just a set selected out of a plethora of puzzles imposed by our limited knowledge of DNA polymerases of eukaryotes. One point is clear, however: future research must involve an enormous amount of effort and work. Nothing is more befitting, therefore, for the conclusion of this chapter and book than the eternal words of Jerome Klapka Jerome in his writing, "It is impossible to enjoy idling thoroughly unless one has plenty of work to do."[13]

REFERENCES

1. Yamaguchi, M., Takahashi, T., Yasuda, K., Shimura, Y., and Matsukage, A., Characterization of messenger RNA for chick embryo DNA polymerase-β and its translation product *in vitro, Eur. J. Biochem.*, 133, 277, 1983.
2. Detera-Wadleigh, S., Karawya, E., and Wilson, S. H., Synthesis of DNA polymerase by *in vitro* translation of calf RNA, *Biochem. Biophys. Res. Commun.*, 122, 420, 1984.
3. Liu, P. K. and Loeb, L. A., Transfection of DNA polymerase-α gene, *Science*, 226, 833, 1984.
4. Wang, T. S-F. Pearson, B. E., Suomalainen, H. A., Mohandas, T., Shapiro, L. J., Schröder, J., and Korn D., Assignment of the gene for human DNA polymerase-α to the X chromosome, *Proc. Natl. Acad. Sci. U.S.A.*, 82, 5270, 1985.
5. Wang, T. S-F., Hu, S-Z., and Korn, D., DNA primase from KB cells. Characterization of a primase activity tightly associated with immunoaffinity-purified polymerase-α, *J. Biol. Chem.*, 259, 1854, 1984.
6. Wahl, A. F., Kowalski, S. P., Harwell, L. W., Lord, E. M., and Bambara, R. A., Immunoaffinity purification and properties of a high molecular weight calf thymus α-polymerase, *Biochemistry*, 23, 1895, 1984.
7. Masaki, S., Tanabe, K., and Yoshida, S., Large polypeptides of 10 S DNA polymerase-α from calf thymus: rapid isolation using monocloncal antibody and tryptic peptide mapping analysis, *Nucl. Acids Res.*, 72, 4455, 1984.
8. Chang, L. M. S., Rafter, E., Augl, C., and Bollum, F. J., Purification of a DNA polymerase-DNA primase complex from calf thymus glands, *J. Biol. Chem.*, 259, 14679, 1984.
9. Yamaguchi, M., Matsukage, A., and Takahashi, T., Chick embryo DNA polymerase-γ. Purification and structural analysis of nearly homogeneous enzyme, *J. Biol. Chem.*, 255, 7002, 1980.
10. Matsukage, A., Yamaguchi, M., Tanabe, K., Taguchi, Y. N., Nishizawa, M., and Takahashi, T., DNA chain elongation mechanism of DNA polymerases-α, -β and -γ, in *New Approaches in Eukaryotic DNA Replication*, de Recondo, A. M., Ed., Plenum Press, New York, 1983, 81.
11. Ottiger, H. P. and Hübscher, V., Mammalian DNA polymerase-α holoenzymes with possible functions at the leading and lagging strand of the replication fork, *Proc. Natl. Acad. Sci. U.S.A.*, 81, 3993, 1984.
12. Subramani, S. and Berg, P., Homologous and nonhomologous recombination in monkey cells, *Mol. Cell. Biol.*, 3, 1040, 1983.
13. Jerome, K. J., On being idle, in *Idle Thoughts of an Idle Fellow*, 1889.

INDEX

A

E

F

G

H

I

K

L

M

N

O

P

R

S

T

U

V

X

Y